BIOCHEMICAL SOCIETY SYMPOSIA

No. 72

LIPIDS, RAFTS AND TRAFFIC

BIOCHEMICAL SOCIETY SYMPOSIUM No. 72
held at BioScience2004, Glasgow, July 2004

Lipids, Rafts and Traffic

ORGANIZED AND EDITED BY
R.A.J. McILHINNEY AND N.M. HOOPER

PORTLAND PRESS

Published by Portland Press,
59 Portland Place, London W1B 1QW, U.K.
on behalf of The Biochemical Society
Tel: (+44) 20 7580 5530; email: editorial@portlandpress.com
http://www.portlandpress.com

ISBN 1 85578 162 X
ISSN (print) 0067-8964
ISSN (online) 1744-1439

British Library Cataloguing in Publication Data
A catalogue record for this book is available from the British Library

Typeset by Portland Press Ltd
Printed in Great Britain by Cambridge University Press

Contents

Preface vii

Abbreviations xi

1 The role of microtubules in transport between the endoplasmic reticulum and Golgi apparatus in mammalian cells
By K.J. Palmer, P. Watson and D.J. Stephens 1

2 Kinases regulating Golgi apparatus structure and function
By C. Preisinger and F.A. Barr 15

3 Phosphoinositides and membrane traffic at the *trans*-Golgi network
By R.R. Choudhury, N. Hyvola and M. Lowe 31

4 N-glycans as apical targeting signals in polarized epithelial cells
By P. Urquhart, S. Pang and N.M. Hooper 39

5 Defective kidney anion-exchanger 1 (AE1, Band 3) trafficking in dominant distal renal tubular acidosis (dRTA)
By A.M. Toye 47

6 Regulation of the clathrin-coated vesicle by reversible phosphorylation
By A. Flett, S. Semerdjieva, A.P. Jackson and E. Smythe 65

7 Single-molecule live-cell imaging of clathrin-based endocytosis
By T. Kirchhausen, W. Boll, A. van Oijen and M. Ehrlich 71

8 Membrane traffic to and from lysosomes
By J.P. Luzio, P.R. Pryor, S.R. Gray, M. Gratian, R.C. Piper and N.A. Bright 77

9 Dynamin I phosphorylation and the control of synaptic vesicle endocytosis
By K.J. Smillie and M.A. Cousin 87

10 Early endocytic Rabs: functional prediction to functional characterization
By J.C. Simpson and A.T. Jones 99

11 The 'lipid raft' microdomain proteins reggie-1 and reggie-2 (flotillins) are scaffolds for protein interaction and signalling
By C.A.O. Stuermer and H. Plattner 109

12 Spatial control of actin-based motility through plasmalemmal PtdIns(4,5)P_2-rich raft assemblies
By T. Golub and P. Caroni 119

13 The conflicting role of brain cholesterol in Alzheimer's disease: lessons from the brain plasminogen system
By M.D. Ledesma and C.G. Dotti 129

14 Endocytosis and retrograde axonal traffic in motor neurons
By K. Deinhardt and G. Schiavo 139

15 Consequences of lipid raft association on G-protein-coupled receptor function
By A. Becher and R.A.J. McIlhinney 151

16 Functional role of lipid rafts in CD20 activity?
By E. Janas, R. Priest and R. Malhotra 165

17 Biophysics (and sociology) of ceramides
By F.M. Goñi, F.-X. Contreras, L.-R. Montes, J. Sot and A. Alonso 177

18 Reversible — through calmodulin — electrostatic interactions between basic residues on proteins and acidic lipids in the plasma membrane
By S. McLaughlin, G. Hangyás-Mihályné, I. Zaitseva and U. Golebiewska 189

19 Translocation of the cell-penetrating Tat peptide across artificial bilayers and into living cells
By P. Curnow, H. Mellor, D.J. Stephens, M. Lorch and P.J. Booth 199

20 Aggregation and fibrillization of prions in lipid membranes
By J. Kazlauskaite and T.J.T. Pinheiro 211

21 BAR domains and membrane curvature: bringing your curves to the BAR
By J.L. Gallop and H.T. McMahon 223

Subject index 233

Preface

The Annual Symposium of the Biochemical Society *Lipids, Rafts and Traffic* took place over 2 days at the BioScience2004 meeting in Glasgow, in July 2004. This book brings together many of the presentations from that meeting and provides a unique description of the endocytosis and exocytosis of proteins and lipids, focusing on the roles and mechanisms of action of lipid rafts and clathrin-coated vesicles in these processes. The roles of lipids and rafts in cell signalling and in disease states are also described.

The book begins with a discussion from David Stephens and colleagues of our current understanding of the mechanisms that link membrane traffic in the early secretory pathway to the microtubule cytoskeleton. In addition to a description of the roles of the different proteins involved in this process, the relative merits of the hypotheses to describe how directionality of vesicle movement is generated are contrasted, namely the 'tug-of war' and 'co-ordinated motor' theories. This is followed by a chapter from Christian Preisinger and Francis Barr describing the role of kinases, such as Cdk1–cyclin and the MAP kinases, in the regulation of Golgi function and re-organization during the cell cycle. In addition, the role of the mammalian Ste protein kinase family and protein kinase D in the establishment and maintenance of cell polarity and migration is examined.

In Chapter 3, Martin Lowe and colleagues turn our attention to phosphoinositide regulation of the formation of cargo vesicles at the *trans*-Golgi network with particular emphasis on PtdIns(4)P, which is enriched in the Golgi apparatus. They describe how low levels of another phosphoinositide, PtdIns(4,5)P_2, and its associated phosphatase OCLR1 are found on Golgi membranes and suggest mechanisms by which PtdIns(4,5)P_2 might play a role in membrane traffic. The next two chapters deal with the trafficking of proteins in polarized epithelial cells. Paula Urquhart and colleagues (Chapter 4) describe how N-glycans, rather than the glycosylphosphatidylinositol (GPI) anchor, are responsible for the apical targeting of membrane dipeptidase, while Ashley Toye (Chapter 5) shows that although some mutations in the kidney anion exchanger cause the protein to remain in the endoplasmic reticulum, other mutations misdirect the protein from the basolateral to the apical membrane due to the removal of a basolateral targeting motif.

Liz Smythe focuses our attention in Chapter 6 on clathrin-mediated endocytosis, and the roles that dynamin I has in constricting the neck of the vesicle and the adaptor protein AP-2 has in stimulating cargo sequestration into coated pits.

In Chapter 7, our view of clathrin-mediated endocytosis is challenged by Tom Kirchhausen and colleagues, who summarize their data obtained from live imaging of fluorescently tagged clathrin or its adaptor protein AP-2. Analysis of the results suggests that there are no preferred nucleation sites for endocytosis and that assembly of coated pits aborts unless cargo is captured. They propose a new model of endocytosis, termed the 'exploratory model'. Continuing the theme of vesicle fusion or formation, in the next chapter Paul Luzio and colleagues describe how, using live imaging of cells pre-loaded with Rhodamine–dextran, and then exposed to Oregon Green–dextran, both 'kissing and running' and 'explosive' fusion events contribute to the maturation of endosomes to lysosomes, but not vesicle trafficking between the organelles. Using a cell-free assay system, they describe how lysosomes fuse with endosomes to form a hybrid organelle and that different SNAREs, VAMP 7 (vesicle-associated membrane protein 7) and VAMP 8, are required for heterotypic endosome–lysosome fusion and homotypic endosome fusion respectively.

The next two chapters deal with endocytosis, with reviews by Karen Smillie and Michael Cousin on the role of dynamin I phosphorylation in the control of synaptic vesicle endocytosis, and by Jeremy Simpson and Arwyn Jones on the role of the Rab proteins in endocytosis. The former describes the identification of cyclin-dependent kinase and calcineurin as the proteins important in regulating thc phosphorylation of dynamin, and discusses the possible consequences of these processes, with respect to changing the interaction of dynamin I with other proteins and lipids. The latter concentrates on Rab5, Rab21 and Rab22, which localize to the early endocytic pathway, and examines their roles in this process, and highlights the difficulties in dissecting the functions of these widely expressed proteins.

The evidence that the reggie proteins 1 and 2 (otherwise known as the flotillins) form intracellular non-caveolar lipid rafts, contained within the inner leaflet of the lipid bilayer, is presented by Claudia Stuermer and Helmut Plattner in Chapter 11. These rafts are thought to function as scaffolds for the assembly of signalling complexes containing kinases and other membrane proteins.

In Chapter 12, Tamara Golub and Pico Caroni describe the role of plasmalemmal rafts and actin regulation in neurons, with particular focus on PtdIns(4,5)P_2-rich rafts that accumulate at the leading edge of motile lamellipoda and promote actin assembly. In Chapter 13, Maria Dolores Ledesma and Carlos Dotti describe work showing that the low level of plasmin observed in brains from Alzheimer's disease patients is due to decreased binding of its precursor plasminogen to rafts, resulting in reduced metabolism of amyloid-β peptide. They also challenge the currently accepted stance concerning cholesterol and Alzheimer's disease by claiming that increasing cholesterol levels actually blocks amyloid-β production. Endocytosis and retrograde transport in neurons are the subject of the next chapter by Katrin Deinhardt and Giampietro Schiavo. They describe how tetanus and botulinum toxins bind to different arrays of gangliosides, GPI-anchored and transmembrane proteins in order to enter cells, and that tetanus toxin is internalized via clathrin-coated pits. Once internalized, the endocytic vesicles are transported retrogradely along the axon in a microtubule -dependent manner. Anja Becher and Jeff McIlhinney discuss the consequences of lipid raft association on G-protein-coupled receptor function in Chapter 15, and Eva Janas and colleagues describe that the B-cell antigen CD20, which is involved in store-operated calcium entry, does not normally reside in rafts, but that on cross-linking with therapeutic antibodies moves into rafts and that this affects its activity (Chapter 16).

The biophysics (and sociology!) of ceramides and how short- and long-chain ceramides differently affect the curvature of the membrane and flip-flop transverse motion across the bilayer are the subject of Chapter 17 by Felix Goñi and colleagues. The theme of proteins binding to the surface of membranes is the focus of Chapter 18 by Stuart McLaughlin and colleagues, who describe how proteins with clusters of basic residues interact with acidic phospholipid head-groups in a reversible manner mediated through calmodulin, thereby affording a mechanism of regulation. In Chapter 19, Paula Booth and colleagues describe the translocation of the cell-penetrating Tat peptide across artificial bilayers and into living cells. Jurate Kazlauskaite and Teresa Pinheiro then describe experiments using recombinant prion protein and lipid vesicles to investigate the role of lipids in the conversion and pathogenesis of the prion protein. In the final chapter, Jennifer Gallop and Harvey McMahon describe how endophilin 1 and other proteins containing BAR domains, through binding selectively to curved membranes, may be involved in mediating clathrin-coated pit formation.

Finally, we thank all the authors for their scholarly contributions, and Mike Cunningham and his colleagues at Portland Press for their assistance in editing and compiling this book.

Jeff McIlhinney and Nigel Hooper

Abbreviations

A568	Alexa Fluor 568
Aβ	β-amyloid peptide
AAK1	adaptor-associated kinase 1
AD	Alzheimer's disease
ADCC	antibody-dependent cell cytoxicity
AE1	anion-exchanger 1
AKAP	A-kinase-anchoring protein
ANTS	8-aminonaphthalene-1,3,6-trisulphonic acid
AP	adaptor protein
AP180	assembly protein 180
ApoE4	apolipoprotein E4
APP	amyloid precursor protein
APR	array of pre-synaptic receptors
AQP-2	aquaporin-2
ARAP1	ARF-GAP, Rho-GAP, ankyrin repeat and PH domain-containing protein 1
ARF	ADP-ribosylation factor
ARH	autosomal recessive hypercholesterolaemia
Arp	actin-related protein
AT_1	angiotensin II type 1 receptor
AT_2	angiotensin II type 2 receptor
BAPTA	1,2-bis[2-aminophenoxy]ethane-*N*,*N*,*N*′,*N*′-tetra-acetic acid
BAR	bin, amphiphysin and Rvs161/167
BCR	B-cell receptor
BDNF	brain-derived neurotrophic factor
BicD	Bicaudal D
BIG2	brefeldin-A-inhibited guanine nucleotide-exchange factor 2
BoNT	botulinum neurotoxin
BRAP	breast-cancer-associated protein
BSE	bovine spongiform encephalopathy
CaM	calmodulin
CAP	c-Cbl-associated protein
CAP23	cortical cytoskeleton-associated protein of 23 kDa
CCPA	2-chloro-N^6-cyclopentyladenosine
CCR5	chemokine receptor 5
Cdc	cell-division control
Cdk	cyclin-dependent kinase

CEACAM1-L	long isoform of carcinoembryonic antigen-related cell-adhesion molecule 1
CFTR	cystic fibrosis transmembrane conductance regulator
CGN	*cis*-Golgi network
CHO	Chinese-hamster ovary
CIN	Cbl-interacting protein
CJD	Creutzfeldt–Jakob disease
CK	casein kinase
CLEM	correlative live cell and electron microscopy
CMV	cytomegalovirus
CNT	clostridial neurotoxin family
COPI	coatamer protein I
COPII	coatamer protein II
CPP	cell-penetrating peptide
cryo-EM	cryo-electron microscopy
CSF	cerebrospinal fluid
CWD	chronic wasting disease
DAG	diacylglycerol
DAPI	4,6-diamidino-2-phenylindole
DEPE	dielaidoyl PE
DHC	dynein heavy chain
7-DHC	7-dehydrocholesterol
DIC	dynein intermediate chain
DIV	days *in vitro*
DLC	dynein light chain
DLIC	dynein light intermediate chain
DMPC	dimyristoyl phosphatidylcholine
dn	dominant-negative
DOPC	1,2-dioleoyl-*sn*-3-glycerophosphocholine
DPPC	dipalmitoyl phosphatidylcholine
DPX	p-xylenebispyridinium bromide
DRM	detergent-resistant membrane
dRTA	distal renal tubular acidosis
DSC	differential scanning calorimetry
dynactin	dynein activator
EEA1	early endosome antigen 1
EGF	epidermal growth factor
EGFP	enhanced green fluorescent protein
EH	eps1S homology
EHNA	erythro-9-(2-hydroxy-3-nonyl) adenine
EM	electron microscopy
eps15	EGF receptor pathway substrate 15
ER	endoplasmic reticulum
ERAD	ER-associated degradation
ERK	extracellular-signal-regulated kinase
ERM	ezrin, radixin and moesin
ESCRT	endosomal sorting complexes required for transport

F-actin	filamentous actin
FAK	focal adhesion kinase
FAPP	four-phosphate-adapter protein
FcϵRIβ	β-subunit of the high-affinity IgE receptor
FCS	fluorescence correlation spectroscopy
FGF	fibroblast growth factor
FM 4-64	*N*-(3-triethylammoniumpropyl)-4-{6-[4-(diethylamino)phenyl]hexatrienyl} pyridinium dibromide
Fmoc	fluoren-9-ylmethoxycarbonyl
FRET	fluorescence resonance energy transfer
FSH	follicle-stimulating hormone
GAP	GTPase-activating protein
GAP-43	growth-associated protein 43
GAPDH	glyceraldehyde-3-phosphate dehydrogenase
GDI	GDP-dissociation inhibitor
GDNF	glial-derived neurotrophic factor
GED	GTPase effector domain
GEF	guanine nucleotide-exchange factor
GFP	green fluorescent protein
GFRα1	GDNF-family receptor α1
GGA	Golgi-localized, γ-ear-containing, ARF-binding protein
GluT	glucose transporter
GM130	Golgi matrix protein of 130 kDa
GnRH	gonadotropin-releasing hormone
GPI	glycosylphosphatidylinositol
GPBP	Goodpasture antigen-binding protein
GPCR	G-protein-coupled receptor
GRASP	Golgi reassembly stacking protein
GRK5	G-protein-coupled receptor kinase 5
GST	glutathione S-transferase
HA	haemagglutinin
H_C	non-toxic fragment of tetanus neurotoxin
H_{CC}	C-terminal domain of H_C
H_{CN}	N-terminal domain of H_C
HEK-293	human embryonic kidney
Hsc70	heat-shock cognate 70
HTm4	haematopoietic transmembrane-4 protein
IAF	5-iodoacetamidofluorescein
IDE	insulin-degrading enzyme
IgCAMs	Ig superfamily cell-adhesion molecules
IPTG	isopropyl β-D-thiogalactoside
IrreC-rst	irregular chiasm C roughest
JNK	c-Jun N-terminal kinase
KAP	kinesin-associated protein
KDELR	KDEL (Lys-Asp-Glu-Leu) receptor
kAE1	kidney isoform of AE1
LAT	linker for activation of T-cells

LCa	light chain A
LDL	low-density lipoprotein
LH	luteinizing hormone
LIM	Lin-11, Isl-1 and Mec-3
Lis1	lissencephaly 1
LM	light microscopy
LUV	large unilamellar vesicle
MacMARCKS	macrophage myristoylated alanine-rich C kinase substrate
MAL	myelin and lymphocyte protein
MAPK	mitogen-activated protein kinase
MARCKS	myristoylated alanine-rich C kinase substrate
MDCK	Madin–Darby canine kidney
MDP	membrane dipeptidase
MEK	MAPK/ERK kinase
MHCI	major histocompatibility complex class I
MLK	mixed-lineage kinase
MMP9	matrix metalloproteinase 9
MN	motor neuron
MPR	mannose 6-phosphate receptor
MS4A	membrane-spanning 4A
MST4	mammalian Ste20 4
MT	microtubule
MVB	multivesicular body
N-BAR	N-terminal domain of BAR
NCAM	neural cell-adhesion molecule
NGF	nerve growth factor
NLS	nuclear localization signal
NMDA	*N*-methyl-D-aspartate
NMJ	neuromuscular junction
NPC	Niemann–Pick type C
NRK	normal rat kidney
NSF	*N*-ethylmaleimide-sensitive factor
Nudel	NudE-like
N-WASP	neural Wiskott–Aldrich syndrome protein
OCRL	oculocerebrorenal syndrome of Lowe
OSBP	oxysterol-binding protein
$p75^{NTR}$	p75 low-affinity NGF receptor
PA	phosphatidic acid
PAI	plasminogen-activator inhibitor
PAK	p21-activated kinase
PC	phosphatidylcholine
PE	phosphatidylethanolamine
PEG	poly(ethylene glycol)
PH	pleckstrin homology
PI3K	phosphoinositide 3-kinase
PI4K	phosphoinositide 4-kinase
PI5K	phosphoinositide 5-kinase

PI4P5K	phosphoinositide 4-phosphate 5-kinase
PICK-1	protein interacting with protein kinase C
PIPK1γ	phosphoinositide phosphate kinase 1γ
PIPLC	phosphatidylinositol-specific phospholipase C
PITP	phosphatidylinositol-transfer protein
PKA	protein kinase A (or cAMP-dependent protein kinase)
PKC	protein kinase C
PKD	protein kinase D
PLC	phospholipase C
PLD	phospholipase D
Plk	polo-like kinase
POPC	1-palmitoyl-2-oleoyl phosphatidylcholine
POPG	1-palmitoyl-2-oleoyl phosphatidylglycerol
PP	protein phosphatase
PRD	proline-rich domain
PRK	PKC-related kinase
PrP	prion protein
α-PrP	PrP with high α-helix content
β-PrP	PrP with high β-sheet content
PrP^{*}	PrP with altered conformation
PrP^{c}	cellular PrP
PrP^{Sc}	disease-specific PrP
PS	phosphatidylserine
PSKH1	protein serine kinase H1
PTB	phosphotyrosine-binding domain
PTD	protein transduction domain
PX	phox homology
RabF	Rab-specific motif
RabSF	Rab-specificity-determining region
REP	Rab-escort protein
Rh	rhodamine
RNAi	RNA interference
ROCKII	Rho-associated, coiled-coil-containing protein kinase 2
R-SNARE	arginine-containing SNARE
SAO	Southeast Asian ovalocytosis
SAV	sucrase-isomaltase-carrying vesicle
SERCA	sarcoplasmic/endoplasmic-reticulum Ca^{2+}-ATPase
SH3	Src homology 3
SHP-2	Src-homology-2-domain-containing tyrosine phosphatase
siRNA	small intefering RNA
SNAP	soluble NSF-attachment protein
SNARE	soluble *N*-ethylmaleimide-sensitive fusion protein attachment protein receptor
SnS	sticks and stones
Snx1	sorting nexin 1
SOCC	store-operated calcium channel
SV	synaptic vesicle

SV40	simian virus 40
SVE	SV endocytosis
SYBL1	synaptobrevin-like 1
Tat	*trans*-activating transcription factor
TCR	T-cell receptor
TeNT	tetanus neurotoxin
tER	transitional ER
TGN	*trans*-Golgi network
TI-VAMP	toxin-insensitive VAMP
TM	transmembrane
$T_{1/2}$	midpoint transition temperature
$\Delta T_{1/2}$	endotherm width at mid-height
T_m	melting temperature
tPA	tissue-type plasminogen activator
TRPC1	transient receptor potential canonical 1
TSE	transmissible spongiform encephalopathy
t-SNARE	target SNARE
uPA	urokinase-type plasminogen activator
VAMP	vesicle-associated membrane protein
VIP36	vesicular integral protein 36
Vps	vacuolar protein sorting
VSV-G	G glycoprotein of the vesicular stomatitis virus
VTC	vesicular tubular cluster
WASP	Wiskott–Aldrich syndrome protein
WIPI49	WD40 repeat protein interacting with phosphoinositides of 49 kDa
YFP	yellow fluorescent protein
Ypt	yeast protein transport
YSK1	yeast Sps1/Ste20-related kinase 1
ZO-1	zona occludens 1 protein

Biochem. Soc. Symp. **72**, 1–13
(Printed in Great Britain)

1

The role of microtubules in transport between the endoplasmic reticulum and Golgi apparatus in mammalian cells

Krysten J. Palmer, Peter Watson and David J. Stephens[1]

Department of Biochemistry, University of Bristol, School of Medical Sciences, University Walk, Bristol BS8 1TD, U.K.

Abstract

The organization of intracellular compartments and the transfer of components between them are central to the correct functioning of mammalian cells. Proteins and lipids are transferred between compartments by the formation, movement and subsequent specific fusion of transport intermediates. These vesicles and membrane clusters must be coupled to the cytoskeleton and to motor proteins that drive motility. Anterograde ER (endoplasmic reticulum)-to-Golgi transport, and the converse step of retrograde traffic from the Golgi to the ER, are now known to involve coupling of membranes to the microtubule cytoskeleton. Here we shall discuss our current understanding of the mechanisms that link membrane traffic in the early secretory pathway to the microtubule cytoskeleton in mammalian cells. Recent data have also provided molecular detail of functional co-ordination of motor proteins to specify directionality, as well as mechanisms for regulating motor activity by protein phosphorylation.

Introduction

The intracellular organization of most cells is highly dependent upon the presence of a functional cytoskeleton. Actin, microtubule and intermediate filaments provide frameworks for generating tension, are plastic to allow highly organized changes in cell or organelle shape [1], and can be used for the positioning of intracellular organelles and the movement of intracellular components between

[1]To whom correspondence should be addressed (email david.stephens@bristol.ac.uk).

organelles [2–5]. From ground-breaking experiments using yeast genetics and biochemistry, we now know an extraordinary amount about the machinery involved in the transport of protein and lipid between the ER and Golgi [6,7]. Partly as a consequence of the very different organization of yeast and mammalian cells (for example, see [8]), little is known about the mechanisms used by animal cells to co-ordinate membrane traffic using the microtubule cytoskeleton. Key differences have been found in the way in which plants, animals and lower eukaryotes utilize cytoskeletal components. In the context of the secretory pathway, plants are much more dependent upon the actin cytoskeleton than on microtubules [9,10]. In mammalian cells, both have a significant role to play [5,11]. However, an intact microtubule network is not required for transport through the secretory pathway, but it does greatly enhance its efficiency [12,13]. Furthermore, it is likely to be of much more significance in highly organized cell types, such as neurons [14]. This review will focus on the role of the microtubule network in the control of intracellular membrane traffic in mammalian cells.

The organization of the early secretory pathway

Intracellular organization is determined in part by the polarity of the microtubule cytoskeleton. Rapidly growing plus-ends are generally located in the cell periphery with the more slowly growing minus-ends at a central, juxtanuclear location (Figure 1). One area that has been studied extensively is the positioning of the ER and Golgi apparatus in cells [11,15–18]. The ER is a large membrane-bound organelle that consists of an extensive network of interconnecting tubules and cisternae that spread peripherally along microtubules throughout the cytoplasm [19]. This membrane system is subdivided into regions that are covered with ribosomes (referred to as rough ER) and regions devoid of ribosomes (referred to as smooth ER). The rough ER is the site of translocation of secretory cargo across the membrane, the integration of transmembrane proteins into the membrane, the synthesis, modification and degradation of proteins, and lipid biogenesis. The smooth ER is primarily involved in calcium storage/release. Specialized regions of the smooth ER known as transitional elements, tER (transitional ER) or ER exit sites also exist, and are the sites of secretory cargo exit from the ER [20,21]. Proteins are concentrated into ER exit sites by the action of the COPII (coatamer protein II) coat complex (reviewed in [22]). Subsequent budding of cargo-containing vesicles is followed rapidly by their uncoating and accumulation into a VTC (vesicular tubular cluster) that acts as the transport intermediate for traffic to the Golgi [20,23–25]. This scheme is illustrated in Figure 1. The network of ER membranes is distributed throughout the cell [26]. ER membranes interact with microtubules in a number of ways and are pulled towards the cell periphery by the plus-ended motor, kinesin [27]. Microtubule tips have also been shown to push and pull the ER into networks [15,16,28]. One functional consequence of this is that there can be large distances (tens of microns) across which transport intermediates might have to travel, and it is in this respect that the microtubule cytoskeleton becomes central to the efficiency of membrane traffic between the ER and Golgi.

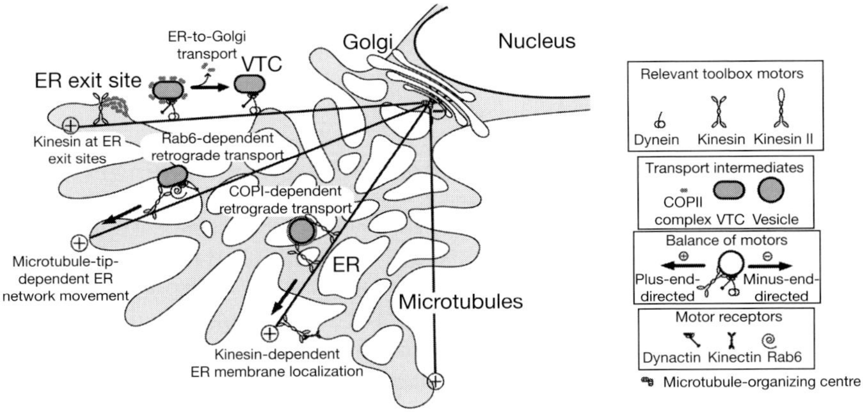

Figure 1 Schematic representation of microtubule-dependent membrane traffic pathways between the ER and Golgi. The anterograde transport of secreted proteins from the ER to the Golgi is initiated at ER exit sites to which kinesin is recruited [46]; subsequent transport of VTCs is dependent on the function of the dynein and dynactin complexes [25,50]. Two pathways of retrograde transport have been isolated, COPI-dependent [32,45] and Rab6-dependent [33,34], both of which almost certainly utilize the plus-end-directed motors of the kinesin family [45,106]. Rab6 recruits dynactin to membranes, which may control motility in this pathway [69]. Kinesin II has been implicated in transport of membranes, probably in the retrograde direction [47]. Microtubules and both plus- and minus-end-directed motors are also involved in the distribution and maintenance of the ER network throughout the cell [15] and Golgi apparatus at its characteristic juxtanuclear location [18]. Directionality of cargo transport is probably determined by a balance of motors attached to the cargo, with interacting proteins determining which motor is activated, allowing movement in the appropriate direction [4,91,94].

The Golgi occupies a central location in most mammalian cells, lying at the microtubule-organizing centre in mammalian cells and both its positioning and orientation (as it is a polarized structure) are controlled by the cytoskeleton [11,18,29]. The Golgi apparatus is a highly dynamic organelle that occupies a central position in the eukaryotic secretory pathway [18,30]. Nocodazole causes the disruption of microtubules and consequently the fragmentation of the Golgi apparatus [29]. These Golgi elements remain polarized and, following washout of nocodazole, translocate along microtubules towards minus-ends located at the centre of the cell [31]. The role of microtubules in Golgi organization and function is dealt with extensively in an excellent recent review [11] and is not considered further here.

Machinery molecules involved in the transport of cargo from the ER to the Golgi [such as SNAREs (soluble *N*-ethylmaleimide-sensitive fusion protein attachment protein receptor)] are recycled back to the ER [6] in a COPI (coatamer protein I)-dependent manner [32]. Other molecules utilize a second, COPI-independent, pathway [33,34]. This retrieval mechanism is also used to direct 'ER-resident' proteins back to the ER should they have exited with cargo. An ever-increasing amount of data suggests that both the anterograde (Golgi-directed) and this retrograde (retrieval) pathway use the microtubule cytoskeleton [25,35].

Microtubule-based motors: kinesin, dynein and dynactin

Molecular motor proteins have the specialized task of converting chemical energy from ATP hydrolysis into force and motion, which is used to drive a wide variety of cellular processes, including the unidirectional transport of molecules and organelles along cytoskeletal tracks. Three superfamilies of motor proteins have so far been identified: kinesins, the majority of which are plus-end-directed microtubule motors, dyneins, which are minus-end-directed microtubule motors, and myosins, which are actin-dependent motors [36,37]. Recent genomic and functional studies suggest that motor proteins in several organisms (fungi, parasites, plants and animals) stem from one of five cargo-carrying motors, the proposed 'toolbox motors' [36]. These proteins, which include three microtubule plus-end-directed kinesins (conventional kinesin, kinesin II and Unc104/KIF1A), cytoplasmic dynein and myosin V, have evolved through gene duplication, alternative splicing and the addition of associated subunits to provide different force-generating functions required with increasing cellular complexity [36]. This review will specifically detail those involved in the transport of membranes within the early secretory pathway of mammalian cells, and the reader is referred to other reviews for further details of motor protein mechanics and function (e.g. [14,38–40]).

Conventional kinesin [41], also referred to as kinesin I or KIF5, is found in many species of animals, and filamentous and fission fungi, but not in budding yeast [42]. This motor protein exists as a homodimer of heavy chains with two flanking light chains that appear to be absent in filamentous fungi [43]. The heavy chains are organized into distinct domains: the N-terminal motor domain, which contains the nucleotide- and microtubule-binding sites, the extended coiled-coil domain interrupted by a central hinge region, and the C-terminal globular tail domain thought to promote light chain and/or cargo binding (see [40] for a review of kinesin structure). The light chains have also been implicated in cargo binding [40], although one study suggests that they are unnecessary for this function [44]. Kinesin II (KIF3) is assembled from two distinct motor subunits that heterodimerize through charge interactions in an extended region of the coiled-coil stalk domain [40]. There is also an additional accessory subunit known as KAP (kinesin-associated protein) that is bound to the motor's C-terminal tail, and for this reason, the motor is also referred to as heterotrimeric kinesin II [40]. Kinesin is centrally involved in ER membrane movement [15] and in the transport of proteins from the Golgi to the ER [45], and has also been localized to ER exit sites [46] (Figure 1). Kinesin II has been localized by immunofluorescence and immunoelectron microscopy to KDEL (Lys-Asp-Glu-Leu)-receptor-containing transport intermediates in *Xenopus* cells [47].

Cytoplasmic dynein is a large multiprotein complex of 1.2 MDa, which drives motility towards the minus-ends of microtubules [48]. The complex comprises two identical heavy chains (DHCs), four light intermediate chains (DLICs), two intermediate chains (DICs) and a number of light chains (DLCs) with molecular masses of between 10 and 13 kDa (see [49] for a review). In humans, there are two DHCs (CD1 and CD2) which intriguingly can both be localized to the Golgi apparatus [50,51]. Multiple isoforms of these subunits exist, giving rise to multiple species of dynein, differing in their subunit composition [52]. This isoform diversity supports a key role for these subunits in dynein function, such as regulation of activity or cargo binding. The combinatorial presence or absence of these accessory subunits is likely to be central to the definition of cytoplasmic dynein function in cells.

Dynein-based motility is enhanced by a 1.2 MDa multiprotein protein complex called dynactin ('dynein activator') [53]. Dynactin comprises ten distinct polypeptide subunits, and contributes to cytoplasmic dynein function by acting as an 'adaptor', expanding the number and types of cargo dynein can transport and by enhancing motor processivity [54]. Attachment to dynein occurs through direct association between the N-terminus of DIC to the central region of the dynactin subunit $p150^{glued}$ [55]. The $p150^{glued}$, $p50^{dynamitin}$ and p24 subunits form a structurally distinct side-arm that binds strongly to microtubules via $p150^{glued}$ [56]. The remaining dynactin subunits form a filament-like structure that comprises the actin-related protein Arp1, conventional actin, α- and β-actin-capping proteins, and p62, p27 and p25 [57]. This filament, referred to as the Arp1 filament is thought to mediate attachment to membranes through interactions with spectrin [58]. Disrupting the dynein–dynactin interaction with specific inhibitory antibodies disrupts dynein-based motility [59]. Overexpression of individual dynactin components causes specific changes in cellular function that are well documented [60]. For example, overexpression of $p50^{dynamitin}$ disrupts microtubule organization [60] and mitosis [61], and causes Golgi fragmentation, as well as disrupting the localization of endosomes and lysosomes in interphase cells [59–61]. Overexpression of $p150^{glued}$ [60] results in dissociation of the $p150^{glued}$-containing side-arm from the remainder of the dynactin complex, uncoupling microtubule binding and cargo-attachment functions.

Recruitment to membranes

Clearly, the central point of co-ordination of membrane traffic by the cytoskeleton is the means of attachment of motors and their regulators, such as dynactin, to membranes [62]. Although motor proteins might associate with several membranous organelles and mediate a wide range of intracellular transport steps, the nature and identity of the motor–cargo interactions remains particularly elusive. In particular, the specificity of motor binding in both space and time is a highly complex process that must be tightly regulated. It is likely that, in many cases, both the recruitment and activity of motors will be coordinated. Associated accessory subunits can often mediate cargo interactions and cargo-bound 'receptor' proteins such as dynactin [62] or kinectin [63] could also mediate motor-protein attachment. This area is also of increasing interest in terms of target discovery in the pharmaceutical industry [64].

Rab proteins act as major co-ordinators of intracellular membrane traffic [65]. Considerable evidence now exists that Rabs are primary recruiters of motors to membranes [66]. Rab6 localizes to the cytoplasmic face of the Golgi, and is involved in both endosome-to-Golgi [67] and Golgi-to-ER transport [33,34,68]. Rab6 has been shown to regulate recruitment of dynactin to membranes (Figure 1), while other Golgi-localized Rabs do not [69]. The interaction between Rab6 and the dynein–dynactin complexes is regulated by BicD (Bicaudal D). BicD co-localizes and can be co-immunoprecipitated with dynein and dynactin, and has been shown to interact with the dynactin subunit $p50^{dynamitin}$ [70]. Furthermore, overexpression of BicD increases the recruitment of dynactin and dynein to Rab6-containing vesicles [71]. Overexpression of a C-terminal fragment of BicD that can bind to Rab6 (but not to the motor complex) inhibits the movement of Rab6-labelled vesicles [71]. BicD has been shown to bind only to the active form of Rab6, providing the basis for a molecular switch for the transport of activated Rab6-decorated vesicles [71].

Other polypeptides that associate with dynactin have also been shown to play a role in membrane traffic. Lis1 (lissencephaly 1) was initially identified as a non-catalytic subunit of brain platelet-activating factor acetylhydrolase [72], but has since been shown to interact with cytoplasmic dynein and dynactin [73,74]. Nudel (NudE-like), a mammalian homologue of the *Aspergillus* dynein regulator NudE, is functionally involved in the transport of proteins that contribute to the inactivation of the spindle checkpoint in mitosis [75,76]. Lis1 and Nudel appear to regulate cytoplasmic dynein in neuronal migration and mitosis through a direct interaction [77]. Expression of Nudel mutants unable to bind to either Lis1 or DHC1, or depletion of Nudel using RNA interference, cause fragmentation/dispersion of the Golgi cisternae, lysosomes and endosomes [77]. Dispersion of the typical juxtanuclear localization of the ER–Golgi intermediate compartment was also observed [77]. Other modulators of dynein function that may have a role in transport within the early secretory pathway have also been identified. ZW10 interacts with $p50^{dynamitin}$, and has recently been shown to play a role in membrane trafficking between the ER and the Golgi [78]. Specifically, ZW10 binds to the ER-localized t-SNARE (target SNARE), syntaxin 18, and its overexpression results in disorganization of the early secretory pathway and disruption of anterograde and retrograde transport [78].

Evidence for the role of motors in membrane traffic between the ER and Golgi

In addition to the role of the cytoskeleton in the positioning of the ER and Golgi, the movement of membrane-bound structures (vesicles and tubules) between the ER and Golgi has received much interest of late. There is now considerable evidence for a direct role for the microtubule cytoskeleton in the co-ordination of traffic between the ER and Golgi [11,79], and in the generation and maintenance of the steady-state localization of components of the pathway [18,29]. Live-cell imaging of the COPII complex has also shown that depolymerization of microtubules with nocodazole results in an enlargement of the size of COPII-coated ER exit sites and an inhibition of their mobility [80]. COPII-coated ER exit sites are positioned along microtubules in skeletal muscle fibres [81], and are reorganized upon muscle cell differentiation [82]. This microtubule-dependent reorganization results in the localization of ER exit sites to the minus-ends of microtubules in myotubes. Thus there is a clear interdependence of the sites of VTC formation and the microtubule cytoskeleton. A clear correlation also exists between ER exit sites and the microtubule cytoskeleton in fibroblast and epithelial cells (D.J. Stephens, unpublished observations; and Figure 2). COPII-coated ER exit sites accumulate in the juxtanuclear area (in direct proximity to Golgi membranes) following expression of Sar1p–GTP [83,84], and also intriguingly as cells proceed through the cell cycle [85].

Visualization of ER-to-Golgi transport by electron microscopy [23] and time-lapse microscopy of live cells [25] strongly implicates a role for microtubules in ER-to-Golgi transport. Live-cell imaging of GFP (green fluorescent protein)-tagged cargo molecules has shown that translocation of VTCs occurs along curvilinear tracks at speeds consistent with microtubule-based motility [24,25]. Inhibition of dynein-mediated transport by overexpression of the $p50^{dynamitin}$ subunit of dynactin inhibits the transport of these structures [25] by blocking recruitment of dynein to these membranes [50]. Somewhat surprisingly, kinesin is recruited to ER exit sites upon incubation with a GTP-restricted mutant of Sar1p [46], a small GTPase that initiates COPII complex assembly at ER exit sites [22]. Given the minus-end-directed destina-

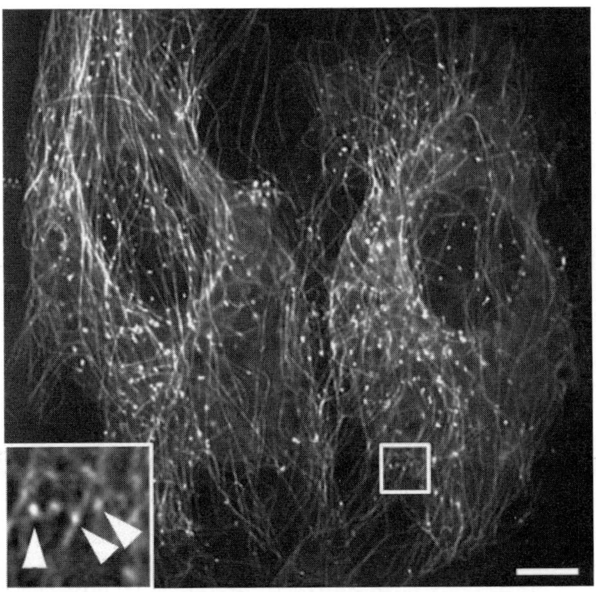

Figure 2 COPII-coated ER exit sites associate with microtubules in mammalian cells. PtK2 cells were transiently transfected with plasmids encoding YFP (yellow fluorescent protein)–Sec23Ap [107] and YFP–α-tubulin [108], and subsequently imaged using a spinning disk confocal microscope. Imaging of these two structures using the same fluorescent tag is possible due to their completely distinct intracellular morphology and localization. Note the clear association of punctate ER exit sites with microtubules. Scale bar, 10 μm.

tion of ER-to-Golgi transport carriers, this recruitment of kinesin may reflect a role in positioning of ER exit sites rather than a direct role in secretory cargo export. The role of motors in these processes is illustrated by Figure 1. The role of kinesins with respect to COPII [46] may be in the initial deformation of the ER membrane or organization of ER export by direct association of export sites with microtubules. Alternatively, or in addition, the recruitment of kinesin at this point may be related to its later function in retrograde (Golgi-to-ER) transport [45]. This would be consistent with a model in which both plus- and minus-end-directed motors are present on VTCs, with directionality of transport being controlled by regulation of these two motor populations.

Our working hypothesis is that a direct interaction between COPII-coated ER exit sites and microtubules co-ordinates the initial stages of ER-to-Golgi transport, establishing this functional link at an early stage prior to microtubule-based transport of membranes to the Golgi [86]. There are two principal modes that could be envisaged for a mechanism of ER exit site capture by microtubules. First, that the ends of microtubules associate with ER exit sites either during growth or shrinkage. Alternatively, that ER exit sites associate laterally with microtubules independent of changes in length. This second mechanism might act independently of the presence of dynactin at plus-ends, and may depend on recruitment of dynactin to membranes prior to its binding to microtubules. Intriguingly, a pool of p150^{glued} has recently been shown to be involved in capture of membranes during reformation of the Golgi apparatus following removal of brefeldin A from treated cells [86]. Brefeldin A causes a redistribution of proteins that are normally localized within Golgi cisternae to the ER [87]. This effect is reversible, allowing reformation of the Golgi by an ER-to-Golgi transport process. Vaughan et al. [86] observed that membrane clusters containing Golgi enzymes transiently associated with p150^{glued} at the plus-ends of microtubules during Golgi reformation after brefeldin A washout. This process was also shown to be regulated by phosphorylation of p150^{glued}, suggesting a mechanism for controlling capture and/or motility of VTCs. One can envisage that this could easily be linked to some quality-control mechanism operating to ensure productive VTC formation.

Directionality and regulation

Directionality is a vital concept in intracellular transport. Membranes (vesicles and/or tubules) must be linked to the microtubule, motor proteins must be recruited and their activity regulated. The regulation of motors is particularly important in the context of directionality of membrane traffic. A considerable amount of recent data has revealed the presence of multiple motors on intracellular structures, with the regulation of the balance between plus- and minus-end-directed motors resulting in directed transport [88–90]. Time-lapse-imaging experiments have revealed that many cargo proteins move bidirectionally along cytoskeletal tracks and frequently change their direction (e.g. see [24,25]). Yet, since motor proteins only move in a unidirectional manner either towards the plus- or minus-ends of microtubules, how is bidirectional transport achieved? What are the controlling factors that determine direction? Two models have been proposed: the 'tug-of-war' scenario and the 'motor co-ordination' model [89,91]. The first model, as the name suggests, involves a struggle between opposing motors bound to the same cargo molecule. The winning motor type determines the direction of transport, and regulation could involve altering force production, enzymic activity or changing the type and number of active motor proteins bound to the cargo. The second model relies on motor protein co-ordination [89], and in this instance, opposite-polarity motors are co-ordinated in such a way that they do not interfere with each other's function. Regulating the 'state' (on or off) that the motor proteins are in determines directionality simply by keeping one set of motors 'on' for longer. Data exist in favour of both of these models, and it is possible that both operate under different scenarios in different cell types. The co-ordination model is supported by work on *Drosophila* embryos in which the transport of lipid droplets was monitored following perturbation of minus-end-directed movement [89]. Somewhat surprisingly, plus-end-directed movement was also found to be inhibited. In the tug-of-war model, one would expect this mode to predominate, but instead this provides evidence in support of co-ordination of minus- and plus-end transport. Indeed, directionality determinants have been identified in this system [89,92,93]: dynactin has a role to play [89], as do other newly identified proteins such as klarsicht and halo [92,93]. Together these factors prevent the onset of a tug-of-war scenario through regulation of individual motor activity. Although much of this work relates to *Drosophila* motor function, much of it is directly relevant to mammalian systems.

In further support of motor co-ordination, GFP–dynein has been shown to remain bound to cargo even when that cargo switches direction [90]. Further work using *Xenopus* melanophores has provided the first physical evidence that dynactin, known to link cytoplasmic dynein to membranes, also connects kinesin II to cargo molecules [94]. This dual interaction was shown to occur through the direct binding of the $p150^{glued}$ dynactin subunit to either DIC or the accessory protein KAP [94]. Binding was also found to be exclusive and competitive, with both DIC and KAP associating to similar regions on the $p150^{glued}$ protein (residues 600–811). This suggests that the DIC/KAP-binding site on $p150^{glued}$ is probably central to the control of directionality. Since binding is competitive, these data could be seen to support the tug-of-war model, but in fact can equally be integrated into a motor co-ordination model. One simply needs to invoke the presence of additional directionality determinant acting in concert with $p150^{glued}$. It is likely that many of the mechanisms co-ordinating the activity of opposite polarity motors are applicable in other systems controlling organelle positioning as well as motility of transport intermediates in the early secretory pathway.

Regulation by phosphorylation

Mechanisms that are likely to regulate the activity of microtubule-based motor proteins include conformational changes, protein phosphorylation and isoform variability. Both the heavy and light chains of kinesin are phosphorylated *in vivo* on serine residues [95,96]; these phosphorylation events appear to regulate membrane association [97] and motor activity [98], but their precise significance remains unclear. Dynein and dynactin are also subject to regulation by phosphorylation [99–103]. Similarly, experiments using dynactin-null mutants of *Neurospora* have revealed dynactin-dependent phosphorylation of the DLCs, thereby regulating dynein ATPase activity [103]. Other *in vitro* assays have suggested that increased phosphorylation of DICs and DHCs results in an inhibition of ATPase activity [104]. Both DICs and $p150^{glued}$ have been found to be subject to cell-cycle regulation, becoming hyperphosphorylated during metaphase in HeLa cells [99]. This is likely to control the association of these complexes with membranes [102]. DIC phosphorylation has been implicated in the regulation of the all-important dynein–dynactin interaction [105]. Phosphorylation of DIC Ser^{84} has been shown (by site-directed mutagenesis) to diminish $p150^{glued}$ binding and thereby reduce the affinity of dynein for dynactin [105]. More recently, binding of the dynactin subunit $p150^{glued}$ to elongating microtubule plus-ends has been shown to be subject to regulation by phosphorylation [86]. A phospho-sensitive monoclonal antibody identified Ser^{19} as the site of $p150^{glued}$ phosphorylation. Stimulation of PKA (protein kinase A) with forskolin treatment promoted $p150^{glued}$ phosphorylation in COS-7 cells and reduce microtubule binding [86]. Phosphorylation site mutants further reinforced this with a Ser^{19}→Glu point mutant, mimicking phosphorylated $p150^{glued}$, showing diminished microtubule association, while the Ser^{19}→Ala mutant promoted microtubule binding. Live-cell imaging also revealed that a transient association of $p150^{glued}$-labelled microtubules with membranes occurred prior to minus-end-directed transport [86]. This led to a model in which phosphorylation of $p150^{glued}$ controls its association with microtubules: the non-phosphorylated form preventing motility by allowing microtubule binding (acting as a brake) with phosphorylation triggering release from the microtubule and subsequent dynein-mediated minus-end motility [86]. Furthermore, these data implicate dynactin (and specifically $p150^{glued}$) in a 'search and capture' mechanism [86], directing the initial stages of coupling of membranes to microtubules prior to minus-end-directed movement of Golgi-directed membranes. These data strongly suggest that dynactin is directly coupled to the membranes of the early secretory pathway and that its phosphorylation controls the efficient transport of intermediates between the ER and Golgi. The mechanisms defining recruitment of these motor complexes and accessory proteins to membranes of the early secretory pathway remain unknown.

Conclusions and perspectives

Our understanding of the molecular mechanisms that integrate membrane traffic with the cytoskeleton is increasing steadily. More is now known about the mechanisms coupling individual membranes to microtubules, but many questions still remain unanswered. Are there independent cargo adaptors for each membrane/vesicle type? How and when are these adaptors (such as dynactin) and motors recruited? How initiation of motility is coupled with the formation of productive transport carriers remains unclear. Do quality-control mechanisms ensure that all relevant cargo and machinery are included prior to the initiation of transport? Our knowledge of how motors are regulated (for example, by phosphorylation or subunit composition) now needs to be coupled with our knowledge of the organization and directionality of individual membrane traffic events.

We thank Jon Lane for helpful discussions and critical reading of the manuscript. Work in the Stephens laboratory is funded by the UK Medical Research Council through a Non-Clinical Research Career Development Fellowship and by the Wellcome Trust.

References

1. Sheetz, M.P. (2001) Nat. Rev. Mol. Cell Biol. **2**, 392–396
2. Helfand, B.T., Chang, L. and Goldman, R.D. (2003) Annu. Rev. Cell Dev. Biol. **19**, 445–467
3. Lane, J. and Allan, V. (1998) Biochim. Biophys. Acta **1376**, 27–55
4. Rogers, S.L. and Gelfand, V.I. (2000) Curr. Opin. Cell Biol. **12**, 57–62
5. Kamal, A. and Goldstein, L.S. (2000) Curr. Opin. Cell Biol. **12**, 503–508
6. Bonifacino, J.S. and Glick, B.S. (2004) Cell **116**, 153–166
7. Schekman, R. (2004) Nat. Cell Biol. **6**, 483–486
8. Rossanese, O.W., Soderholm, J., Bevis, B.J., Sears, I.B., O'Connor, J., Williamson, E.K. and Glick, B.S. (1999) J. Cell Biol. **145**, 69–81
9. Brandizzi, F., Snapp, E.L., Roberts, A.G., Lippincott-Schwartz, J. and Hawes, C. (2002) Plant Cell **14**, 1293–1309
10. Nebenfuhr, A., Frohlick, J.A. and Staehelin, L.A. (2000) Plant Physiol. **124**, 135–151
11. Allan, V.J., Thompson, H.M. and McNiven, M.A. (2002) Nat. Cell Biol. **4**, E236–E242
12. Stults, N.L., Fechheimer, M. and Cummings, R.D. (1989) J. Biol. Chem. **264**, 19956–19966
13. Kelly, R.B. (1990) Cell **61**, 5–7
14. Holzbaur, E.L. (2004) Trends Cell Biol. **14**, 233–240
15. Waterman-Storer, C.M. and Salmon, E.D. (1998) Curr. Biol. **8**, 798–806
16. Lane, J.D. and Allan, V.J. (1999) Mol. Biol. Cell **10**, 1909–1922
17. Allan, V.J. (1998) Methods Enzymol. **298**, 339–353
18. Rios, R.M. and Bornens, M. (2003) Curr. Opin. Cell Biol. **15**, 60–66
19. Voeltz, G.K., Rolls, M.M. and Rapoport, T.A. (2002) EMBO Rep. **3**, 944–950
20. Bannykh, S.I., Rowe, T. and Balch, W.E. (1996) J. Cell Biol. **135**, 19–35
21. Orci, L., Ravazzola, M., Meda, P., Holcomb, C., Moore, H.P., Hicke, L. and Schekman, R. (1991) Proc. Natl. Acad. Sci. U.S.A. **88**, 8611–8615
22. Barlowe, C. (2002) Curr. Opin. Cell Biol. **14**, 417–422
23. Saraste, J. and Svensson, K. (1991) J. Cell Sci. **100**, 415–430
24. Scales, S.J., Pepperkok, R. and Kreis, T.E. (1997) Cell **90**, 1137–1148
25. Presley, J.F., Cole, N.B., Schroer, T.A., Hirschberg, K., Zaal, K.J. and Lippincott-Schwartz, J. (1997) Nature (London) **389**, 81–85
26. Terasaki, M., Chen, L.B. and Fujiwara, K. (1986) J. Cell Biol. **103**, 1557–1568
27. Feiguin, F., Ferreira, A., Kosik, K.S. and Caceres, A. (1994) J. Cell Biol. **127**, 1021–1039
28. Allan, V. and Vale, R. (1994) J. Cell Sci. **107**, 1885–1897
29. Thyberg, J. and Moskalewski, S. (1985) Exp. Cell Res. **159**, 1–16
30. Lippincott-Schwartz, J., Roberts, T.H. and Hirschberg, K. (2000) Annu. Rev. Cell Dev. Biol. **16**, 557–589
31. Ho, W.C., Allan, V.J., van Meer, G., Berger, E.G. and Kreis, T.E. (1989) Eur. J. Cell Biol. **48**, 250–263
32. Letourneur, F., Gaynor, E.C., Hennecke, S., Demolliere, C., Duden, R., Emr, S.D., Riezman, H. and Cosson, P. (1994) Cell **79**, 1199–1207

33. Girod, A., Storrie, B., Simpson, J.C., Johannes, L., Goud, B., Roberts, L.M., Lord, J.M., Nilsson, T. and Pepperkok, R. (1999) Nat. Cell Biol. **1**, 423–430
34. White, J., Johannes, L., Mallard, F., Girod, A., Grill, S., Reinsch, S., Keller, P., Tzschaschel, B., Echard, A., Goud, B. and Stelzer, E.H. (1999) J. Cell Biol. **147**, 743–760
35. Lippincott-Schwartz, J. and Cole, N.B. (1995) Biochem. Soc. Trans. **23**, 544–548
36. Vale, R.D. (2003) Cell **112**, 467–480
37. Kamal, A. and Goldstein, L.S. (2002) Curr. Opin. Cell Biol. **14**, 63–68
38. Wozniak, M.J., Milner, R. and Allan, V. (2004) Traffic **5**, 400–410
39. Reilein, A.R., Rogers, S.L., Tuma, M.C. and Gelfand, V.I. (2001) Int. Rev. Cytol. **204**, 179–238
40. Hirokawa, N. (1998) Science **279**, 519–526
41. Vale, R.D., Reese, T.S. and Sheetz, M.P. (1985) Cell **42**, 39–50
42. Moore, J.D. and Endow, S.A. (1996) Bioessays **18**, 207–219
43. Bloom, G.S. and Endow, S.A. (1995) Protein Profile **2**, 1105–1171
44. Skoufias, D.A., Cole, D.G., Wedaman, K.P. and Scholey, J.M. (1994) J. Biol. Chem. **269**, 1477–1485
45. Lippincott-Schwartz, J., Cole, N.B., Marotta, A., Conrad, P.A. and Bloom, G.S. (1995) J. Cell Biol. **128**, 293–306
46. Aridor, M., Fish, K.N., Bannykh, S., Weissman, J., Roberts, T.H., Lippincott-Schwartz, J. and Balch, W.E. (2001) J. Cell Biol. **152**, 213–229
47. Le Bot, N., Antony, C., White, J., Karsenti, E. and Vernos, I. (1998) J. Cell Biol. **143**, 1559–1573
48. Schroer, T.A., Steuer, E.R. and Sheetz, M.P. (1989) Cell **56**, 937–946
49. Vallee, R.B., Vaughan, K.T. and Echeverri, C.J. (1995) Cold Spring Harb. Symp. Quant. Biol. **60**, 803–811
50. Roghi, C. and Allan, V.J. (1999) J. Cell Sci. **112**, 4673–4685
51. Tai, A.W., Chuang, J.Z. and Sung, C.H. (1998) J. Biol. Chem. **273**, 19639–19649
52. King, S.J., Bonilla, M., Rodgers, M.E. and Schroer, T.A. (2002) Protein Sci. **11**, 1239–1250
53. Gill, S.R., Schroer, T.A., Szilak, I., Steuer, E.R., Sheetz, M.P. and Cleveland, D.W. (1991) J. Cell Biol. **115**, 1639–1650
54. King, S.J. and Schroer, T.A. (2000) Nat. Cell Biol. **2**, 20–24
55. Vaughan, K.T. and Vallee, R.B. (1995) J. Cell Biol. **131**, 1507–1516
56. Holleran, E.A., Karki, S. and Holzbaur, E.L. (1998) Int. Rev. Cytol. **182**, 69–109
57. Schroer, T.A. (1996) Semin. Cell Dev. Biol. **7**, 321–328
58. Holleran, E.A., Tokito, M.K., Karki, S. and Holzbaur, E.L. (1996) J. Cell Biol. **135**, 1815–1829
59. Burkhardt, J.K., Echeverri, C.J., Nilsson, T. and Vallee, R.B. (1997) J. Cell Biol. **139**, 469–484
60. Quintyne, N.J., Gill, S.R., Eckley, D.M., Crego, C.L., Compton, D.A. and Schroer, T.A. (1999) J. Cell Biol. **147**, 321–334
61. Echeverri, C.J., Paschal, B.M., Vaughan, K.T. and Vallee, R.B. (1996) J. Cell Biol. **132**, 617–633
62. Karcher, R.L., Deacon, S.W. and Gelfand, V.I. (2002) Trends Cell Biol. **12**, 21–27
63. Blocker, A., Severin, F.F., Burkhardt, J.K., Bingham, J.B., Yu, H., Olivo, J.C., Schroer, T.A., Hyman, A.A. and Griffiths, G. (1997) J. Cell Biol. **137**, 113–129
64. Phelps, M.A., Foraker, A.B. and Swaan, P.W. (2003) Drug Discovery Today **8**, 494–502
65. Zerial, M. and McBride, H. (2001) Nat. Rev. Mol. Cell Biol. **2**, 107–117
66. Seabra, M.C. and Coudrier, E. (2004) Traffic **5**, 393–399
67. Mallard, F., Tang, B.L., Galli, T., Tenza, D., Saint-Pol, A., Yue, X., Antony, C., Hong, W., Goud, B. and Johannes, L. (2002) J. Cell Biol. **156**, 653–664
68. Martinez, O., Schmidt, A., Salamero, J., Hoflack, B., Roa, M. and Goud, B. (1994) J. Cell Biol. **127**, 1575–1588
69. Short, B., Preisinger, C., Schaletzky, J., Kopajtich, R. and Barr, F.A. (2002) Curr. Biol. **12**, 1792–1795
70. Hoogenraad, C.C., Akhmanova, A., Howell, S.A., Dortland, B.R., De Zeeuw, C.I., Willemsen, R., Visser, P., Grosveld, F. and Galjart, N. (2001) EMBO J. **20**, 4041–4054

71. Matanis, T., Akhmanova, A., Wulf, P., Del Nery, E., Weide, T., Stepanova, T., Galjart, N., Grosveld, F., Goud, B., De Zeeuw, C.I. et al. (2002) Nat. Cell Biol. **4**, 986–992
72. Hattori, M., Arai, H. and Inoue, K. (1993) J. Biol. Chem. **268**, 18748–18753
73. Smith, D.S., Niethammer, M., Ayala, R., Zhou, Y., Gambello, M.J., Wynshaw-Boris, A. and Tsai, L.H. (2000) Nat. Cell Biol. **2**, 767–775
74. Faulkner, N.E., Dujardin, D.L., Tai, C.Y., Vaughan, K.T., O'Connell, C.B., Wang, Y. and Vallee, R.B. (2000) Nat. Cell Biol. **2**, 784–791
75. Howell, B.J., McEwen, B.F., Canman, J.C., Hoffman, D.B., Farrar, E.M., Rieder, C.L. and Salmon, E.D. (2001) J. Cell Biol. **155**, 1159–1172
76. Yan, X., Li, F., Liang, Y., Shen, Y., Zhao, X., Huang, Q. and Zhu, X. (2003) Mol. Cell. Biol. **23**, 1239–1250
77. Liang, Y., Yu, W., Li, Y., Yang, Z., Yan, X., Huang, Q. and Zhu, X. (2004) J. Cell Biol. **164**, 557–566
78. Hirose, H., Arasaki, K., Dohmae, N., Takio, K., Hatsuzawa, K., Nagahama, M., Tani, K., Yamamoto, A., Tohyama, M. and Tagaya, M. (2004) EMBO J. **23**, 1267–1278
79. Mizuno, M. and Singer, S.J. (1994) J. Cell Sci. **107**, 1321–1331
80. Stephens, D.J., Lin-Marq, N., Pagano, A., Pepperkok, R. and Paccaud, J.P. (2000) J. Cell Sci. **113**, 2177–2185
81. Ralston, E., Ploug, T., Kalhovde, J. and Lomo, T. (2001) J. Neurosci. **21**, 875–883
82. Lu, Z., Joseph, D., Bugnard, E., Zaal, K.J. and Ralston, E. (2001) Mol. Biol. Cell **12**, 795–808
83. Sato, K. and Nakano, A. (2004) J. Biol. Chem. **279**, 1330–1335
84. Stephens, D.J. and Pepperkok, R. (2004) J. Cell Sci. **117**, 3635–3644
85. Stephens, D.J. (2003) EMBO Rep. **4**, 210–217
86. Vaughan, P.S., Miura, P., Henderson, M., Byrne, B. and Vaughan, K.T. (2002) J. Cell Biol. **158**, 305–319
87. Lippincott-Schwartz, J., Yuan, L.C., Bonifacino, J.S. and Klausner, R.D. (1989) Cell **56**, 801–813
88. Vale, R.D., Malik, F. and Brown, D. (1992) J. Cell Biol. **119**, 1589–1596
89. Gross, S.P., Welte, M.A., Block, S.M. and Wieschaus, E.F. (2002) J. Cell Biol. **156**, 715–724
90. Ma, S. and Chisholm, R.L. (2002) J. Cell Sci. **115**, 1453–1460
91. Gross, S.P. (2003) Curr. Biol. **13**, R320–R322
92. Welte, M.A., Gross, S.P., Postner, M., Block, S.M. and Wieschaus, E.F. (1998) Cell **92**, 547–557
93. Gross, S.P., Guo, Y., Martinez, J.E. and Welte, M.A. (2003) Curr. Biol. **13**, 1660–1668
94. Deacon, S.W., Serpinskaya, A.S., Vaughan, P.S., Lopez Fanarraga, M., Vernos, I., Vaughan, K.T. and Gelfand, V.I. (2003) J. Cell Biol. **160**, 297–301
95. Hollenbeck, P.J. (1993) J. Neurochem. **60**, 2265–2275
96. Sato-Yoshitake, R., Yorifuji, H., Inagaki, M. and Hirokawa, N. (1992) J. Biol. Chem. **267**, 23930–23936
97. Lee, K.D. and Hollenbeck, P.J. (1995) J. Biol. Chem. **270**, 5600–5605
98. McIlvain, Jr, J.M., Burkhardt, J.K., Hamm-Alvarez, S., Argon, Y. and Sheetz, M.P. (1994) J. Biol. Chem. **269**, 19176–19182
99. Huang, C.Y., Chang, C.P., Huang, C.L. and Ferrell, Jr, J.E. (1999) J. Biol. Chem. **274**, 14262–14269
100. Gill, S.R., Cleveland, D.W. and Schroer, T.A. (1994) Mol. Biol. Cell **5**, 645–654
101. Dillman, 3rd, J.F. and Pfister, K.K. (1994) J. Cell Biol. **127**, 1671–1681
102. Niclas, J., Allan, V.J. and Vale, R.D. (1996) J. Cell Biol. **133**, 585–593
103. Kumar, S., Lee, I.H. and Plamann, M. (2000) J. Biol. Chem. **275**, 31798–31804
104. Runnegar, M.T., Wei, X. and Hamm-Alvarez, S.F. (1999) Biochem. J. **342**, 1–6
105. Vaughan, P.S., Leszyk, J.D. and Vaughan, K.T. (2001) J. Biol. Chem. **276**, 26171–26179
106. Lippincott-Schwartz, J., Donaldson, J.G., Schweizer, A., Berger, E.G., Hauri, H.P., Yuan, L.C. and Klausner, R.D. (1990) Cell **60**, 821–836
107. Toomre, D., Keller, P., White, J., Olivo, J.C. and Simons, K. (1999) J. Cell Sci. **112**, 21–33

Biochem. Soc. Symp. **72**, 15–30
(Printed in Great Britain)

2

Kinases regulating Golgi apparatus structure and function

Christian Preisinger and Francis A. Barr[1]

Max-Planck Institute of Biochemistry, Am Klopferspitz 18, Martinsried 82152, Germany

Abstract

Protein kinases control Golgi function in both mitotic and interphase cells. In mitosis, phosphorylation of structural proteins by Cdk1 (cyclin-dependent kinase 1)–cyclin B, Polo-like and mitogen-activated protein kinases underlie changes in Golgi reorganization during cell division. While in interphase, signalling pathways that are associated with the Golgi control secretory function through a variety of mechanisms. Some of these, notably those involving protein kinase D and Ste20 family kinases, are also relevant for the establishment and maintenance of cell polarization and migration.

Introduction

Cell growth and division requires the synthesis and delivery of membranes and proteins to numerous destinations within the cell, and thus depends on the function of the secretory pathway. Membrane lipid and secretory protein biosynthesis is compartmentalized to the ER (endoplasmic reticulum) of eukaryotic cells, and the Golgi receives the secretory output of the ER for further post-translational modification and sorting of proteins and lipids [1]. The enzymes involved in post-translational modifications are spread throughout the Golgi, related to the order in which they act upon their substrates. Sorting functions, meanwhile, are compartmentalized at the CGN and TGN (*cis*- and *trans*-Golgi network respectively). ER and CGN sorting functions are generally associated with quality-control mechanisms for the correct assembly and folding of proteins [2], while TGN sorting is required for separation of cargo molecules destined for different cellular compartments, such as the plasma membrane and endocytic pathways [3]. Here we will discuss the kinases responsible for controlling Golgi function in mitotic and interphase

[1]To whom correspondence should be addressed (email barr@biochem.mpg.de).

mammalian cells (see Table 1), introducing relevant substrates and the effects that phosphorylation has on their function where appropriate.

M-phase regulation of Golgi structure

Cdk1 (cyclin-dependent kinase 1)–cyclin B — the master regulator of cell division

The mitotic kinase Cdk1–cyclin B exerts a key regulatory influence on both secretory-protein transport and Golgi structure in mitotic cells [4,5]. In mitosis, protein transport is blocked at the level of membrane fusion events required for delivery of secretory cargo to the Golgi from the ER, and a variety of target molecules for Cdk1–cyclin B have been identified at the CGN that may contribute to this block. Experiments using the mouse FT210 cell line with a temperature-sensitive allele of Cdk1 suggest that this kinase is one of the key regulators not only of mitotic entry, but also of Golgi fragmentation [6]. At the restrictive temperature, these cells arrest in G_2 phase with a normal ribbon-like array of Golgi stacks, although Plk1 (polo-like kinase 1) is partially activated [6,7], and then enter mitosis and disassemble the Golgi synchronously when placed at the permissive temperature for Cdk1 activity [6]. One of the first Golgi mitotic phosphoproteins to be identified as a Cdk1–cyclin B target was the small GTPase Rab1 [8], which is required for the recruitment of the cytosolic vesicle tethering factor p115 to ER-to-Golgi transport vesicles [9]. The second was another Rab1-binding partner: the large coiled-coil protein GM130 (Golgi matrix protein of 130 kDa) [10], a member of the large family of golgins that comprise the structural Golgi matrix [11]. The network of interactions between Rab1, p115 and GM130 is thought to play a pivotal role in the process of *cis*-Golgi recognition by ER-derived vesicles prior to the membrane-fusion step [12] and cisternal stacking [13,14]. Additionally, recent work implicates yet another Rab1 interacting protein, the transmembrane golgin, golgin-84, as a mitotic phosphoprotein with a function in Golgi organization and ER to Golgi traffic [15]. Phosphorylation of a serine residue in the N-terminus of GM130 prevents binding of p115 to the *cis*-Golgi cisternae, and, while Rab1 phosphorylation interferes with its cycle of membrane association, it has not been established if this has any effect on its interaction with either p115 or GM130 [8,10]. Regulation of interactions between these Rab1-binding proteins at the onset of mitosis might therefore explain the block of ER to Golgi traffic during cell division [16]. Interestingly, the major phosphoprotein of Golgi membranes in mitosis is another GM130-interacting protein, GRASP65 (Golgi reassembly stacking protein of 65 kDa), which is required for the reassembly of the Golgi on mitotic exit [13]. Cdk1–cyclin B and at least one other mitotic kinase phosphorylate GRASP65 [17], and this has been proposed to control a pathway leading to the establishment of stacked Golgi cisternae perhaps by modulating GRASP65 self-interaction [18].

Table 1 Golgi-associated kinases. AKAP, A-kinase-anchoring protein, DAG,diacylglycerol; PSKH1, protein serine kinase H1; PRK, PKC-related kinase

Protein kinase	Interaction partners	Substrates at the Golgi	Proposed functions at the Golgi	References
Cdk1	Cyclins B1 and B2	Rab1, GM130, GRASP65	Mitotic regulation	[8,12]
Plk1	–	GRASP65	Mitotic regulation	[17,18,21]
Plk3/Fnk/PRK	Giantin	–	Mitotic regulation, apoptosis	[25,26]
ERK1/2	–	GRASP55	Mitotic regulation	[19]
MEK1/MKK1	–	–	Mitotic regulation	[30–32]
CK2, CK2-like	AP-1	p115	Vesicle traffic, cargo sorting at the TGN	[55,58,59]
PKA	AKAP350, BIG2	KDELR	Vesicle traffic	[48,49,90]
PKCι/λ	Rab2	–	Vesicle traffic	[51,52,54]
PKD	Gβγ, DAG	–	Vesicle fission, cargo sorting at the TGN	[65,67–69]
PSKH1	–	–	–	[91,92]
Src	–	–	Ras activation, Golgi structure, KDELR transport	[60–62]
YSK1	GM130	14-3-3ζ	Cell migration	[75]
MST4	GM130	–	Cell migration	[75]
ROCKII	Citron-N	–	Golgi structure, actin reorganization	[86]
PAK4	Cdc42	–	Actin reorganization	[87,88]

Plks as modulators of Golgi disassembly

Several studies have proposed the involvement of both the mammalian Plk1 and the MAPK (mitogen-activated protein kinase) pathways in the phosphorylation of structural Golgi proteins [17,19]. Significantly, the two substrates identified are both members of the GRASP family, GRASP65 at the *cis*- and GRASP55 at the medial-Golgi [13,20]. A screen for substrates of Plk1 in mitosis identified GRASP65, and a subsequent report demonstrated that a dominant-negative form of Plk1 or depletion of Plk1 from mitotic extracts could inhibit Golgi fragmentation *in vitro* [21]. Introduction of various GRASP65 constructs to an *in vitro* mitotic Golgi fragmentation assay significantly reduced the initial phase of Golgi dispersal [21]. However, Plk1 alone was not sufficient to initiate mitotic Golgi disassembly, suggesting that Plk1 acts through additional components absent from this assay. Consideration of the complex regulatory mechanisms of Plks [22] suggests that one such component could be Cdk1–cyclin B. A key step in Plk activation is the binding of phosphorylated docking partners to its phosphopeptide-binding polo box domain, which shows a strong preference for a motif related to that phosphorylated by Cdk1–cyclin B [23,24]. Whether or not Plk1 has any docking partners at the Golgi is currently unknown, but since spatial regulation through docking proteins is thought to be critical for the action of Plk1 at other structures in mitosis, it seems possible that such proteins exist at the Golgi.

Another recent report has investigated the relative roles of Plk1 and Cdk1–cyclin B further [18]. Despite the fact that GM130 and GRASP65 are *in vitro* substrates of Cdk1–cyclin B and Plk1, micro-injection of a recombinant Cdk1–cyclin B complex into cells caused a dramatic increase in Golgi dispersal, whereas Plk1 did not stimulate fragmentation at all. In contrast with this, both kinases exhibit the same potential to unstack Golgi cisternae *in vitro* [18], and thus may be promoting an initial step prior to complete mitotic Golgi disassembly. A possible explanation for this comes from the observation that GRASP65 can form oligomers under control or interphasic conditions, and that this interaction is abolished by treatment with mitotic cytosol, Cdk1–cyclin B or Plk1 [18]. Thus, in this *in vitro* assay, Plk1 appears to be a modulator of Golgi disassembly, possibly through control of cisternal unstacking, while Cdk1–cyclin B is absolutely required for cisternal unstacking and further fragmentation.

Other Plks have been proposed to exert some control over Golgi structure. Plk3, a serine/threonine protein kinase closely related to Plk1, has been found to localize to the Golgi through an interaction with the 400-kDa coiled-coil golgin, giantin [25]. Overexpression of wild-type and constitutively active forms of Plk3 can induce Golgi fragmentation in interphase cells, whereas a kinase-dead mutant does not have any effect. Plk3 also interacts with ERK (extracellular-signal-regulated kinase) 2 and MEK1 (MAPK/ERK kinase 1), and these kinases appear to be needed upstream of Plk3 for its activation [26]. However, it is unclear if this reflects a role of Plk3 in controlling Golgi structure during mitosis or some other cellular function, since Plk3 has been reported to function in the control of apoptosis [27], a condition where the Golgi is also broken down into fragments [28]. Recently, the non-Plk family kinase MLK3 (mixed-lineage kinase) has been found to phosphorylate golgin-160 and to regulate its cleavage by caspases, suggesting that such regulation of the Golgi does occur during apoptosis [29].

MAPKs and the initiation of Golgi fragmentation

As mentioned previously, MAPKs can phosphorylate Golgi structural proteins of the GRASP family [19]. Work from the Malhotra group has promoted the idea that, besides Cdk1–cyclin B and Plk1, MEK1 is a key regulator of early mitotic Golgi breakdown [30,31]. In a recent publication, the activation of MEK1 by Raf1 was found to be required for mitotic Golgi dispersal in prophase [32]. Consistent with this, activated MEK1 was present on Golgi membranes in prophase, but not at later stages of mitosis. The regulatory pathway upstream of MEK1 may involve Raf1, since the auto-inhibitory domain of Raf1 causes a decrease in MEK1 activity, as well as a block in mitotic Golgi dissociation that could be reversed by the application of constitutively active MEK1 [32]. The same effects were also observed following the application of Raf1 kinase inhibitory protein, which inhibits Raf1-dependent MEK1 activation, and also Golgi complex fragmentation. Since the specific MEK1 antagonist PD98059, which binds to inactive MEK1 and prevents activation by upstream regulators such as Raf1, and depletion of Raf1 by siRNA (small interfering RNA) reduce the level of Cdk1–cyclin B activity (although not Plk1 activity) and prevent entry of cells into mitosis, these results therefore suggest that MEK1 may play a general role in mitotic entry, as well as a specific role in the initial stages of Golgi fragmentation.

In contrast with these observations, other publications give a different view regarding the necessity of MEK activity for Golgi disassembly in mitosis [12,33]. Depletion or inactivation of MEK1 had no effect on Golgi fragmentation in an *in vitro* assay, while micro-injection of Cdk1–cyclin B complexes into mammalian cells caused the same amount of Golgi fragmentation, both in the presence or absence of the MEK1 inhibitors U0126 and PD98059. It was also reported that inactivation of MEK1 had no effect on the entry of mammalian cells into mitosis or on Golgi fragmentation [12,33], and it was therefore concluded that MEK1 activity was not required for efficient Golgi fragmentation at the onset of mitosis. Exactly why these differences were observed is unclear. The simplest explanation amenable to further study is that they may be due to technical differences, such as the concentrations of drugs or the cell types used. Another possibility is that MEK2 or other kinases take over some of the functions of MEK1, as proposed in an analysis of *mek1*$^{-/-}$ mice that did not show defects prior to mid-gestation [34]. Further investigation will be needed to clarify these issues.

A direct link suggesting a role for MAPK signalling in Golgi fragmentation is the phosphorylation of GRASP55 [19]. ERK2, as well as JNK (c-Jun N-terminal kinase), has been reported to phosphorylate the Golgi matrix protein GRASP55 *in vitro*, generating an epitope for the anti-mitosis-specific phosphoprotein monoclonal antibody, MPM2 [19]. GRASP55 phosphorylation was inhibited by the application of U0126 both *in vivo* and *in vitro*, suggesting that MEK1 is acting upstream of ERK2. Furthermore, an antibody specific for ERK singly phosphorylated at Tyr^{185} stained the Golgi as well as the nucleus during late G_2 and early prophase, similar to activated MEK1 [32,35]. Overexpression of an activated mutant of MEK1, which in turn activates ERK1/2, caused a partial disruption of the Golgi. Surprisingly, this effect did not appear to require the kinase activity of ERK, but instead the presence of the phosphorylated tyrosine residue [35]. These authors therefore proposed that tyrosine phosphorylated ERK might have a structural role in organization of protein complexes that are important for protein trafficking and Golgi structure. How these findings relate to those of the Malhotra group is unclear [30,31], since cleavage of MEK1 by the anthrax lethal factor uncoupling it from its cognate MAPK substrates ERK1/2 [36,37] does not prevent Golgi fragmentation [12,31]. Additional substrates must therefore be crucial for the action of MEK1.

An organelle checkpoint in mitotic entry and Golgi fragmentation

In order to ensure high-fidelity inheritance of the genome, cells have evolved a mechanism, termed the spindle checkpoint, that ensures that progression through mitosis is coupled with attachment of chromosomes to the mitotic spindle and subsequent segregation. Sutterlin et al. [38] asked the provocative question of whether such a mechanism exists for organelles, specifically the Golgi. Micro-injection of an antibody against either GRASP65 or a C-terminal fragment of GRASP65 was able to firstly block Golgi fragmentation, and secondly to delay entry of the cells into mitosis [38]. These authors hypothesized that these effects were due to competing out of binding partners for GRASP65. In the first case, this may be self-interaction of GRASP65 itself, due to its role in Golgi structure [13,18], while, in the second case, it may be an unknown partner that contributes to the control of mitotic entry. Why should the Golgi serve as a trigger for proper entry into mitosis? One possibility is that such a checkpoint mechanism is needed to ensure that both daughter cells receive a complete Golgi after cytokinesis. On the other hand, the Golgi might serve as a scaffold for proteins that are required for regulating mitotic entry, which could be transported by Golgi fragments to their sites of action, or released from the Golgi and thus become activated. In this context, it is interesting to note that the kinase Myt1, a key negative regulator of Cdk1, localizes to the ER and Golgi in interphase mammalian cells [39,40].

Multiple kinases co-operate to determine the fate of the Golgi in mitosis

As a result of these findings, the models for Golgi fragmentation have to be revised to incorporate multiple kinases acting on multiple substrates (Figure 1). In summary, it appears that Cdk1–cyclin B is required to sustain Golgi fragmentation, while Plk1 and MEK1 may be required for the initial breakdown of the Golgi, perhaps at the cisternal unstacking stage [4]. However, it seems that none of these kinases is acting alone, and co-operation between them is crucial for regulating the Golgi during mitosis. Whether or not they are all essential regulators or whether they simply control the rate of disassembly requires further investigation. The initial fragmentation stage may also feed into some form of organelle checkpoint pathway, and one tempting speculation is that Plk1 is active in this process, since it plays an important role in the complex regulatory

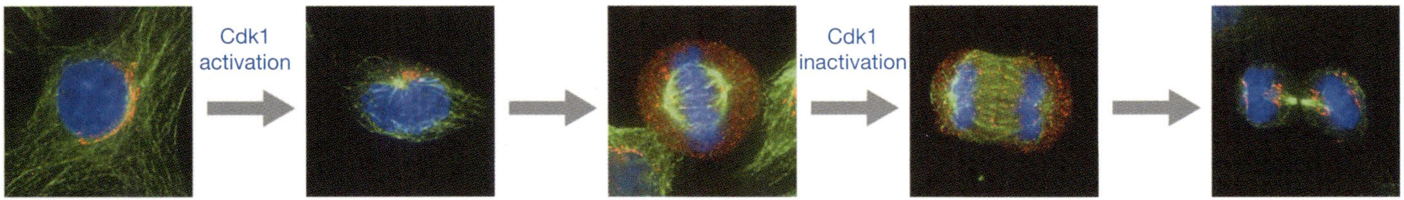

Figure 1 M-phase regulation of the Golgi apparatus. In interphase mammalian cells, the Golgi is found collected in the perinuclear region around the centrosome. The ribbon-like structure characteristic of interphase cells is formed from an end-to-end array of Golgi stacks. At the onset of mitosis in G_2 and prophase, this ribbon-like organization is lost, and Golgi cisternae start to unstack and fragment, probably due to the concerted action of Cdk1–cyclin B, Plk1 and MEK1. At this time, exocytic protein transport is stopped. This fragmentation process continues until metaphase, by which time the Golgi cisternae have been reduced to dispersed small vesicles and some larger tubulovesicular clusters. At the metaphase-to-anaphase transition, Cdk1–cyclin B is inactivated through cyclin B degradation; however, Plk1 remains active. During anaphase and into telophase, as Plk1 is inactivated and degraded, functional Golgi stacks reassemble and exocytic protein transport resumes. Kinases implicated in the various stages of Golgi disassembly are listed at the appropriate points. The images show cells at the indicated stages of the cell cycle stained for the Golgi matrix protein GM130 (red), microtubules (green) and DNA (blue).

network controlling activation of Cdk1–cyclin B, through the phosphatase Cdc25 (cell-division control 25) and direct phosphorylation of cyclin B [22].

Kinases regulating interphase Golgi function

PKA (protein kinase A)

PKA or cAMP-dependent protein kinase has been reported to play a number of roles in the regulation of protein traffic, and here we will focus on a few studies related to the Golgi. It has been known for some time that a pool of the PKA regulatory subunit (RII) can be detected at the Golgi, both at the CGN and TGN [41,42], and this latter pool has recently been proposed to play a role in traffic between endosomes and the Golgi [43]. PKA inhibitors, such as H89, can inhibit secretory protein transport at the level of the Golgi [44], while PKA activators stimulate transport from the TGN to the apical, but not basolateral, plasma membrane in polarized epithelial cells [45]. This inhibition may be at the level of vesicle budding from the TGN [46], although there is some evidence that PKA can regulate the recruitment of the small GTPase ARF1 (ADP-ribosylation factor 1) to Golgi membranes [47], which would indicate that it could play a general role in transport throughout the Golgi.

PKA also seems to control the traffic of specific proteins such as the KDELR [KDEL (Lys-Asp-Glu-Leu) receptor] by direct phosphorylation [48]. Mutation of the phosphorylated Ser^{209} to alanine reduces the efficiency of KDELR recycling between the ER and the Golgi, and results in the secretion of KDEL-bearing reporter proteins into the culture medium [48]. Significantly, the cytoplasmic tail peptide of the KDELR can bind to the COPI (coatamer protein I) coat more efficiently when this residue is mutated to aspartate to mimic phosphorylation [48]. This is consistent with the observations that PKA can regulate recruitment of ARF1, another component of the COPI coat [47]. Intriguingly, BIG2 (brefeldin-A-inhibited guanine-nucleotide-exchange factor 2), one of the Golgi-associated brefeldin-A-sensitive ARF exchange factors, can interact with various PKA regulatory subunits [49], providing another link between PKA and regulation of Golgi-associated coat proteins.

PKC (protein kinase C)

The small GTPase Rab2 plays an important role in the regulation of transport between the Golgi and ER [50], most likely by promoting the formation of COPI-coated transport intermediates enriched in recycling cargo such as p53/58 and lacking anterograde cargo [51]. One of its effector proteins in this context is PKCι/λ, which is recruited to vesicular tubular transport intermediates through a direct interaction with the first 19 amino acids of Rab2 [52]. Inhibition of PKCι/λ activity in an *in vitro* vesicle-formation assay with a pseudosubstrate peptide leads to the accumulation of dilated vesicle structures containing recycling cargo, Rab2, and the vesicle coat COPI [53]. Another binding partner and one of the substrates for PKCι/λ is GAPDH (glyceraldehyde-3-phosphate dehydrogenase) [54]. Rab2 can inhibit this phosphorylation event [52], and, in addition, appears to regulate the membrane recruitment of

GAPDH [54]. GAPDH is also associated with the cytoskeleton, and it has therefore been proposed that, by modulating its membrane recruitment and phosphorylation state, Rab2 controls microtubule dynamics important for retrograde transport events in the early secretory pathway [54].

CKs (casein kinases)

As discussed above, the tethering of ER to Golgi transport vesicles is mediated by the cytosolic factor p115 in conjunction with Rab1 and GM130. p115 is phosphorylated in its C-terminal domain by CK2 or a CK2-like enzyme, and this promotes its membrane binding and interaction with GM130 [55]. This regulation by CK2 is also needed for the reassembly of the Golgi at the end of mitosis, at least *in vitro*, where it enhances the p115-dependent interaction of the Golgi matrix proteins GM130 and giantin [55]. Whether or not this CK2-like enzyme is an essential component of the transport machinery and how it is regulated is unknown.

In addition to these functions, involvement of CK-like activities has also been reported in protein-sorting events at the TGN. Perhaps the best understood example is the role of CK2 in MPR (mannose 6-phosphate receptor) sorting, and we will therefore focus on this here. Two types of sorting adaptor, AP-1 (adaptor protein 1) and the GGAs (Golgi-localized, γ-ear-containing, ARF-binding proteins) [56,57], co-localize in clathrin-coated buds enriched in the MPR at the TGN. Biochemical studies indicate that purified AP-1 contains an associated CK2 activity that phosphorylates GGA1 and GGA3 and this inhibits their MPR-binding activity [58]. This CK2 phosphorylation event may be important for the transfer of the MPR from GGAs to AP-1 and hence efficient sorting into vesicles [58]. An additional layer of regulation comes from CK2 phosphorylation of the MPR cytoplasmic domain, which is involved in the high-affinity binding of AP-1 to membranes [59].

Src family kinases

Ras GTPases are important regulators of cell growth and differentiation, typically thought to act through activation of the protein kinase Raf1 at the plasma membrane. It is therefore of considerable interest that a pool of activated H- and N-Ras is generated at the Golgi in mitogenic signalling [60]. Activation of Ras at the Golgi requires the tyrosine kinase Src, a small pool of which can also be found at the Golgi [60]. Src appears to be mediating these effects through activation of phospholipase Cγ, and downstream of this translocation of the Ras exchange factor, RasGRP1, to the Golgi [61]. The pool of activated Ras at the Golgi appears to be biologically significant, since, in assays for Ras activity, it was able to drive transformation of mouse fibroblasts, mediate radio-resistance of RIE-1 cells, and promote neuronal differentiation of PC12 cells [61]. Exactly why activation of Ras occurs at the Golgi, and the biological functions mediated by endogenous Ras at the Golgi still remain to be determined, but are clearly exciting areas for future research.

In addition to these effects on Ras signalling, Src family kinases appear to be important for other Golgi functions. In cells deleted for the three Src family kinases (Src, Fyn and Yes), the morphology of the Golgi is significantly altered, cisternae become swollen and the long-range organization of Golgi stacks is lost [62]. This phenotype is specifically due to the loss of Src, since it can be rescued by the expression of Src alone [62]. Furthermore, while expression of activated Src does not alter anterograde secretory traffic, it does cause a redistribution of the KDELR [62], which is recycling between the ER and Golgi. Together, these findings suggest that events at the Golgi can be modulated by extracellular signals, via pathways involving Src and Ras activation.

Golgi function in cell polarization

PKD (protein kinase D) is a key regulator at the TGN

Trimeric G-protein βγ-subunits have been known for some time to regulate vesicle formation events at the TGN [63,64], and, subsequently, it was shown that this is in part through the direct activation of PKD [65]. The group of Malhotra then went on to demonstrate that binding of PKD to diacylglycerol through its first cysteine-rich domain is necessary for its recruitment to the TGN [66,67], and for the scission of TGN-derived transport carriers destined for the plasma membrane in non-polarized cells [67,68].

The sorting and transport of proteins and lipids to specific plasma membrane domains is an important aspect of cell polarization, and a recent report has identified PKD as a key component required for efficient transport to the basolateral plasma membrane of polarized epithelial cells [69]. Whereas PKD1 had been shown previously to be required for protein transport from the TGN to the plasma membrane in non-polarized cells [65,68], the roles of the other PKD family members PKD2 and PKD3 in protein transport from the TGN to the cell surface had not been evaluated. Surprisingly, in polarized MDCK (Madin–Darby canine kidney) cells transport to the apical membrane measured using a variety of marker proteins was not influenced by expression of either wild-type or kinase-dead dominant-negative forms of PKD1 or PKD2. However, the dominant-negative forms of PKD1 and PKD2, but not PKD3, caused a dramatic decrease in the transport of the exogenous marker VSV-G (G glycoprotein of the vesicular stomatitis virus), as well as the endogenous markers β1-integrin and E-cadherin to the basolateral plasma membrane [69]. Interestingly, PKD2 caused an accumulation of basolateral cargo in the TGN, while PKD1 caused it to accumulate in cytoplasmic vesicles. A possible explanation for this observation is the different effects that these two PKDs have on localization of the exocyst complex member Sec6, which is important for basolateral traffic at the stage of docking with the plasma membrane [70,71]. Overexpression of dominant-negative PKD1 caused an accumulation of Sec6 in the TGN, whereas this was not observed with PKD2 [69]. This suggests that PKD1 and 2 are involved in regulating transport of a subset of proteins with basolateral sorting information from the TGN to the basolateral membrane. PKD3 also appears to have a function in transport out of the TGN, but the relevant pathway in polarized cells has not yet been identified. The PKD family members therefore appear to function at the TGN in signalling events needed for vesicle scission and docking, cargo sorting and transport pathways relevant for establishment and maintenance of the polarized cell phenotype.

PKD and cell motility

Directional cell movement is a complex process that requires the delivery of membrane to the leading edge of the cell, and this is thought to involve polarization of the Golgi in the direction of cell migration [72]. The delivery of membrane to the leading edge of the cell from the Golgi has been proposed to promote a retrograde flow at the plasma membrane in migrating cells. A recent study has investigated the role of TGN to plasma membrane traffic in this membrane-delivery event, using dominant-negative PKD1 to block transport out of the TGN [73] (Figure 2). As found previously, dominant-negative PKD1 causes tubulation of the TGN, as well as a block in VSV-G transport, and, furthermore, cell motility was reduced by a factor of two. However, if mutant forms of PKD1 unable to bind to membranes were used, VSV-G transport and cell motility were not affected [73]. This block in transport results in a cessation of retrograde membrane flow at the leading edge, which the authors postulate underlies the motility defect. Whether or not this reflects a role of PKD1 in polarized membrane delivery to the leading edge, perhaps similar to the function in basolateral delivery in polarized epithelial cells [69], was not determined. A transport block exerted by a dominant-negative form of the GTPase Arf1 had similar effects on migration [73], and this is expected to inhibit transport prior to sorting and exit from the TGN [74].

Ste20 kinases and signalling at the Golgi in cell migration

Polarization of the Golgi accompanies changes in cell polarity and migration in response to extracellular signals [3,72], and we have recently investigated the role of Golgi-associated signalling pathways in these events in human cells [75]. Kinases of the Ste20 family were first identified in yeast as mutants unable to undergo cell polarization and hence mating. Two members of the human Ste20 family, YSK1 (yeast Sps1/Ste20-related kinase 1) and MST4 (mammalian Ste20 4), localize to the Golgi, via an interaction with the Golgi matrix protein GM130 [75] (Figure 2A). This is reminiscent of the assembly of signalling complexes at the plasma membrane, since GM130 is part of a complex with GRASP65 that contains a membrane-targeting PDZ-like domain [76]. Furthermore, the interaction of YSK1 and MST4 with GM130 leads to activation of these kinases. A variety of experiments interfering with the function of these kinases revealed that they are required for both cell migration and polarization of the Golgi in the direction of migration in cultured fibroblasts [75]. In the case of YSK1, 14-3-3ζ was identified as a specific substrate localized at the Golgi [75] (Figure 2B). The family of 14-3-3 proteins regulate the activity and localization of many different proteins involved in diverse cellular processes [77]. In the context of YSK1 signalling at the Golgi, 14-3-3 functions may include the quality control of multimeric membrane protein complexes at the *cis*-Golgi [78,79], phosphorylation-dependent integrin regulation [80,81], and the control of cell-polarity complexes including Par3/Baz [82]. It may also be relevant for the MAPK-controlled Golgi fragmentation, since 14-3-3 pro-

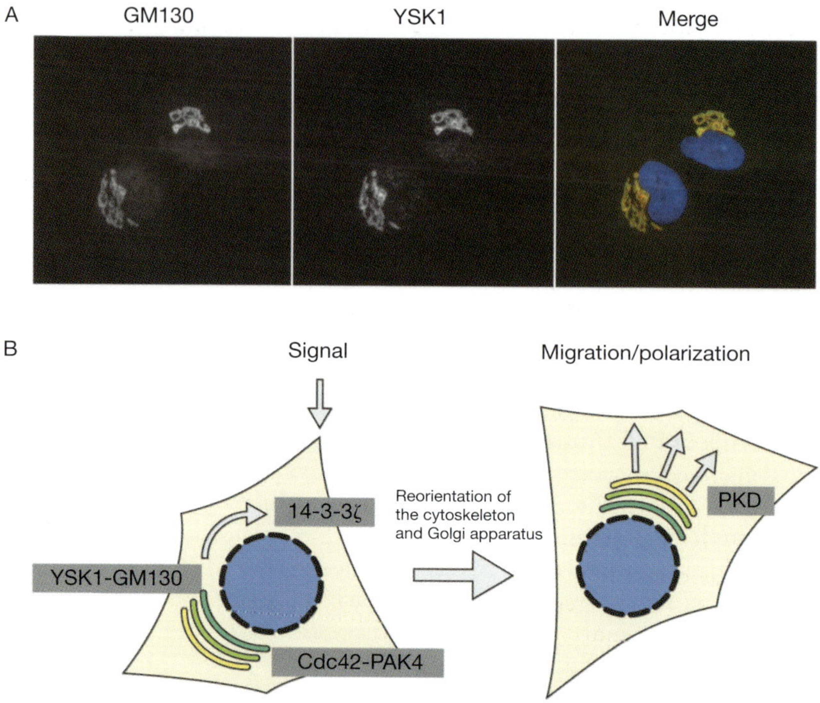

Figure 2 Golgi-associated kinases, cell polarization and migration. (A) Golgi association of the Ste20 family kinase YSK1 (red), which co-localizes with the *cis*-Golgi matrix protein GM130 (green). DNA stained with DAPI (4,6-diamidino-2-phenylindole) is shown in blue in the merged image. (**B**) Cells receive directional inputs from the extracellular environment that lead to the activation of signalling pathways that are important for cell polarization and migration. At the Golgi, the Ste20 family kinases YSK1 and MST4 (only YSK1 is depicted for simplicity) bind to and are activated by the Golgi matrix protein GM130. Activated YSK1 phosphorylates 14-3-3ζ, and possibly other substrates important for cell migration and polarization [75]. PAK4 is recruited to the Golgi by activated Cdc42 and regulates the actin cytoskeleton [87]. At the TGN, PKD family members contribute to the polarized delivery of specific cargo molecules to subdomains of the plasma membrane. In migrating cells, this delivery is required to establish a retrograde membrane flux at the cell surface needed for normal cell migration and polarization [73].

teins interact with Raf1 [83,84], which has been implicated in the control of MEK1 at the Golgi [32]. Interestingly, 14-3-3 proteins can be regulated by dimerization [85], and 14-3-3ζ is phosphorylated by YSK1 in the region reported to be important for dimerization, providing a potential regulatory mechanism. Although this is a promising new area of kinase function at the Golgi, exactly which signalling pathways lie upstream of these Golgi-associated Ste20 kinases and their physiological functions, for example in all or only specific types of cell polarization and migration, remain to be uncovered.

Kinase regulation of actin at the Golgi

Evidence for a link between the Golgi and regulation of the actin cytoskeleton comes from studies on Citron-N in hippocampal neurons [86], and the PAK (p21-activated kinase)-related kinase PAK4 [87,88]. PAK4 is a binding partner for the Rho-family GTPase Cdc42, and activated mutants of Cdc42 trigger the recruitment of PAK4 to the Golgi [87]. Activated PAK4 can be detected at the Golgi with phospho-specific antibodies, and it causes the formation of filopodia by inducing actin polymerization [87] and transforms NIH3T3 cells [88]. A dominant-negative form of PAK4 has the opposite effect and can block Ras-induced cell transformation [88]. PAK4 therefore provides an exciting link between the Golgi, cell growth and morphogenesis.

The Citron gene encodes a kinase required for the control of cytokinesis in some tissues, and a brain-specific alternative transcript Citron-N. Citron-N is a Rho-binding protein localized at the Golgi of cultured hippocampal neurons [86]. Knockdown of Citron-N using antisense oligonucleotides or expression of a dominant-negative form unable to bind Rho causes dispersal of the Golgi, while expression of the wild-type protein protects it from changes induced by actin-depolymerizing drugs [86]. An explanation for these results comes from the findings that the C-terminal domain of Citron-N is a scaffold for a number of proteins linked with regulation of actin including the Rho kinase ROCKII (Rho-associated, coiled-coil-containing protein kinase 2), LIM (Lin-11, Isl-1 and Mec-3) kinase, profillin IIa, and $p116^{Rip}$ [86,89]. Citron-N therefore appears to provide a link between the Golgi and actin cytoskeleton in neurons.

Outlook and future directions

As we have discussed, there is mounting evidence for the regulation of protein-transport events at the Golgi and Golgi structure by protein kinases in both interphase and mitotic cells. However, as the recent discoveries related to PKD and Ste20 kinase function in cell polarization and migration show, there are probably still many aspects of kinase function at the Golgi to be uncovered. One of the key areas for further investigation is the need for the systematic identification of which substrates are being modified. This is clearly necessary for a fuller understanding of how kinases regulate Golgi structure and function. Another outstanding issue reflecting the current convergence of cell and developmental biology concerns the limitations of cell-biological approaches to kinase function. It is likely that Golgi-associated signalling pathways might have specific functions in specific tissues, and this is obviously an exciting area for future research.

We thank Graham Warren, Jonathon Pines and Martin Lowe for their comments and useful discussion, and Ben Short and Evelyn Fuchs for carefully reading the manuscript. We apologize to our colleagues whose work could not be discussed for lack of space. The Max-Planck Society and the EMBO Young Investigator Programme support research in the laboratory of F.A.B.

References

1. Farquhar, M.G. and Palade, G.E. (1998) Trends Cell Biol. **8**, 2–10
2. Ellgaard, L. and Helenius, A. (2003) Nat. Rev. Mol. Cell Biol. **4**, 181–191
3. Nelson, W.J. and Yeaman, C. (2001) Trends Cell Biol. **11**, 483–486
4. Colanzi, A., Suetterlin, C. and Malhotra, V. (2003) Curr. Opin. Cell Biol. **15**, 462–467
5. Shorter, J. and Warren, G. (2002) Annu. Rev. Cell Dev. Biol. **18**, 379–420
6. Misteli, T. and Warren, G. (1995) J. Cell Sci. **108**, 2715–2727
7. Liu, X., Yan, S., Zhou, T., Terada, Y. and Erikson, R.L. (2004) Oncogene **23**, 763–776
8. Bailly, E., McCaffrey, M., Touchot, N., Zahraoui, A., Goud, B. and Bornens, M. (1991) Nature (London) **350**, 715–718
9. Allan, B.B., Moyer, B.D. and Balch, W.E. (2000) Science **289**, 444–448
10. Nakamura, N., Lowe, M., Levine, T.P., Rabouille, C. and Warren, G. (1997) Cell **89**, 445–455
11. Barr, F.A. and Short, B. (2003) Curr. Opin. Cell Biol. **15**, 405–413
12. Lowe, M., Rabouille, C., Nakamura, N., Watson, R., Jackman, M., Jamsa, E., Rahman, D., Pappin, D.J. and Warren, G. (1998) Cell **94**, 783–793
13. Barr, F.A., Puype, M., Vandekerckhove, J. and Warren, G. (1997) Cell **91**, 253–262
14. Shorter, J. and Warren, G. (1999) J. Cell Biol. **146**, 57–70
15. Diao, A., Rahman, D., Pappin, D.J., Lucocq, J. and Lowe, M. (2003) J. Cell Biol. **160**, 201–212
16. Lowe, M., Nakamura, N. and Warren, G. (1998) Trends Cell Biol. **8**, 40–44
17. Lin, C.Y., Madsen, M.L., Yarm, F.R., Jang, Y.J., Liu, X. and Erikson, R.L. (2000) Proc. Natl. Acad. Sci. U.S.A. **97**, 12589–12594
18. Wang, Y., Seemann, J., Pypaert, M., Shorter, J. and Warren, G. (2003) EMBO J. **22**, 3279–3290
19. Jesch, S.A., Lewis, T.S., Ahn, N.G. and Linstedt, A.D. (2001) Mol. Biol. Cell **12**, 1811–1817
20. Shorter, J., Watson, R., Giannakou, M.E., Clarke, M., Warren, G. and Barr, F.A. (1999) EMBO J. **18**, 4949–4960
21. Sutterlin, C., Lin, C.Y., Feng, Y., Ferris, D.K., Erikson, R.L. and Malhotra, V. (2001) Proc. Natl. Acad. Sci. U.S.A. **98**, 9128–9132
22. Barr, F.A., Sillјé, H.W. and Nigg, E.A. (2004) Nat. Rev. Mol. Cell Biol. **5**, 429–440
23. Elia, A.E., Cantley, L.C. and Yaffe, M.B. (2003) Science **299**, 1228–1231
24. Elia, A.E., Rellos, P., Haire, L.F., Chao, J.W., Ivins, F.J., Hoepker, K., Mohammad, D., Cantley, L.C., Smerdon, S.J. and Yaffe, M.B. (2003) Cell **115**, 83–95
25. Ruan, Q., Wang, Q., Xie, S., Fang, Y., Darzynkiewicz, Z., Guan, K., Jhanwar-Uniyal, M. and Dai, W. (2004) Exp. Cell Res. **294**, 51–59
26. Xie, S., Wang, Q., Ruan, Q., Liu, T., Jhanwar-Uniyal, M., Guan, K. and Dai, W. (2004) Oncogene **23**, 3822–3829
27. Xie, S., Wu, H., Wang, Q., Cogswell, J.P., Husain, I., Conn, C., Stambrook, P., Jhanwar-Uniyal, M. and Dai, W. (2001) J. Biol. Chem. **276**, 43305–43312
28. Lane, J.D., Lucocq, J., Pryde, J., Barr, F.A., Woodman, P.G., Allan, V.J. and Lowe, M. (2002) J. Cell Biol. **156**, 495–509
29. Cha, H., Smith, B.L., Gallo, K., Machamer, C.E. and Shapiro, P. (2004) J. Cell Sci. **117**, 751–760
30. Acharya, U., Mallabiabarrena, A., Acharya, J.K. and Malhotra, V. (1998) Cell **92**, 183–192
31. Colanzi, A., Deerinck, T.J., Ellisman, M.H. and Malhotra, V. (2000) J. Cell Biol. **149**, 331–339
32. Colanzi, A., Sutterlin, C. and Malhotra, V. (2003) J. Cell Biol. **161**, 27–32
33. Draviam, V.M., Orrechia, S., Lowe, M., Pardi, R. and Pines, J. (2001) J. Cell Biol. **152**, 945–958
34. Giroux, S., Tremblay, M., Bernard, D., Cardin-Girard, J.F., Aubry, S., Larouche, L., Rousseau, S., Huot, J., Landry, J., Jeannotte, L. and Charron, J. (1999) Curr. Biol. **9**, 369–372
35. Cha, H. and Shapiro, P. (2001) J. Cell Biol. **153**, 1355–1367
36. Duesbery, N.S., Webb, C.P., Leppla, S.H., Gordon, V.M., Klimpel, K.R., Copeland, T.D., Ahn, N.G., Oskarsson, M.K., Fukasawa, K., Paull, K.D. and Vande Woude, G.F. (1998) Science **280**, 734–737
37. Bardwell, A.J., Abdollahi, M. and Bardwell, L. (2004) Biochem. J. **378**, 569–577

38. Sutterlin, C., Hsu, P., Mallabiabarrena, A. and Malhotra, V. (2002) Cell **109**, 359–369
39. Liu, F., Stanton, J.J., Wu, Z. and Piwnica-Worms, H. (1997) Mol. Cell. Biol. **17**, 571–583
40. Liu, F., Rothblum-Oviatt, C., Ryan, C.E. and Piwnica-Worms, H. (1999) Mol. Cell. Biol. **19**, 5113–5123
41. Griffiths, G., Hollinshead, R., Hemmings, B.A. and Nigg, E.A. (1990) J. Cell Sci. **96**, 691–703
42. Martin, M.E., Hidalgo, J., Vega, F.M. and Velasco, A. (1999) J. Cell Sci. **112**, 3869–3878
43. Birkeli, K.A., Llorente, A., Torgersen, M.L., Keryer, G., Tasken, K. and Sandvig, K. (2003) J. Biol. Chem. **278**, 1991–1997
44. Muniz, M., Alonso, M., Hidalgo, J. and Velasco, A. (1996) J. Biol. Chem. **271**, 30935–30941
45. Pimplikar, S.W., Ikonen, E. and Simons, K. (1994) J. Cell Biol. **125**, 1025–1035
46. Muniz, M., Martin, M.E., Hidalgo, J. and Velasco, A. (1997) Proc. Natl. Acad. Sci. U.S.A. **94**, 14461–14466
47. Martin, M.E., Hidalgo, J., Rosa, J.L., Crottet, P. and Velasco, A. (2000) J. Biol. Chem. **275**, 19050–19059
48. Cabrera, M., Muniz, M., Hidalgo, J., Vega, L., Martin, M.E. and Velasco, A. (2003) Mol. Biol. Cell **14**, 4114–4125
49. Li, H., Adamik, R., Pacheco-Rodriguez, G., Moss, J. and Vaughan, M. (2003) Proc. Natl. Acad. Sci. U.S.A. **100**, 1627–1632
50. Tisdale, E.J. and Balch, W.E. (1996) J. Biol. Chem. **271**, 29372–29379
51. Tisdale, E.J. (2000) Traffic **1**, 702–712
52. Tisdale, E.J. (2003) J. Biol. Chem. **278**, 52524–52530
53. Tisdale, E.J., Wang, J., Silver, R.B. and Artalejo, C.R. (2003) J. Biol. Chem. **278**, 38015–38021
54. Tisdale, E.J. (2002) J. Biol. Chem. **277**, 3334–3341
55. Dirac-Svejstrup, A.B., Shorter, J., Waters, M.G. and Warren, G. (2000) J. Cell Biol. **150**, 475–488
56. Robinson, M.S. and Bonifacino, J.S. (2001) Curr. Opin. Cell Biol. **13**, 444–453
57. Bonifacino, J.S. (2004) Nat. Rev. Mol. Cell Biol. **5**, 23–32
58. Doray, B., Ghosh, P., Griffith, J., Geuze, H.J. and Kornfeld, S. (2002) Science **297**, 1700–1703
59. Mauxion, F., Le Borgne, R., Munier-Lehmann, H. and Hoflack, B. (1996) J. Biol. Chem. **271**, 2171–2178
60. Chiu, V.K., Bivona, T., Hach, A., Sajous, J.B., Silletti, J., Wiener, H., Johnson, 2nd, R.L., Cox, A.D. and Philips, M.R. (2002) Nat. Cell Biol. **4**, 343–350
61. Bivona, T.G., Perez De Castro, I., Ahearn, I.M., Grana, T.M., Chiu, V.K., Lockyer, P.J., Cullen, P.J., Pellicer, A., Cox, A.D. and Philips, M.R. (2003) Nature (London) **424**, 694–698
62. Bard, F., Mazelin, L., Pechoux-Longin, C., Malhotra, V. and Jurdic, P. (2003) J. Biol. Chem. **278**, 46601–46606
63. Barr, F.A., Leyte, A., Mollner, S., Pfeuffer, T., Tooze, S.A. and Huttner, W.B. (1991) FEBS Lett. **294**, 239–243
64. Jamora, C., Takizawa, P.A., Zaarour, R.F., Denesvre, C., Faulkner, D.J. and Malhotra, V. (1997) Cell **91**, 617–626
65. Jamora, C., Yamanouye, N., Van Lint, J., Laudenslager, J., Vandenheede, J.R., Faulkner, D.J. and Malhotra, V. (1999) Cell **98**, 59–68
66. Maeda, Y., Beznoussenko, G.V., Van Lint, J., Mironov, A.A. and Malhotra, V. (2001) EMBO J. **20**, 5982–5990
67. Baron, C.L. and Malhotra, V. (2002) Science **295**, 325–328
68. Liljedahl, M., Maeda, Y., Colanzi, A., Ayala, I., Van Lint, J. and Malhotra, V. (2001) Cell **104**, 409–420
69. Yeaman, C., Ayala, M.I., Wright, J.R., Bard, F., Bossard, C., Ang, A., Maeda, Y., Seufferlein, T., Mellman, I., Nelson, W.J. and Malhotra, V. (2004) Nat. Cell Biol. **6**, 106–112
70. Lipschutz, J.H., Guo, W., O'Brien, L.E., Nguyen, Y.H., Novick, P. and Mostov, K.E. (2000) Mol. Biol. Cell **11**, 4259–4275
71. Novick, P. and Guo, W. (2002) Trends Cell Biol. **12**, 247–249

72. Ridley, A.J., Schwartz, M.A., Burridge, K., Firtel, R.A., Ginsberg, M.H., Borisy, G., Parsons, J.T. and Horwitz, A.R. (2003) Science **302**, 1704–1709
73. Prigozhina, N.L. and Waterman-Storer, C.M. (2004) Curr. Biol. **14**, 88–98
74. Ward, T.H., Polishchuk, R.S., Caplan, S., Hirschberg, K. and Lippincott-Schwartz, J. (2001) J. Cell Biol. **155**, 557–570
75. Preisinger, C., Short, B., De Corte, V., Bruyneel, E., Haas, A., Kopajtich, R., Gettemans, J. and Barr, F.A. (2004) J. Cell Biol. **164**, 1009–1020
76. Barr, F.A., Nakamura, N. and Warren, G. (1998) EMBO J. **17**, 3258–3268
77. Dougherty, M.K. and Morrison, D.K. (2004) J. Cell Sci. **117**, 1875–1884
78. O'Kelly, I., Butler, M.H., Zilberberg, N. and Goldstein, S.A. (2002) Cell **111**, 577–588
79. Yuan, H., Michelsen, K. and Schwappach, B. (2003) Curr. Biol. **13**, 638–646
80. Han, D.C., Rodriguez, L.G. and Guan, J.L. (2001) Oncogene **20**, 346–357
81. Santoro, M.M., Gaudino, G. and Marchisio, P.C. (2003) Dev. Cell **5**, 257–271
82. Hurd, T.W., Fan, S., Liu, C.J., Kweon, H.K., Hakansson, K. and Margolis, B. (2003) Curr. Biol. **13**, 2082–2090
83. Fantl, W.J., Muslin, A.J., Kikuchi, A., Martin, J.A., MacNicol, A.M., Gross, R.W. and Williams, L.T. (1994) Nature (London) **371**, 612–614
84. Freed, E., Symons, M., Macdonald, S.G., McCormick, F. and Ruggieri, R. (1994) Science **265**, 1713–1716
85. Tzivion, G., Luo, Z. and Avruch, J. (1998) Nature (London) **394**, 88–92
86. Camera, P., da Silva, J.S., Griffiths, G., Giuffrida, M.G., Ferrara, L., Schubert, V., Imarisio, S., Silengo, L., Dotti, C.G. and Di Cunto, F. (2003) Nat. Cell Biol. **5**, 1071–1078
87. Abo, A., Qu, J., Cammarano, M.S., Dan, C., Fritsch, A., Baud, V., Belisle, B. and Minden, A. (1998) EMBO J. **17**, 6527–6540
88. Callow, M.G., Clairvoyant, F., Zhu, S., Schryver, B., Whyte, D.B., Bischoff, J.R., Jallal, B. and Smeal, T. (2002) J. Biol. Chem. **277**, 550–558
89. Rosso, S., Bollati, F., Bisbal, M., Peretti, D., Sumi, T., Nakamura, T., Quiroga, S., Ferreira, A. and Caceres, A. (2004) Mol. Biol. Cell **15**, 3433–3449
90. Larocca, M.C., Shanks, R.A., Tian, L., Nelson, D.L., Stewart, D.M. and Goldenring, J.R. (2004) Mol. Biol. Cell **15**, 2771–2781
91. Brede, G., Solheim, J., Stang, E. and Prydz, H. (2003) Exp. Cell Res. **291**, 299–312
92. Brede, G., Solheim, J., Troen, G. and Prydz, H. (2000) Genomics **70**, 82–92

Biochem. Soc. Symp. **72**, 31–38
(Printed in Great Britain)

3

Phosphoinositides and membrane traffic at the *trans*-Golgi network

Rawshan R. Choudhury, Noora Hyvola and Martin Lowe[1]

School of Biological Sciences, University of Manchester, The Michael Smith Building, Oxford Road, Manchester, M13 9PT, U.K

Abstract

Cargo proteins moving along the secretory pathway are sorted at the TGN (*trans*-Golgi network) into distinct carriers for delivery to the plasma membrane or endosomes. Recent studies in yeast and mammals have shown that formation of these carriers is regulated by PtdIns(4)*P*. This phosphoinositide is abundant at the TGN and acts to recruit components required for carrier formation to the membrane. Other phosphoinositides are also present on the TGN, but the extent to which they regulate trafficking is less clear. Further characterization of phosphoinositide kinases and phosphatases together with identification of new TGN-associated phosphoinositide-binding proteins will reveal the extent to which different phosphoinositides regulate TGN trafficking, and help define the molecular mechanisms involved.

Introduction

The TGN (*trans*-Golgi network) is the final compartment encountered by cargo proteins as they transit the Golgi apparatus [1,2]. A major function of the TGN is to segregate cargo into distinct carriers for delivery to different destinations. Proteins are constitutively delivered to the plasma membrane in pleiomorphic tubulosaccular structures that extend from the TGN and pinch off before motor-driven translocation to the cell surface, where they undergo fusion. In polarized cells, distinct tubulosaccular carriers mediate delivery of proteins to the apical and basolateral membranes respectively. Proteins destined for delivery to the endosome/lysosome system are instead packaged into TGN-derived clathrin-coated intermediates that translocate to the peripheral cytoplasm where they fuse with endosomes. In regulated secretory cells, a

[1]To whom correspondence should be addressed (email martin.lowe@man.ac.uk)

fourth pathway exists in which cargo is packaged into secretory granules that fuse with the plasma membrane upon stimulation. There is accumulating evidence that formation of these different TGN carriers is regulated by phosphoinositides. These lipids exist in multiple forms that are phosphorylated at different positions within the inositol headgroup. Different phosphoinositide species are specifically recognized by modular binding domains in proteins to facilitate their recruitment and sometimes activation at the appropriate membrane [3]. In this way, phosphoinositides can regulate membrane identity and function. In this review, we will discuss the role of phosphoinositides in regulating trafficking from the TGN, with particular emphasis on PtdIns(4)*P*, which regulates traffic to both the plasma membrane and endosomes (see Figure 1).

PtdIns(4)*P* and trafficking from the TGN to the plasma membrane

The first indication that phosphoinositides could regulate trafficking at the TGN came from analysis of the PITP (phosphatidylinositol-transfer protein) Sec14 in yeast [4]. The ability of Sec14 to deliver PtdIns to Golgi membranes was found to be critical for secretion, although the underlying mechanisms were not known. It now appears that PtdIns can influence trafficking by acting as a precursor for PtdIns(4)*P*. Yeast possess a single Golgi-localized PI4K (phosphoinositide 4-kinase), Pik1, and mutation of this protein reduces cellular PtdIns(4)*P* levels and inhibits TGN to plasma membrane traffic [5–7]. Pik1 overexpression gives the opposite results, and this can be antagonized by Sac1, a phosphatase that hydrolyses PtdIns(4)*P* [as well as PtdIns(3)*P*, PtdIns(5)*P* and PtdIns(3,5)P_2] to PtdIns [8]. Together, these findings indicate a key role for PtdIns(4)*P* in TGN-to-plasma-membrane transport. This is likely to be a direct effect and not due to PtdIns(4)*P* acting as a precursor for PtdIns(4,5)P_2, since the only yeast PI4P5K (phosphoinositide 4-phosphate 5-kinase), Mss4, is located at the plasma membrane and Mss4 mutants have normal secretion [9].

Mammals possess several PI4Ks. The mammalian homologue of Pik1 type-III PI4Kβ is recruited to the Golgi apparatus by the small GTPase ARF (ADP-ribosylation factor), suggesting a role in membrane traffic [10]. Overexpression of PI4Kβ inhibits transport from the TGN to both the apical and basolateral plasma membranes, whereas expression of a kinase-dead form inhibits TGN-to-basolateral delivery while stimulating TGN-to-apical-membrane transport [11]. Interestingly, it has recently been shown that PI4Kβ can recruit Rab11 to the Golgi apparatus, suggesting that it may regulate trafficking not only by modulation of PtdIns(4)*P* levels, but also through Rab11 recruitment [12]. In line with this, expression of a mutant PI4Kβ defective in Rab11 binding inhibits TGN-to-plasma-membrane traffic. The major PI4K at the mammalian Golgi apparatus is type-II PI4Kα [13]. Depletion of this kinase by RNA interference significantly reduces Golgi PtdIns(4)*P* levels, abolishes binding of Golgi-associated PtdIns(4)*P*-binding proteins (see below), and inhibits TGN to cell-surface transport [13]. Another potential regulator of PtdIns(4)*P* synthesis at the TGN is PKD (protein kinase D). This kinase is

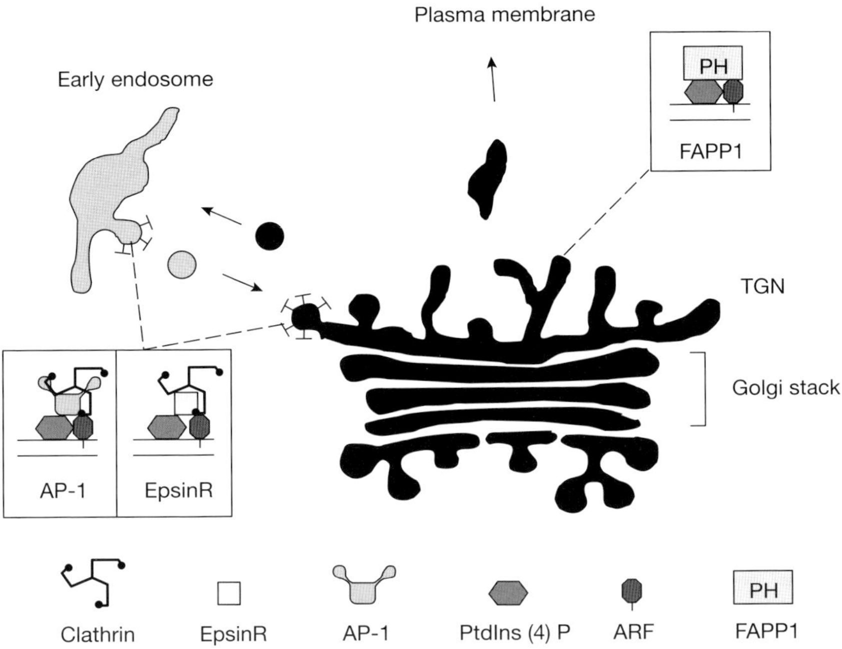

Figure 1 PtdIns(4)*P* and transport-carrier formation at the TGN. Formation of tubulosaccular carriers destined for the plasma membrane is regulated by FAPP1, which is recruited to the membrane via binding of its PH domain to PtdIns(4)*P* and membrane-associated ARF. Once on the membrane, FAPP1 is thought to regulate scission of transport carriers. Other related proteins (FAPP2, OSBP and GPBP) may regulate scission of different carriers destined for the plasma membrane. Budding of clathrin-coated vesicles from the TGN and early endosomes requires the accessory proteins AP-1 and epsinR. Both proteins are recruited to the membrane through association with PtdIns(4)*P* and ARF. AP-1 binds to cytoplasmic tails of cargo proteins for sorting into transport intermediates, and, together with epsinR, promotes assembly of the clathrin coat.

localized to specific domains of the TGN and regulates formation of constitutive TGN to plasma membrane carriers [14]. PKD can associate with a type II PI4K and a PI4P5K, and this association is promoted by PKD activity, suggesting that PKD may regulate levels of PtdIns(4)*P* [and PtdIns(4,5)P_2] during trafficking from the TGN [15].

How does PtdIns(4)*P* regulate TGN to plasma membrane traffic? Several PtdIns(4)*P*-binding proteins have been localized to the Golgi apparatus. FAPP (four-phosphate-adaptor protein) 1 and 2 are recruited to the TGN by a PH (pleckstrin homology) domain that interacts with both PtdIns(4)*P* and ARF [16]. Depletion of FAPPs inhibits delivery of cargo from the TGN to the cell surface, indicating a key role for these proteins in this transport step. Interestingly, overexpression of the FAPP PH domain appears to inhibit fission of TGN tubules, suggesting that FAPPs may regulate the formation of

TGN-derived carriers. They may do this directly, perhaps acting in a manner analogous to vesicle coat proteins, or indirectly, by influencing the availability of PtdIns(4)*P* for other enzymes or fission-promoting proteins. OSBP (oxysterol-binding protein) and GPBP (Goodpasture antigen-binding protein) also contain PtdIns(4)*P*-binding PH domains and are localized to the Golgi apparatus [17,18]. The function of these proteins is unclear, but a role in carrier formation appears likely. Together with FAPPs, they may comprise a family that regulates the formation of distinct classes of TGN-derived carriers.

PtdIns(4)*P* and trafficking from the TGN to endosomes

PtdIns(4)*P* also regulates transport from the TGN to endosomes. The first indication of this came when Pik1 mutant strains were found to have a reduced rate of delivery of proteins from the TGN to the pre-vacuolar compartment and the vacuole (the yeast equivalent of endosomes and the lysosome in animals) [6,7]. Further evidence came from studies on the yeast synaptojanin-like protein Sjl3/Inp53, which has both Sac1-like and PtdIns(4,5)P_2 5-phosphatase domains [19,20]. Both activities are required for proper delivery of proteins from the TGN to the pre-vacuolar compartment and vacuole, suggesting that PtdIns(4)*P* is involved. Sjl3/Inp53 binds to clathrin, further supporting a role for this protein in clathrin-mediated trafficking at the TGN. PtdIns(4)*P* effectors that regulate clathrin budding at the TGN have recently been identified. The tetrameric clathrin adaptor AP-1 (adaptor protein 1) mediates cargo selection and clathrin-coat assembly at the TGN, and is targeted to the membrane by association with both PtdIns(4)*P* and ARF [13]. Similarly, the AP-1 accessory protein epsinR also has a dual requirement for PtdIns(4)*P* and ARF for membrane binding [21–24]. Thus PtdIns(4)*P* plays a direct role in budding of clathrin-coated vesicles from the TGN. Recent studies suggest it may perform a similar function on endosomes, since both AP-1 and epsinR have been localized to clathrin buds there, and both are required for efficient retrograde trafficking from endosomes to the TGN [25–30].

PtdIns(4,5)P_2 at the TGN

Labelling of yeast with a PH domain specific for PtdIns(4,5)P_2 suggests that this phosphoinositide is present at the Golgi apparatus at very low levels, consistent with the lack of a Golgi-associated PI4P5K in these cells [31]. Similarly, mammalian cells also appear to have low steady-state levels of Golgi-associated PtdIns(4,5)P_2 [32]. However, PI4P5K activity has been measured in Golgi membranes isolated from mammalian cells, and, interestingly, this activity is stimulated by ARF, suggesting a role in membrane traffic [33,34]. A role for PtdIns(4,5)P_2 in Golgi trafficking is supported further by the finding that inhibition of TGN to cell-surface transport upon PI4KIIα depletion is rescued by shuttling PtdIns(4,5)P_2 into the cells [13]. The mechanism by which PtdIns(4,5)P_2 regulates this step is unknown, despite a number of potential PtdIns(4,5)P_2 effectors having been identified. PtdIns(4,5)P_2 is a potent stimulator of PLD (phospholipase D), a

Golgi-associated enzyme that catalyses the production of PA (phosphatidic acid) [35]. PA may promote membrane fission by physically inducing membrane curvature. Alternatively, it may be converted into DAG (diacylglycerol), which stimulates TGN-to-plasma-membrane trafficking through activation of PKD [36]. PA is also an activator of PI4P5K activity, giving rise to a potential positive-feedback loop for PtdIns(4,5)P_2 synthesis [37,38]. This, combined with the ability of PtdIns(4,5)P_2 to regulate members of the ARF GAP (GTPase-activating protein) and GEF (guanine nucleotide-exchange factor) families and thereby regulate ARF activation [39], might allow for a very rapid, but tightly controlled, production of PtdIns(4,5)P_2 at localized sites on the TGN. Other potential PtdIns(4,5)P_2 effectors include dynamin 2, which has been implicated in budding of TGN-derived clathrin-coated vesicles, and Golgi spectrin, which may regulate ARF-mediated trafficking and/or Golgi architecture via the actin cytoskeleton [10,40]. Further characterization of these proteins and identification of new Golgi-associated PtdIns(4,5)P_2-binding proteins will be required to understand the role of this phosphoinositide in trafficking at the TGN.

Several phosphoinositide 5-phosphatases have been described at the mammalian Golgi apparatus, but only one has a marked preference for PtdIns(4,5)P_2. Mutations in this protein, called OCRL1, are responsible for the rare X-linked disorder OCRL (oculocerebrorenal syndrome of Lowe), characterized by defects in the brain, kidney and eye [41,42]. OCRL1 is present at the TGN [43, 44], and one study also suggests a lysosomal localization [45]. Cells from OCRL patients have elevated levels of PtdIns(4,5)P_2, suggesting that changes in phosphoinositide levels contribute to the disease state [45,46]. One mechanism may be through alteration of the actin cytoskeleton, which is perturbed in OCRL patient cells [47]. Another possibility is that OCRL1 regulates membrane traffic at the TGN through modulation of PtdIns(4)P and/or PtdIns(4,5)P_2 levels. Alternatively, its main function may be to ‘mop up’ excess PtdIns(4,5)P_2 to maintain the relatively high and low levels of PtdIns(4)P and PtdIns(4,5)P_2 respectively at the TGN.

Other phosphoinositides

Phosphoinositides phosphorylated at the 3-position have not been detected at the Golgi apparatus using specific probes. This contrasts with endosomes, which are abundant in both PtdIns(3)P and PtdIns(3,5)P_2, and contain many proteins that specifically bind to these lipids [48–50]. However, studies using the PI3K (phosphoinositide 3-kinase) inhibitor wortmannin, as well as purified PI3K, support a role for this kinase in carrier formation and/or cargo sorting at the TGN [51–53]. More recently, the wortmannin-resistant PI3K-C2α has been localized to TGN-derived clathrin-coated transport intermediates, and overexpression of this protein has been shown to disrupt trafficking between the TGN and endosomes [54]. Interestingly, binding to clathrin activates PI3K-C2α, suggesting that localized PtdIns(3)P or PtdIns(3,4,5)P_3 production contributes to clathrin bud formation. A potential regulator of this process is WIPI49 (WD40 repeat protein interacting with phosphoinositides of 49 kDa), a PtdIns(3)P-binding protein localized to the

TGN and endosomes [55]. WIPI49 is enriched in clathrin-coated vesicles and is required for efficient TGN to endosome trafficking. PtdIns(3,4,5)P_3 may influence trafficking by activating the ARF-GAP activity of ARAP1 (ARF-GAP, Rho-GAP, ankyrin repeat and PH domain-containing protein 1), a Golgi-associated multifunctional signalling protein, or through activation of other members of the ARF-GAP and -GEF families [56,57].

Conclusions and future perspectives

Different phosphoinositides are enriched in different membrane compartments. PtdIns(4)P is abundant at the Golgi apparatus, and a number of PtdIns(4)P-binding proteins have been identified there. Several of these are localized to the TGN and are important for trafficking to both the plasma membrane and endosomes. Given the abundance of PtdIns(4)P at the Golgi apparatus, it is likely that other PtdIns(4)P effectors will be discovered there, and it will be interesting to find out what these proteins do with respect to membrane traffic. The situation with PtdIns(4,5)P_2 and 3-phosphorylated phosphoinositides is less clear. These are very abundant at the plasma membrane and endosomes respectively, and have clearly defined roles at these compartments. At the Golgi apparatus, they are much less abundant and, although a number of regulators and one or two potential effector proteins have been identified, the extent to which they mediate trafficking there is unclear. Further studies are required to resolve this issue. This will require the identification of proteins that bind to these phosphoinositides with clearly defined roles in trafficking. Only then will we fully understand the diverse roles that phosphoinositides play in membrane traffic at the Golgi apparatus.

M.L. is the recipient of a senior research fellowship from the Medical Research Council. R.C. and N.H. are supported by a project grant from the Biotechnology and Biological Sciences Research Council (BBSRC) and a Wellcome Trust PhD studentship respectively.

References

1. Griffiths, G. and Simons, K. (1986) Science **234**, 438–443
2. Gleeson, P.A., Lock, J.G., Luke, M.R. and Stow, J.L. (2004) Traffic **5**, 315–326
3. Cullen, P.J., Cozier, G.E., Banting, G. and Mellor, H. (2001) Curr. Biol. **11**, R882–R893
4. Bankaitis, V.A., Aitken, J.R., Cleves, A.E. and Dowhan, W. (1990) Nature (London) **347**, 561–562
5. Hama, H., Schnieders, E.A., Thorner, J., Takemoto, J.Y. and DeWald, D.B. (1999) J. Biol. Chem. **274**, 34294–34300
6. Walch-Solimena, C. and Novick, P. (1999) Nat. Cell Biol. **1**, 523–525
7. Audhya, A., Foti, M. and Emr, S.D. (2000) Mol. Biol. Cell **11**, 2673–2689
8. Schorr, M., Then, A., Tahirovic, S., Hug, N. and Mayinger, P. (2001) Curr. Biol. **11**, 1421–1426
9. Homma, K., Terui, S., Minemura, M., Qadota, H., Anraku, Y., Kanaho, Y. and Ohya, Y. (1998) J. Biol. Chem. **273**, 15779–15786
10. Godi, A., Santone, I., Pertile, P., Marra, P., Di Tullio, G., Luini, A., Corda, D. and De Matteis, M.A. (1999) Biochem. Soc. Trans. **27**, 638–642

11. Bruns, J.R., Ellis, M.A., Jeromin, A. and Weisz, O.A. (2002) J. Biol. Chem. **277**, 2012–2018
12. de Graaf, P., Zwart, W.T., van Dijken, R.A., Deneka, M., Schulz, T.K., Geijsen, N., Coffer, P.J., Gadella, B.M., Verkleij, A.J., van der Sluijs, P. and van Bergen en Henegouwen, P.M. (2004) Mol. Biol. Cell **15**, 2038–2047
13. Wang, Y.J., Wang, J., Sun, H.Q., Martinez, M., Sun, Y.X., Macia, E., Kirchhausen, T., Albanesi, J.P., Roth, M.G. and Yin, H.L. (2003) Cell **114**, 299–310
14. Liljedahl, M., Maeda, Y., Colanzi, A., Ayala, I., Van Lint, J. and Malhotra, V. (2001) Cell **104**, 409–420
15. Nishikawa, K., Toker, A., Wong, K., Marignani, P.A., Johannes, F.J. and Cantley, L.C. (1998) J. Biol. Chem. **273**, 23126–23133
16. Godi, A., Di Campli, A., Konstantakopoulos, A., Di Tullio, G., Alessi, D.R., Kular, G.S., Daniele, T., Marra, P., Lucocq, J.M. and De Matteis, M.A. (2004) Nat. Cell Biol. **6**, 393–404
17. Levine, T.P. and Munro, S. (2002) Curr. Biol. **12**, 695–704
18. Raya, A., Revert-Ros, F., Martinez-Martinez, P., Navarro, S., Rosello, E., Vieites, B., Granero, F., Forteza, J. and Saus, J. (2000) J. Biol. Chem. **275**, 40392–40399
19. Stolz, L.E., Huynh, C.V., Thorner, J. and York, J.D. (1998) Genetics **148**, 1715–1729
20. Ha, S.A., Torabinejad, J., DeWald, D.B., Wenk, M.R., Lucast, L., De Camilli, P., Newitt, R.A., Aebersold, R. and Nothwehr, S.F. (2003) Mol. Biol. Cell **14**, 1319–1333
21. Hirst, J., Motley, A., Harasaki, K., Peak Chew, S.Y. and Robinson, M.S. (2003) Mol. Biol. Cell **14**, 625–641
22. Kalthoff, C., Groos, S., Kohl, R., Mahrhold, S. and Ungewickell, E.J. (2002) Mol. Biol. Cell **13**, 4060–4073
23. Mills, I.G., Praefcke, G.J., Vallis, Y., Peter, B.J., Olesen, L.E., Gallop, J.L., Butler, P.J., Evans, P.R. and McMahon, H.T. (2003) J. Cell Biol. **160**, 213–222
24. Wasiak, S., Legendre-Guillemin, V., Puertollano, R., Blondeau, F., Girard, M., de Heuvel, E., Boismenu, D., Bell, A.W., Bonifacino, J.S. and McPherson, P.S. (2002) J. Cell Biol. **158**, 855–862
25. Crump, C.M., Xiang, Y., Thomas, L., Gu, F., Austin, C., Tooze, S.A. and Thomas, G. (2001) EMBO J. **20**, 2191–2201
26. Mallard, F., Antony, C., Tenza, D., Salamero, J., Goud, B. and Johannes, L. (1998) J. Cell Biol. **143**, 973–990
27. Meyer, C., Zizioli, D., Lausmann, S., Eskelinen, E.L., Hamann, J., Saftig, P., von Figura, K. and Schu, P. (2000) EMBO J. **19**, 2193–2203
28. Saint-Pol, A., Yelamos, B., Amessou, M., Mills, I.G., Dugast, M., Tenza, D., Schu, P., Antony, C., McMahon, H.T., Lamaze, C. and Johannes, L. (2004) Dev. Cell **6**, 525–538
29. Stoorvogel, W., Oorschot, V. and Geuze, H.J. (1996) J. Cell Biol. **132**, 21–33
30. Valdivia, R.H., Baggott, D., Chuang, J.S. and Schekman, R.W. (2002) Dev. Cell **2**, 283–294
31. Stefan, C.J., Audhya, A. and Emr, S.D. (2002) Mol. Biol. Cell **13**, 542–557
32. Watt, S.A., Kular, G., Fleming, I.N., Downes, C.P. and Lucocq, J.M. (2002) Biochem. J. **363**, 657–666
33. Godi, A., Pertile, P., Meyers, R., Marra, P., Di Tullio, G., Iurisci, C., Luini, A., Corda, D. and De Matteis, M.A. (1999) Nat. Cell Biol. **1**, 280–287
34. Jones, D.H., Morris, J.B., Morgan, C.P., Kondo, H., Irvine, R.F. and Cockcroft, S. (2000) J. Biol. Chem. **275**, 13962–13966
35. Freyberg, Z., Siddhanta, A. and Shields, D. (2003) Trends Cell Biol. **13**, 540–546
36. Baron, C.L. and Malhotra, V. (2002) Science **295**, 325–328
37. Jenkins, G.H., Fisette, P.L. and Anderson, R.A. (1994) J. Biol. Chem. **269**, 11547–11554
38. Siddhanta, A., Backer, J.M. and Shields, D. (2000) J. Biol. Chem. **275**, 12023–12031
39. Donaldson, J.G. and Jackson, C.L. (2000) Curr. Opin. Cell Biol. **12**, 475–482
40. Jones, S.M., Howell, K.E., Henley, J.R., Cao, H. and McNiven, M.A. (1998) Science **279**, 573–577

41. Attree, O., Olivos, I.M., Okabe, I., Bailey, L.C., Nelson, D.L., Lewis, R.A., McInnes, R.R. and Nussbaum, R.L. (1992) Nature (London) **358**, 239–242
42. Lowe, C.U., Terrey, M. and Mac, L.E. (1952) AMA Am. J. Dis. Child. **83**, 164–184
43. Olivos-Glander, I.M., Janne, P.A. and Nussbaum, R.L. (1995) Am. J. Hum. Genet. **57**, 817–823
44. Dressman, M.A., Olivos-Glander, I.M., Nussbaum, R.L. and Suchy, S.F. (2000) J. Histochem. Cytochem. **48**, 179–190
45. Zhang, X., Hartz, P.A., Philip, E., Racusen, L.C. and Majerus, P.W. (1998) J. Biol. Chem. **273**, 1574–1582
46. Wenk, M.R., Lucast, L., Di Paolo, G., Romanelli, A.J., Suchy, S.F., Nussbaum, R.L., Cline, G.W., Shulman, G.I., McMurray, W. and De Camilli, P. (2003) Nat. Biotechnol. **21**, 813–817
47. Suchy, S.F. and Nussbaum, R.L. (2002) Am. J. Hum. Genet. **71**, 1420–1427
48. Birkeland, H.C. and Stenmark, H. (2004) Curr. Top. Microbiol. Immunol. **282**, 89–115
49. Dove, S.K., Piper, R.C., McEwen, R.K., Yu, J.W., King, M.C., Hughes, D.C., Thuring, J., Holmes, A.B., Cooke, F.T., Michell, R.H. et al. (2004) EMBO J. **23**, 1922–1933
50. Friant, S., Pecheur, E.I., Eugster, A., Michel, F., Lefkir, Y., Nourrisson, D. and Letourneur, F. (2003) Dev. Cell **5**, 499–511
51. Jones, S.M., Alb, Jr, J.G., Phillips, S.E., Bankaitis, V.A. and Howell, K.E. (1998) J. Biol. Chem. **273**, 10349–10354
52. Davidson, H.W. (1995) J. Cell Biol. **130**, 797–805
53. Gaffet, P., Jones, A.T. and Clague, M.J. (1997) J. Biol. Chem. **272**, 24170–24175
54. Gaidarov, I., Smith, M.E., Domin, J. and Keen, J.H. (2001) Mol. Cell **7**, 443–449
55. Jeffries, T.R., Dove, S.K., Michell, R.H. and Parker, P.J. (2004) Mol. Biol. Cell **15**, 2652–2663
56. Miura, K., Jacques, K.M., Stauffer, S., Kubosaki, A., Zhu, K., Hirsch, D.S., Resau, J., Zheng, Y. and Randazzo, P.A. (2002) Mol. Cell **9**, 109–119
57. Randazzo, P.A., Nie, Z., Miura, K. and Hsu, V.W. (2000) Sci STKE **2000**, RE1

Biochem. Soc. Symp. **72**, 39–45
(Printed in Great Britain)

4

N-glycans as apical targeting signals in polarized epithelial cells

Paula Urquhart[1], Susan Pang and Nigel M. Hooper

School of Biochemistry and Microbiology, University of Leeds, Leeds LS2 9JT, U.K.

Abstract

MDCK (Madin–Darby canine kidney) cells represent a good model of polarized epithelium to investigate the signals involved in the apical targeting of proteins. As reported previously, GPI (glycosylphosphatidylinositol) anchors mediate the apical sorting of proteins in polarized epithelial cells through their interaction with lipid rafts. However, using a naturally N-glycosylated and GPI-anchored protein, we found that the GPI anchor does not influence the targeting of the protein. It is, in fact, the N-glycans that signal the protein to the apical surface. In the present review, the role of N-glycans and GPI anchors as apical signals is discussed along with the putative mechanisms involved.

Introduction

Epithelial cells line the surface of all organs and cavities of the body, having a pleiotropic role in nutrient uptake, host defence and homoeostasis. They are polarized, i.e. they have two distinct domains: the apical and basolateral domains. The apical surface faces the lumen of the tissue, while the basolateral surface faces neighbouring cells and the extracellular matrix. Tight junctions separate the two domains, so that diffusion of proteins and lipids between the two surfaces cannot occur. As a result of this polarization, the cell must target newly synthesized proteins to the correct surface to be able to function properly. Several disease states result from the mistargeting and trafficking of proteins in epithelial cells, e.g. sitosterolaemia, distal renal tubular acidosis and cystic fibrosis [1–3]. It is therefore physiologically important to investigate the signals and mechanisms that are involved in polarized protein transport.

The signals involved in basolateral targeting are well studied. Many proteins targeted to the basolateral membrane contain tyrosine- or dileucine-

[1]To whom correspondence should be addressed (email P.Urquhart@leeds.ac.uk).

based amino acid motifs in their cytoplasmic domains which are recognized by distinct molecular subunits of adaptor complexes [4]. However, the signals for apical targeting are not so well defined.

Apical targeting signals

Studies involving chimaeric proteins in which the anchoring domain of a GPI (glycosylphosphatidylinositol)-anchored protein was fused to the ectodomain of a basolaterally sorted protein, resulting in the apical delivery of the chimaeric GPI-anchored protein, also appeared to support the role of lipid rafts in apical targeting [5–7]. However, the replacement of the cytosolic and transmembrane domains of a basolaterally sorted protein with a GPI anchor may have inadvertently removed the dominant basolateral targeting signals to leave a previously masked apical sorting signal in the ectodomain. Because other studies investigating apical signalling and targeting have used fusion proteins, i.e. normally soluble proteins with an engineered GPI anchor, the true nature of the targeting signal has been altered. Additionally, evidence supporting a role for lipid rafts and GPI-anchored proteins in targeting proteins to the apical membrane [8] has been superseded by the report that every study on the targeting of GPI-anchored proteins has used glycosylated proteins, suggesting a role for N-glycans in signalling [9]. Therefore we decided to investigate the apical targeting signals of a naturally GPI-anchored, N-glycosylated protein.

MDP (membrane dipeptidase), glycans and apical targeting

MDP is an ectoenzyme involved in the hydrolysis of dipeptides and is found in the brush-border region of the proximal tubule of the kidney. Native porcine MDP possesses a GPI anchor and two N-glycans at positions Asn^{41} and Asn^{263}. Using constructs of MDP lacking either the N-glycans (MDPΔgly) or the GPI anchor (MDPΔGPI), or both of these post-translational modifications (MDP ΔGPI,Δgly), we show for the first time on an endogenously N-glycosylated GPI-anchored protein, that it is the N-glycans, not the GPI anchor, that determines apical targeting in polarized cells [10]. As all of the constructs of MDP were enzymically active and were either secreted from the cells or present at the cell surface, the lack of the GPI anchor or the N-linked glycans did not drastically interfere with the folding or trafficking of the protein (Table 1).

Evaluation by immunofluorescence microscopy and activity assays of polarized MDCK (Madin–Darby canine kidney) cells grown on filters revealed that wild-type MDP was targeted predominantly to the apical surface (60%). Removal of the N-glycans changed the surface expression to 80% at the basolateral surface. The same pattern was found when trafficking of the non-GPI-anchored mutants were assessed.

In order to identify if lipid raft association accounted for the difference in targeting, MDCK cells expressing the GPI-anchored MDP constructs (wtMDP and MDPΔgly), were subjected to buoyant sucrose-density-gradient centrifuga-

Table 1 Properties of wtMDP, MDPΔgly, MDP ΔGPI and MDP ΔGPI,Δgly mutants. Also listed are the specific activities (nmol of D-phenylalanine/μg of MDP per min), GPI-anchor status, lipid raft association and percentage apical release for each construct. See Pang et al. [10] for details. Diagonal striped boxes represent 16-amino-acid signal peptide, lollipops represent N-glycans and hatched boxes represent GPI anchor.

Construct	Schematic diagram	Enzymic specific activity (nmol of D-phenylalanine/μg of MDP per min)	GPI-anchored	Secreted	Lipid-raft -associated	Apical release (%)
wtMDP	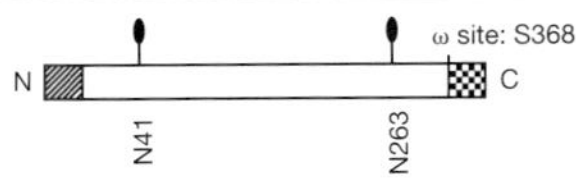	11.6±1.2	Yes	No	Yes	60±2.4
MDPΔgly	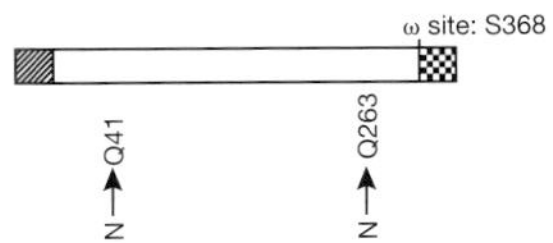	11.0±2.8	Yes	No	Yes	82±1.2
MDPΔGPI	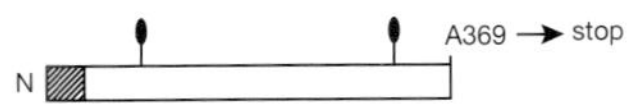	31.2±2.6	No	Yes	No	82±1.0
MDPΔGPI,Δgly	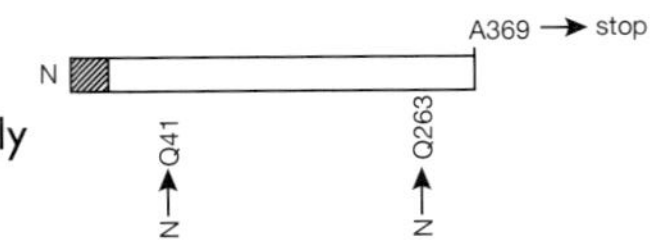	8.1±0.4	No	Yes	No	24±2.0

tion. The position of lipid rafts in the sucrose gradients was determined by immunoblotting with an antibody against caveolin-1, a well-defined marker of lipid rafts [11]. The distribution in the sucrose gradients of the non-raft proteins clathrin and dipeptidyl peptidase IV was also determined. Both of these proteins were detected exclusively in the soluble fractions. The distribution of wtMDP and MDPΔgly in the sucrose gradients was determined by both immunoblotting and activity assays. Both constructs were located predominantly (72±5% and 72±6% respectively) in the lipid raft fractions with negligible enzyme activity or MDP protein present in the soluble fractions of the sucrose gradients, clearly indicating that both constructs were present in lipid rafts, thus indicating that GPI anchoring is not sufficient for apical targeting. This result is in agreement with the observation that GPI anchoring was not sufficient for apical delivery of rat growth hormone, which is an unglycosylated unpolarized secreted protein [9], and that free GPI anchors are delivered in an unpolarized manner to the surface of MDCK cells [12]. Raft association of proteins located on the basolateral surface has been observed previously [13,14], and caveolae, subdomains of detergent-resistant rafts that contain caveolin-1, are enriched on the basolateral surface of MDCK cells [15]. It can be concluded that the GPI anchor on MDP causes the protein to associate with lipid rafts in the secretory pathway, but that targeting of the protein to the apical surface requires sorting information encoded within the N-glycans. In this context, it is interesting to note that the glycoprotein Thy-1 was secreted from the apical surface after removal of its GPI anchor, leading to the conclusion that sorting information lies elsewhere in the protein other than the GPI anchor [16]. However, it was not determined whether or not the N-glycans on Thy-1 fulfilled this role. Consistent with our data, it has also been found that N-linked glycosylation is required for AQP-2 (aquaporin-2) trafficking to the apical surface of MDCK cells, and, as AQP-2 is soluble in Triton X-100 at 4°C, that lipid rafts are not involved in the apical targeting of this protein [17].

The nature of the apical targeting signal in the glycan of MDP

To investigate if it is the nature of the glycan that is important in the apical targeting of MDP, we have preliminary data using glycosylation inhibitors to disrupt the terminal processing of the N-glycans. Deoxymannojirimycin and swainsonine (α-mannosidase I and II inhibitors respectively), and deoxynojirimycin and castanospermine (α-glucosidase I and II inhibitors respectively) were found to have little or no effect on the targeting of wild-type MDP (P. Urquhart and N.M. Hooper, unpublished work). This suggests that it is the core glycan that is functional in the apical signalling of MDP. In relation to this, a study by Potter et al. [18] showed that the apical trafficking of endolyn, a highly glycosylated transmembrane protein, could be disrupted in tunicamycin-treated cells, and, when subjected to site-directed mutagenesis, they identified two N-glycans that were major determinants in its apical delivery. Further studies are required to identify whether the site of glycosylation on the protein is the crucial factor in the apical targeting of MDP. It is also possible that the prion protein is directed to the basolateral surface in MDCK cells as a result of its two N-glycans not being in the correct position on the surface of the protein to target it to the apical domain [19]. Additionally, a study investigating the role of N-glycans in the apical trafficking of the H^+,K^+-ATPase β subunit found the core region sugars to be more important than terminal sugars for the apical sorting of the protein [20].

An alternative hypothesis is that N-glycans play a facilitative role, by providing structural support or preventing aggregation of the proteins, thereby allowing interaction of proteinaceous apical sorting signals with the sorting machinery [21]. In the case of MDP, this hypothesis seems unlikely. The basolaterally targeted unglycosylated constructs were appropriately folded, as shown by their trafficking competence and enzyme activity, and thus any proteinaceous sorting signals would still be available for interaction with the apical sorting machinery.

Distinct pathways for polarized transport

Two distinct pathways are known to exist that transport cargo to the cell surface in polarized cells: the direct and indirect pathways. The direct pathway delivers cargo exiting the TGN (*trans*-Golgi network) straight to either the apical or basolateral surface by vesicles that bud off from the Golgi stack. Conversely, the indirect route involves traversing to the opposing surface before finally delivering the cargo to its intended surface, also known as transcytosis.

Until now, the sorting of proteins into their specific pathway was thought to occur at the TGN; however, a recent study by Polishchuk et al. [22] provides evidence that this might not be the case. These authors have shown, using live-cell time-lapse microscopy, that a GFP (green fluorescent protein)-tagged GPI anchor was delivered first to the basolateral membrane, then internalized and finally delivered to the apical surface, suggesting that apical sorting signals are recognized at the lateral membrane and not the TGN. It remains to be determined whether the N-glycans are required for the internalization of MDP at the basolateral surface for terminal apical trafficking. If N-glycans are required for MDP to be transcytosed in MDCK cells, it follows that the Δgly mutant will remain at the basolateral membrane.

Regulators of apical targeting

Several putative recognition factors for apical targeting have been identified. However, most are associated with lipid rafts. Jacob et al. [23] monitored the exocytic apical pathway of raft-associated SAVs (sucrase-isomaltase-carrying vesicles). This group identified that annexin II was required to traffic the SAVs to the apical surface of cells. Previously, Fiedler et al. [24] demonstrated that another member of the annexin family, annexin XIIIb, was involved in the delivery of influenza virus HA (haemagglutinin) to the apical surface in MDCK cells.

VIP36 [vesicular integral protein 36; MAL (myelin and lymphocyte protein)] is an intracellular lectin that recognizes high-mannose-type glycans [25]. Overexpression of VIP36 was shown to stimulate the secretion of clusterin, a glycoprotein that carries at least one high-mannose glycan and was therefore thought to be involved in transport to the cell surface [26]. However, further studies revealed that it is the overexpression of VIP36 that leads to post-Golgi transport of the protein [27]. VIP17, another member of the MAL family, is also thought to play a role in the sorting and delivery of proteins from the TGN to the apical surface [28]. In agreement with this, another group found that antisense oligonucleotides to VIP17/MAL resulted in diminished apical transport of HA [29]. MAL2, a novel member of the MAL family, was recently identified, and was found to be required for transcytosis in HepG2 cells [30]. Other apical recognition factors in the non-lipid raft pathway remain to be detected.

Conclusions

N-glycans are not a universal apical targeting signal as they do not function as apical sorting determinants for the hepatitis B surface antigen [31], the soluble corticosteroid binding globulin [32] or the transmembrane ectonucleotide pyrophosphatase/phosphodiesterase NPP3 [33]. In the case of the neurotrophin receptor, a juxtamembrane region of the extracellular domain that is rich in O-glycosylated serine/threonine residues, and not the single N-glycan, was responsible for the apical secretion of both transmembrane and secreted forms of the protein [34]. As the GPI-anchored MDP lacks an O-glycosylated region, it is the N-glycans that provide the apical sorting information in this case. Further studies would be required to determine whether it is a particular glycan sequence on the N-glycans that is responsible for the apical targeting. What is becoming apparent from these studies is that there are a variety of sorting signals responsible for the apical targeting of proteins in polarized cells. However, the data presented in the current study clearly show that N-glycans are involved in the apical targeting of an endogenous GPI-anchored protein in MDCK cells, necessitating a re-evaluation of the roles of GPI anchors and lipid rafts in the apical targeting of proteins in polarized cells.

This work was supported by grants from the Biotechnology and Biological Sciences Research Council (BBSRC) of Great Britain and the Wellcome Trust (Bioimaging Facility, University Of Leeds). The monoclonal antibody against the apical marker gp135 (3F2) was a gift from Dr George K. Ojakian, SUNY Downstate Medical Centre, Brooklyn, NY, U.S.A. We thank Dr Gareth Howell for advice and helpful discussions regarding the confocal microscopy.

References

1. Graff, G.A., Cohen, J.C. and Hobbs, H.H., (2004) J. Biol. Chem. **279**, 24881–24888
2. Devonald, M.A., Smith, A.N., Poon, J.P., Ihrke, G. and Karet, F.E. (2003) Nat. Genet. **33**, 125–127

3. Kleizen, B., Braakman, I. and de Jonge, H.R. (2000) Eur. J. Cell Biol. **79**, 544–556
4. Matter, K. (2000) Curr. Biol. **10**, R39–R42
5. Brown, D.A., Crise, B. and Rose, J.K. (1989) Science **245**, 1499–1501
6. Lisanti, M.P., Caras, I.W., Gilbert, T., Hanzel, D. and Rodriguez-Boulan, E. (1990) Proc. Natl. Acad. Sci. U.S.A. **87**, 7419–7423
7. Soole, K.L., Jepson, M.A., Hazlewood, G.P., Gilbert, H.J. and Hirst, B.H. (1995) J. Cell Sci. **108**, 369–377
8. Simons, K. and Ikonen, E. (1997) Nature (London) **387**, 569–572
9. Benting, J.H., Rietveld, A.G. and Simons, K. (1999) J. Cell Biol. **146**, 313–320
10. Pang, S., Urquhart, P. and Hooper, N.M. (2004) J. Cell Sci. **117**, 5079–5086
11. Lisanti, M.P., Tang, Z. and Sargiacomo, M. (1993) J. Cell Biol. **123**, 595–604
12. van't Hoff, W., Rodriguez-Boulan, E. and Menon, A.K. (1995) J. Biol. Chem. **270**, 24150–24155
13. Naim, H.Y., Dodds, D.T., Brewer, C.B. and Roth, M.G. (1995) J. Cell Biol. **129**, 1241–1250
14. Neame, S.J., Uff, C.R., Sheikh, H., Wheatley, S.C. and Isacke, C.M. (1995) J. Cell Sci. **108**, 3127–3135
15. Scheiffele, P., Verkade, P., Fra, A.M., Virta, H., Simons, K. and Ikonen, E. (1998) J. Cell Biol. **140**, 795–806
16. Powell, S.K., Lisanti, M.P. and Rodriguez-Boulan, E.J. (1991) Am. J. Physiol. **260**, C715–C720
17. Hendriks, G., Koudijs, M., van Balkom, B.W.M., Oorschot, V., Klumperman, J., Deen, P.M.T. and van der Sluijs, P. (2004) J. Biol. Chem. **279**, 2975–2983
18. Potter, B.A., Ihrke, G., Bruns, J.R., Weixel, K.M. and Weisz, O.A. (2004) Mol. Biol. Cell **15**, 1407–1416
19. Sarnataro, D., Paladino, S., Campana, V., Grassi, J., Nitsch, L. and Zurzolo, C. (2002) Traffic **3**, 810–821
20. Vagin, O., Turdikulova, S. and Sachs, G. (2004) J. Biol. Chem. **279**, 39026–39034
21. Rodriguez-Boulan, E. and Gonzalez, A. (1999) Trends Cell Biol. **9**, 291–294
22. Polishchuk, R., Di Pentima, A. and Lippincott-Schwartz, J. (2004) Nat. Cell Biol. **6**, 297–306
23. Jacob, R., Heine, M., Eikemyer, J., Freker, N., Zimmer, K.P., Rescher, U., Gerke, V. and Naim, H.Y. (2004) J. Biol. Chem. **279**, 3680–3684
24. Fiedler, K., Lafont, F., Parton, R.G. and Simons, K. (1995) J. Cell Biol. **128**, 1043–1053
25. Hara-Kuge, S., Ohkura, T., Seko, A. and Yamashita, K. (1999) Glycobiology **9**, 833–839
26. Hara-Kuge, S., Ohkura, T., Ideo, H., Shimada, O., Atsumi, S. and Yamashita, K. (2002) J. Biol. Chem. **277**, 16332–16339
27. Fullekrug, J., Scheiffele, P. and Simons, K. (1999) J. Cell Sci. **112**, 2813–2821
28. Puertollano, R., Martin-Belmonte, F., del Carmen de Marco, C., Albar, J.P., Kremer, L. and Alonso, M.A. (1999) J. Cell Biol. **145**, 141–151
29. Cheong, K.H., Zacchetti, D., Schneeberger, E.E. and Simons, K. (1999) Cell Biol. **96**, 6241–6248
30. de Marco, M.C., Martin-Belmonte, F., Kremer, L., Albar, J.P., Corrreas, I., Vaerman, J.P., Marazuela, M., Byrne, J.A. and Alonso, M.A. (2002) J. Cell Biol. **159**, 37–44
31. Marzolo, M.P., Bull, P. and Gonzalez, A. (1997) Proc. Natl. Acad. Sci. U.S.A. **94**, 1834–1839
32. Larsen, J.E., Avvakumov, G.V., Hammond, G.L. and Vogel, L.K. (1999) FEBS Lett. **451**, 19–22
33. Meerson, N.R., Bello, V., Delaunay, J.L., Slimane, T.A., Delautier, D., Lenoir, C., Trugnan, G. and Maurice, M. (2000) J. Cell Sci. **113**, 4193–4202
34. Yeaman, C., Le Gall, A.H., Baldwin, A.N., Monlauzeur, L., Le Bivic, A. and Rodriguez-Boulan, E. (1997) J. Cell Biol. **139**, 929–940

Biochem. Soc. Symp. **72**, 47–63
(Printed in Great Britain)

5

Defective kidney anion-exchanger 1 (AE1, Band 3) trafficking in dominant distal renal tubular acidosis (dRTA)

Ashley M. Toye[1]

Department of Biochemistry, University of Bristol, University Walk, Bristol BS8 1TD, U.K.

Abstract

dRTA (distal renal tubular acidosis) results from the failure of the α-intercalated cells in the distal tubule of the nephron to acidify the urine. A truncated form of AE1 (anion-exchanger 1; Band 3), kAE1 (kidney isoform of AE1), is located in the basolateral membrane of the intercalated cell. Mutations in the AE1 gene cause autosomal dominant and recessive forms of dRTA. All the dominant dRTA mutations investigated cause aberrant trafficking of kAE1, resulting in its intracellular retention or mistargeting to the apical plasma membrane. Therefore the intracellular retention of hetero-oligomers containing wild-type and dRTA mutants, or the mistargeted protein in the apical membrane neutralizing acid secretion, explains dominant dRTA. The kAE1 (Arg^{901}→stop) mutant has been studied in more detail, since the mistargeting kAE1 (Arg^{901}→stop) from the basolateral to the apical membrane is consistent with the removal of a basolateral localization signal. The C-terminal amino acids deleted by the Arg^{901}→stop mutation, contain a tyrosine motif and a type II PDZ interaction domain. The tyrosine residue (Tyr^{904}), but not the PDZ domain, is critical for basolateral localization. In the absence of the N-terminus of kAE1, the C-terminus was not sufficient to localize kAE1 to the basolateral membrane. This suggests that a determinant within the kAE1 N-terminus co-operates with the C-terminus for kAE1 basolateral localization. Interestingly, Tyr^{359}, in the N-terminal domain, and Tyr^{904} in the C-terminus of AE1 are phosphorylated in red blood cells. A potential scheme is suggested where successive phosphorylation of these residues is necessary for correct localization and recycling of kAE1 to the basolateral membrane.

[1]email Ash.M.Toye@bristol.ac.uk

Introduction

The AE1 (anion-exchanger 1; red blood cell Band 3, SLC4A1) is a multi-spanning membrane protein that conducts chloride–bicarbonate exchange in the red blood cell and in the distal nephron of the kidney (reviewed in [1]). The protein consists of an N-terminal cytoplasmic domain (amino acids 1–359), which associates with peripheral membrane proteins and forms a membrane-anchorage site for the red blood cell cytoskeleton, and a C-terminal domain (amino acids 360–911), which carries out anion-exchange and consists of 12–14 TM (transmembrane) segments and includes a short (33-amino-acid) cytoplasmic C-terminal tail. The N-terminal cytoplasmic domain is known to bind ankyrin, protein 4.1, protein 4.2, glycolytic enzymes and haemoglobin [2]. The C-terminal tail harbours a binding site for carbonic anhydrase II [3,4]. Together, AE1 and carbonic anhydrase II are suggested to form a 'metabolon', resulting in highly efficient chloride–bicarbonate exchange [5].

The kidney isoform of AE1 (kAE1) is identical with AE1, but lacks the N-terminal 65 amino acids of AE1 [6]. This N-terminal truncation removes a central β-strand in the structural model of AE1, probably resulting in substantial changes in the N-terminal structure and binding properties of kAE1 [2,7].

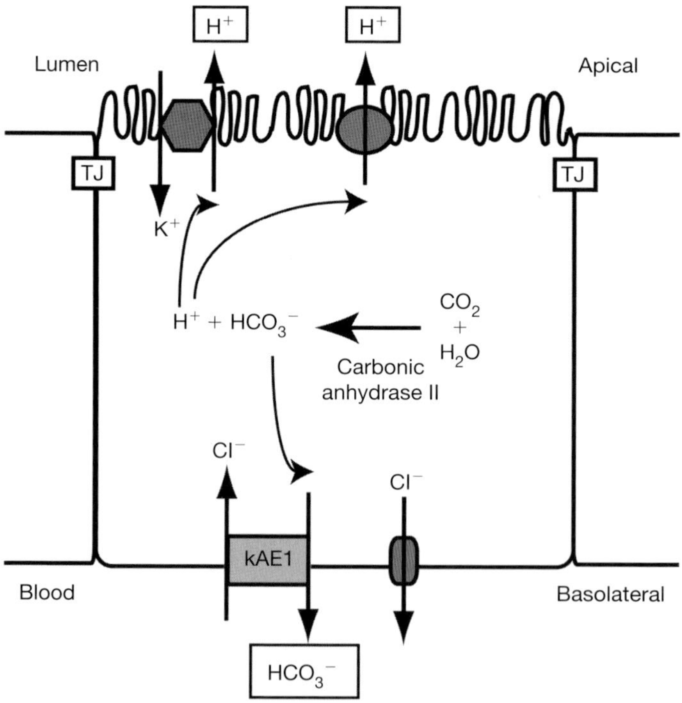

Figure 1 Schematic representation of the α-intercalated cell. Intracellular hydration of CO_2 by carbonic anhydrase II provides protons that are secreted into the kidney lumen by the apical H^+-ATPase and H^+,K^+-ATPase, and the remaining bicarbonate is transported into the blood in exchange for chloride by kAE1 in the basolateral membrane.

Therefore the N-termini of kAE1 and AE1 are likely to interact with different intracellular proteins, but, aside from kanadaptin [8] (whose binding to kAE1 now looks doubtful [9]), no candidate interactors with the N-terminus of kAE1 have been reported. kAE1 is located in the basolateral membrane of kidney α-intercalated cells [10–12] and is required for acid secretion into urine (see Figure 1). The apical H^+-ATPase of the α-intercalated cell pumps protons derived from hydrated CO_2 into the tubular lumen, while kAE1 reclaims the remaining bicarbonate back into the blood in exchange for chloride.

dRTA (distal renal tubular acidosis) is characterized by defective acid secretion by the α-intercalated cells. AE1 gene mutations have been reported for both dominant and recessive forms of dRTA [13–19]. AE1 mutations that cause Band-3-linked hereditary spherocytosis or SAO (Southeast Asian ovalocytosis) in red blood cells, do not generally cause dRTA, unless present in homozygous form as in the Band-3-null individual [20]. In addition to mutations in AE1, mutations in the H^+-ATPase subunits have also been described in recessive dRTA [21–23].

To date, all reported dominantly inherited forms of dRTA result from mutations in the AE1 gene, and reside exclusively in the AE1 membrane domain and C-terminal tail (see Figure 2). The most prevalent, and thus most clinically significant, inherited forms of dRTA are associated with alterations of a conserved arginine at residue 589 (Arg^{589}→His, Arg^{589}→Ser and Arg^{589}→Cys) located at the start of the cytoplasmic loop joining TM spans) 6 and 7 [13,15,19]. The other reported dominant mutations include Ser^{613}→Phe in TM7 [13], Ala^{858}→Asp in the final TM [14] and two truncations of the C-terminus that result in an 11-amino-acid deletion of the C-terminus (Arg^{901}→stop; Band 3 Walton) [19,24] or a 23-amino-acid deletion [25]. Mutations that cause recessive dRTA have been described in Southeast Asia, and include Arg^{602}→His within TM7 [26], Gly^{701}→Asp in the fourth cytoplasmic loop [14,17,18,26], Ser^{773}→Pro in TM11 [26] and deletion of Val^{850} in TM13 [14]. The recessive mutations are often found as compound heterozygotes with SAO [14,17], due to the high prevalence of the latter mutation in this region. When presented in conjunction with SAO, the recessive dRTA mutations have a more severe effect. This is partly due to the presence of functionally inactive SAO causing a pseudodominant phenotype, but also due to interactions that occur between the kAE1 SAO protein and the dRTA proteins exacerbating further the structural and functional characteristics of the dRTA protein [14,27].

The dominant dRTA mutant AE1 proteins all retain substantial anion-transport activity in mutant red blood cells and when the kAE1 proteins are expressed in *Xenopus* oocyte [13–16,24]. In addition, the observation that a single normal kAE1 allele can provide adequate acid secretion by the intercalated cells (as is the case in SAO [13,14]), suggests that dominant dRTA does not result from a simple reduction in anion-transport activity of the mutant kAE1. Alternative mechanisms proposed for the dominant disease include mistargeting of the mutant protein to the apical membrane and/or altered targeting of the normal protein by interaction with the mutant protein [13,15].

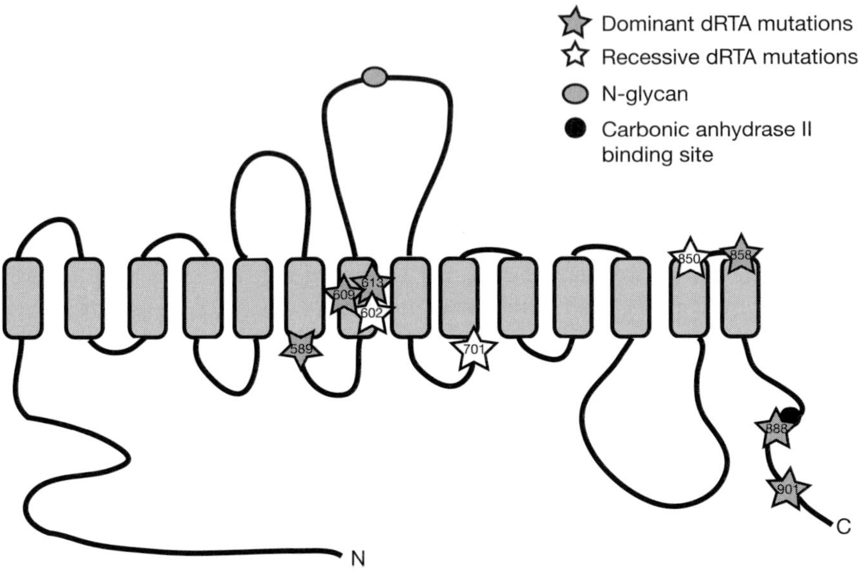

Figure 2 Diagram of the location of dominant and recessive dRTA mutations in relation to the putative topology of kAE1. The approximate locations of the dRTA mutations reported so far are shown. They are clustered in the membrane domain and cytoplasmic C-terminal tail.

Several recent studies have described the successful expression of both wild-type and dominant dRTA mutations in non-polarized [24,28,29] and polarized kidney cells [16,30,31]. This emerging body of work suggests that dominant dRTA mutations cause aberrant targeting of kAE1 in the intercalated cell, and also implicates specific regions of the kAE1 membrane domain for kAE1 trafficking in the kidney. This review will discuss the implications of these findings for our understanding of the dominant dRTA disease mechanism, and also the basolateral trafficking of kAE1 in intercalated cells.

kAE1 expression in mammalian cells

To date, most heterologous expression, particularly for topological studies of AE1, has been carried out using transiently transfected HEK-293 (human embryonic kidney) cells [32–35]. Although HEK-293 cells successfully express both AE1 and kAE1 at the plasma membrane, a major disadvantage is that these cells are unable to polarize into distinct apical and basolateral membrane domains. In addition, post-translational modifications of AE1 and kAE1, such as complex N-glycosylation [29,36] and palmitoylation [37] do not occur in HEK-293 cells (which appears to have no adverse effect on bicarbonate–chloride exchange conducted by AE1). AE1 and kAE1 have also been stably expressed at the cell surface in K562 cells [38,39], which facilitated the identification of the optimum confocal-imaging conditions for visualizing both proteins in mammalian cells [39].

Adair-Kirk et al. [40] were the first to investigate kAE1 trafficking in polarized MDCK (Madin–Darby canine kidney) I cells, finding that the major chicken-specific kidney AE1 isoform (AE1-4) was targeted to the basolateral membrane. AE1-4 basolateral sorting, Golgi recycling from the cell surface and the association with the actin cytoskeleton depended on tyrosine residues in the N-terminus [40]. However, in mammalian cells, this N-terminal sequence is present in erythrocyte AE1, but is absent in kAE1. Therefore there are likely to be species-specific differences in kAE1-targeting mechanisms between human and chicken kidney anion-exchangers.

Human kAE1 has now been expressed transiently in MDCKI [16,31] and LLC-PK1 (porcine kidney) cells [30], and stably in MDCKI cells [24,30]. kAE1 was progressively converted from a core glycosylated form to a complex glycosylated form in both non-polarized and polarized MDCKI cells [30]. Immunofluorescence studies showed expression of kAE1 predominantly at the cell surface in non-polarized MDCKI cells ($51 \pm 6\%$ by biotinylation studies), with variable amounts of intracellular material that co-localized with markers of the secretory pathway [e.g. Golgi and TGN (*trans*-Golgi network)] [24,30]. Importantly, kAE1 also localized to the basolateral membrane in MDCKI cells and LLC-PK1 cells [16,30,31] when these cells were allowed to polarize.

Unexpectedly, overexpression of wild-type kAE1 reduced the transepithelial electrical resistance and increased the permeability of the paracellular barrier of the polarized MDCKI monolayers to a normally impermeant biotinylation reagent [30]. However, the localization of the tight junction marker ZO-1 (zona occludens 1 protein) was unaffected in these cells [30]. The increased permeability of the paracellular barrier occurred with wild-type kAE1 and not the kAE1 ($Arg^{901} \rightarrow$stop) mutant, implicating the involvement of the C-terminal residues of kAE1. It was suggested that kAE1 overexpression may interfere with the actin/spectrin cytoskeleton, therefore affecting the organization of the paracellular permeability barrier provided by epithelial tight junctions [30]. Another possibility is that kAE1 overexpression affects MDCKI epithelial cell–cell adherence and therefore paracellular permeability. This is because the C-terminal tail of kAE1 contains a tyrosine motif [YDEV (Tyr-Asp-Glu-Val), see below], which is similar to a tyrosine motif [YSEV (Tyr-Ser-Glu-Val)] contained in the cytoplasmic domain of CEACAM1-L (long isoform of carcinoembryonic antigen-related cell-adhesion molecule 1) that perturbed desmosomes and disorganized cytokeratin filaments [41].

Mistrafficking of dRTA mutants affecting the putative TM6–7 region of kAE1

Five of the eight dominant dRTA variants reported are located in the region of kAE1 encompassing TMs 6 and 7 (see Figure 2), with three of these mutations affecting the conserved Arg^{589}. The kAE1 ($Arg^{589} \rightarrow$His), kAE1 ($Arg^{589} \rightarrow$Ser) and kAE1 ($Arg^{589} \rightarrow$Cys) proteins were transiently expressed in non-polarized HEK-293 cells, and all were retained within the ER (endoplasmic reticulum) [29]. Importantly, kAE1 (Arg^{589}→His) formed hetero-oligomers with wild-type kAE1 and had a dominant-negative effect on the ability of wild-type protein to reach the cell surface. The mutant did not affect wild-type erythrocyte AE1 trafficking, suggesting that the additional 65 amino acids present in erythrocyte AE1 helped the protein evade this dominant-negative effect [29]. Both kAE1 (Arg^{589}→His) and kAE1 (Ser^{613}→Phe) dRTA mutations were also retained internally when stably expressed in non-polarized and polarized MDCKI cells [30]. Incubation of these cells at 27°C, which corrects CFTR(ΔF508) (cystic fibrosis transmembrane conductance regulator with a deletion of Phe^{508}) mistrafficking in transfected cells [42], did not alter the mistrafficking of Arg^{589}→His or Ser^{613}→Phe [30].

It is unlikely that the Arg^{589}→His and Ser^{613}→Phe dRTA mutations cause gross misfolding of the mutant proteins, since they have significant transport activity in both *Xenopus* oocytes and red blood cells [13,15,16]. In addition, TM6–7 appear not to be essential for AE1 chloride transport [43]. It is possible that localized misfolding in the TM 6 and 7 region increases the aggregation of the mutant proteins or alters their interactions with ER chaperones. This would affect their exit from the ER, and may then lead to their degradation by the ERAD (ER-associated degradation) system.

Recently, a dominant dRTA mutation (Gly^{609}→Arg) [16] and a recessive dRTA mutation (Arg^{602}→His) [26] have been reported. Despite the close proximity of both Arg^{602}→His and Gly^{609}→Arg to the Ser^{613}→Phe dRTA mutation, Arg^{602}→His is reported as being recessive [26] and the Gly^{609}→Arg mutation had both apical and basolateral distribution (termed non-polarized when transiently expressed in MDCKI cells) [16]. These data indicate that, although dRTA mutations are clustered between the TM6–7 region of kAE1, they can still affect kAE1 function and activity in subtly different ways.

Effects of dominant dRTA mutations in the final TM and cytoplasmic C-terminus of kAE1

Three dominant dRTA mutations occur within the final TM and the short cytosolic C-terminal tail of kAE1 (see Figure 2). These are Ala^{858}→Asp within the final TM [14] and two truncations of the cytoplasmic C-terminus, one of 11 amino acids (Band 3 Walton, Arg^{901}→stop) [19,24] and the other of 23 amino acids [25]. Both the Ala^{858}→Asp and the Arg^{901}→stop mutations had significant chloride–chloride exchange activity in red blood cells and *Xenopus* oocytes [14,24], but the effects of the 23-amino-acid truncation mutant have not been investigated.

Topological studies on AE1, position residue 858 within the outer leaflet of the lipid bilayer (near the extracellular face of the plasma membrane) in the final TM [32,35,44]. This residue is suggested to form part of a hydrophobic pore, lining the anion-translocation channel of AE1 [44]. The effect of the Ala^{858}→Asp mutation on kAE1 trafficking in kidney cells has not been studied, but the Ala^{858}→Cys mutant used in the cysteine-scanning mutagenesis experiments had significant transport in HEK-293 cells [35]. In a similar way to the recessive dRTA mutants, compound heterozygotes with SAO and Ala^{858}→Asp had a more severe form of dRTA, and also anaemia [14]. This suggests that hetero-oligomerization of Ala^{858}→Asp with the mutant SAO protein, can exacerbate further the structural/functional effects of the Ala^{858}→Asp mutation in the patients' α-intercalated cells and red blood cells [14,27].

The Arg^{901}→stop form of kAE1 is the most studied dRTA mutation, having been expressed in MDCKI [24,30,31], LLC-PK1 [30] and HEK-293 [28,45] cells. In non-polarized cells, the majority of kAE1 (Arg^{901}→stop) is retained internally, co-localizing with calnexin in both MDCKI and HEK-293 cells [24,28,30]. However, cell-specific differences in the trafficking of the Arg^{901}→stop mutant appear to exist, since the protein is able to exit the ER and mistarget to late endosomes/lysosomes in MDCKI cells, but this does not occur in HEK-293 cells [28,30]. In addition, kAE1 (Arg^{901}→stop) is complex N-glycosylated in MDCKI cells, but the complex N-glycan is larger than that on the wild-type [30]. This was suggested to indicate that the protein traffics differently from normal kAE1, with a longer transit time through the medial Golgi, or increased recycling of the mutant protein. As found for the Arg^{589}→His mutation, kAE1 (Arg^{901}→stop) could form hetero-oligomers with wild-type kAE1, and have a dominant-negative effect on the ability of wild-type protein to reach the cell surface in HEK-293 cells [28]. The Arg^{901}→stop mutation also affects AE1 trafficking in HEK-293 cells, but not in red blood cells or when it was expressed in *Xenopus* oocytes [24]. This highlights differences in AE1 trafficking between the kidney cells, red blood cells and oocytes, and suggests that the C-terminus is not as important for trafficking in red blood cells or oocytes, as it is in the kidney [24]. In addition, the presence of AE1-interacting proteins such as glycophorin A [46] in red blood cells could also help alleviate the trafficking defect of this mutation.

In polarized MDCKI cells stably expressing kAE1 (Arg^{901}→stop), the mutant protein was mistargeted to the apical plasma membrane, as judged by immunofluorescence and domain-selective biotinylation [30]. However, transient expression of kAE1 results in a non-polarized distribution in both polarized MDCKI [31] and polarized LLC-PK1 [30] cells. A potential explanation for the purely apical localization observed in stably transfected cells and the non-polarized distribution in transiently transfected cells could be that transient overexpression of kAE1 saturates the cellular sorting machinery allowing some kAE1 (Arg^{901}→stop) to reside at the basolateral membrane. Alternatively, in polarized cells, kAE1 (Arg^{901}→stop) may be delivered initially to the basolateral plasma membrane, then rapidly internalized prior to delivery to the apical plasma membrane. In cells transiently overexpressing kAE1 (Arg^{901}→stop), there would be more kAE1 (Arg^{901}→stop) detected at the basolateral membrane, whereas, in stable cells, the amount is below the level of detection.

The different localizations of the normal kAE1 and the kAE1 (Arg^{901}→stop) in non-polarized and polarized MDCKI cells, suggest that the C-terminal 11 residues of kAE1 (-RDEYDEVAMPV) contains both a plasma-membrane-targeting determinant (in non-polarized cells) and a basolateral-targeting or -retention determinant (in polarized cells). The removal of the C-terminus determinant would then allow previously cryptic/suppressed signals to predominate in the non-polarized and polarized cells.

These findings would also predict that the 23-amino-acid C-terminal truncation may have a similar effect to Arg^{901}→stop, since the same C-terminal sorting determinant would be absent in this mutation. The 23-amino-acid truncation may also make kAE1 more unstable, since C-terminal truncations of AE1 greater than 20 amino acids are unstable in *Xenopus* oocytes [47] and HEK-293 cells [45]. The 23-amino-acid truncation mutation also affects the binding site for carbonic anhydrase II [48], so is also likely to alter the interaction of kAE1 with carbonic anhydrase II.

Mechanisms leading to dominant dRTA associated with kAE1 mutations

It is clear that kAE1 mutations associated with dominant dRTA affect the targeting of kAE1 in kidney cells. However, these dominant mutations result in disease even though the patients carry a normal AE1 allele as well as the mutant allele. Research to date suggests that dominant dRTA can arise by at least two mechanisms that involve either the intracellular retention of wild-type protein or mistargeting of the mutant protein to the apical plasma membrane (see Figure 3). Similar mechanisms have recently been reported for other dominant diseases, such as familial hypercholesterolaemia and autosomal dominant diabetes insipidus [49–51].

Both the Arg^{589}→His and Ser^{613}→Phe mutations probably result in the intracellular retention of the wild-type kAE1 due to its association with the mutant protein in the form of higher-order hetero-oligomers/aggregates (Figure 3). A similar hetero-oligomerization was observed for the SAO protein with wild-type AE1, where, strikingly, all the wild-type kAE1 forms hetero-oligomers with the mutant SAO protein [27]. The intracellular retention of wild-type kAE1 may then lead to the destruction of normal kAE1 protein by the ERAD pathway or another degradative pathway. In support of this hypothesis, a kidney resection from an Arg^{589}→His patient appeared to have no detectable AE1 immunoreactivity [52].

Both kAE1 (Arg^{901}→stop) and kAE1 (Gly^{609}→Arg) mutations appear to mistarget the protein to the apical plasma membrane [16,30,31] (Figure 3). The presence of the mistargeted bicarbonate–chloride exchanger at the apical membrane would allow bicarbonate to move into the tubular lumen together with protons secreted by the H^+-ATPase, thus negating net acid secretion by the cells. In addition, the kAE1 (Arg^{901}→stop) mutant [and potentially kAE1 (Gly^{609}→Arg)] could hetero-oligomerize with wild-type kAE1 [28], which may then also misroute the normal protein to the apical membrane via the formation of hetero-oligomers. However, these hetero-oligomers would still contain wild-type copies of kAE1, which, hypothetically, should still contain signals to direct the hetero-oligomers to the basolateral membrane. Failure of such hetero-oligomers to be delivered to the basolateral plasma membrane would indicate that, in this case, the apical targeting information in kAE1 (Arg^{901}→stop) protein overrides the basolateral targeting information on the wild-type protein.

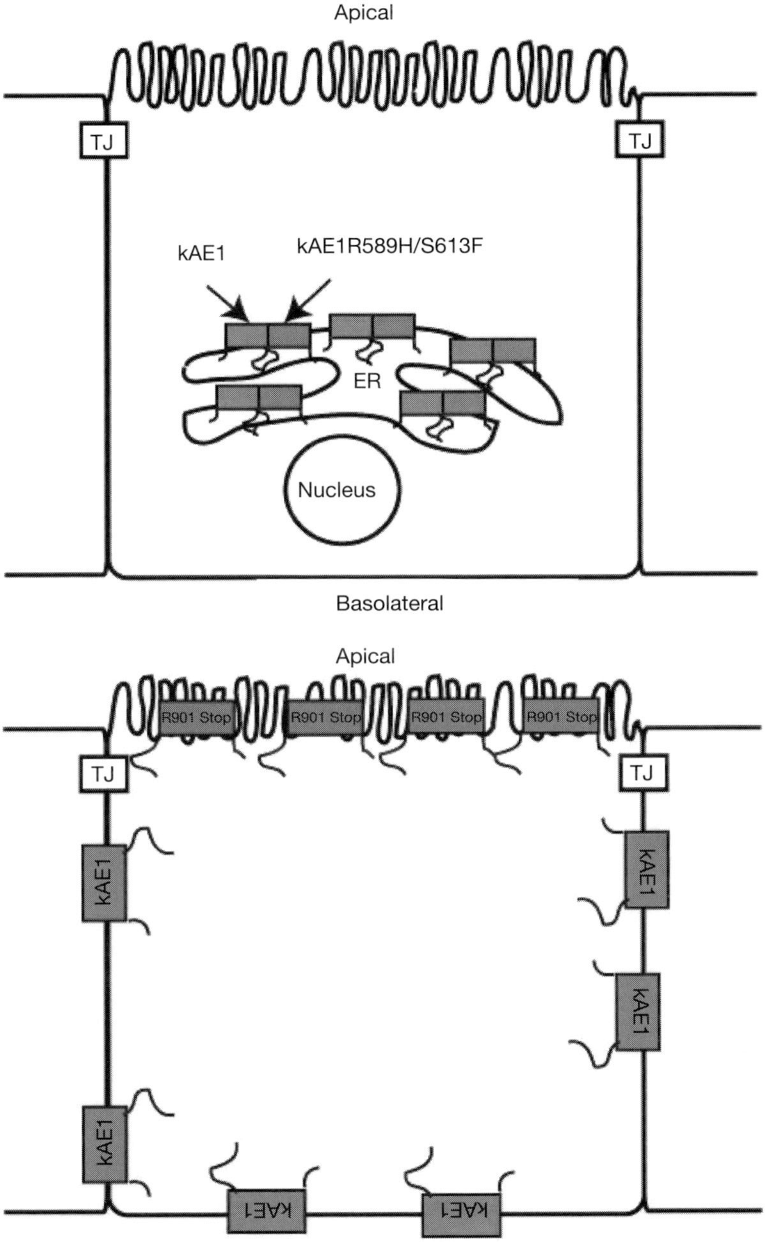

Figure 3 Diagram illustrating the potential molecular mechanisms that cause dRTA. Both Arg[589]→His and Ser[613]→Phe (kAE1R589H/S613F) are retained intracellularly (probably at the ER) and hetero-oligomerize with the wild-type protein, stopping it from reaching the basolateral plasma membrane and resulting in both proteins' degradation. The Arg[901]→stop (R901Stop) mutation is mistargeted to the apical plasma membrane where it neutralizes acid secreted by the cell. TJ, tight junction.

The reverse situation is illustrated by a dominant AQP-2 (aquaporin-2) mutant that has a basolateral-targeting motif in the C-terminus due to a frameshift mutation, and mistargets the normally apical wild-type AQP-2 to the basolateral membrane [53]. Further experiments utilizing polarized cells co-expressing wild-type and mutant kAE1 are needed to address this hypothesis. Initial attempts to investigate co-expression of wild-type and dRTA mutants in polarized cells using transient transfection were unsuccessful [30]. Stable MDCKI cells have now been developed that co-express both wild-type kAE1 and dRTA mutant kAE1 (A.M. Toye, unpublished work). It is hoped that polarization of these stable cells will help to define more conclusively the molecular basis of the disease associated with dRTA kAE1 mutations.

Basolateral-targeting signals in the N- and C-termini of kAE1

Results from investigations of the kAE1 (Arg^{901}→stop) dRTA mutant [30,31], and also the removal of the N-terminus of kAE1 [30], demonstrated that kAE1 has two basolateral determinants, one in the N-terminus and one in the C-terminus. The deletion of either the N- or the C-terminus results in the apical localization of the protein, suggesting that the presence of both determinants is essential for correct basolateral localization of kAE1. In non-polarized MDCKI cells, only the C-terminal determinant was required for plasma-membrane-targeting [30]; this illustrates the greater complexity of the polarized cell system.

The nature of the N-terminal basolateral determinant of human kAE1 has not been defined. Chicken kidney AE1-4 has two basolateral localization determinants in the N-terminus [40], but these are absent in human kAE1. The human kAE1 N-terminal domain may also contain a basolateral retention/sorting signal, or it may have a role in the polarized trafficking pathway before/after the C-terminus becomes involved. Alternatively, both the N- and the C-termini might interact simultaneously with as yet unidentified targeting machinery. The human kAE1 N-terminal domain has several conserved potential targeting motifs that are not present in chicken kAE1, and these are being evaluated at present.

Analysis of the sequence of the C-terminal tail of human AE1 reveals several motifs that appear to be conserved among the three AE1 family members and AE1 orthologues (Figure 4). The sequence -Y^{904}DEV- (of human AE1) within the C-terminus conforms to the consensus motif -YXXΦ- (where Φ is a bulky hydrophobic residue), which binds to medium-chain subunits of adaptor complexes involved in various clathrin-dependent and -independent trafficking events [54–58]. Directly adjacent to the tyrosine-based sorting signal in kAE1 is a type II PDZ binding motif (Ala-Met-Pro-Val), which can bind PICK-1 (protein interacting with protein kinase C 1) [59]. PDZ-binding domain proteins are widely expressed, are often involved in targeting and regulation of membrane proteins [60], and can retain apical or basolateral membrane proteins in their respective domains [61,62]. The C-terminus also contains repeats of di-acidic amino acids (Asp-Glu or Glu-Glu), some of which may also be important in kAE1 trafficking.

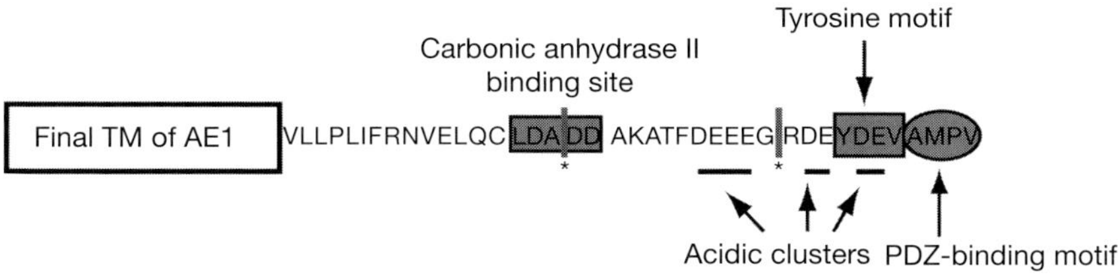

Figure 4 Analysis of the sequence of the C-terminal tail of human AE1. The sequence -YDEV- within the C-terminus conforms to the consensus motif -YXXΦ-, which binds to medium-chain subunits of adaptor complex trafficking events. Adjacent to the tyrosine-based sorting signal in kAE1 is a type II PDZ-binding motif (-AMPV). There are also repeats of di-acidic amino acids (Asp-Glu or Glu-Glu), which may also be important in kAE1 trafficking. The location of the 11-amino-acid and 23-amino-acid truncations which lead to dRTA are indicated by asterisks. The binding site for carbonic anhydrase II (-LDADD-) is also indicated.

The -Y^{904}DEV-, but not the C-terminal PDZ-binding motif, is critical for kAE1 basolateral localization [30]. Substitution of alanine or phenylalanine for Tyr^{904} alters kAE1 trafficking [30,31]. This tyrosine motif may interact directly with adaptor proteins that are necessary for basolateral targeting of kAE1. However, the adaptor complex AP-1B (activator protein 1B), which mediates basolateral sorting of some membrane proteins in epithelial cells [63], is not necessary for kAE1 trafficking to the basolateral membrane [30,31]. Therefore another adaptor complex, such as AP-4 [64] or other unidentified components, may direct kAE1 to the basolateral membrane.

kAE1 can tolerate the substitution of an alanine for valine at the Tyr+3 position in the C-terminal YXXΦ motif. Alanine is present at the Tyr+3 position in the basolateral tyrosine sorting motif of the Na^+–taurocholate co-transporter, which may not require a clathrin-adaptor complex for basolateral targeting [65]. Therefore either the basolateral-targeting machinery that interacts with kAE1 can tolerate the Val^{907}→Ala substitution or kAE1 (Val^{907}→Ala) can be targeted to the basolateral membrane by an alternative mechanism to that used by the wild-type protein.

Although not essential for kidney basolateral targeting, removal of the PDZ-interacting domain increased intracellular kAE1 levels observed in polarized MDCKI cells [30]. Therefore the PDZ-binding domain may still have some role to play in the efficient trafficking to, or retention at, the basolateral membrane. Of interest is the finding that the PDZ-binding domain of the NMDA (*N*-methyl-D-aspartate) receptor has a role in the direct targeting of this neuronal protein to the synapse [66]. Sans et al. [66] suggested two potential pathways for trafficking to the plasma membrane, a direct pathway that utilizes the exocyst complex (possibly used by newly synthesized proteins) and an endosomal route. Hence, if the PDZ-binding domain of kAE1 was involved in a direct route of targeting to the plasma membrane, its removal would still leave the endosomal pathway to predominate (see Figure 5). Therefore the

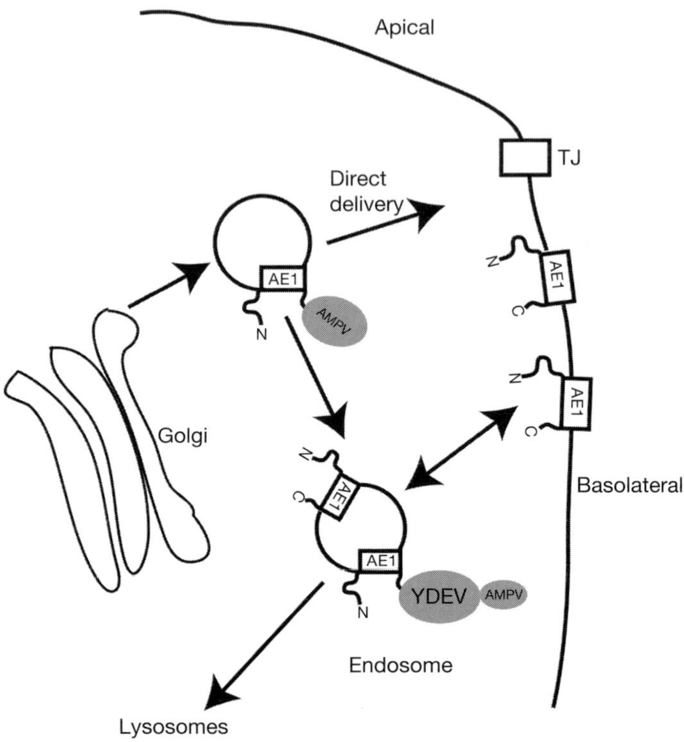

Figure 5 Potential role of the PDZ-binding domain and tyrosine motif in the C-terminus of kAE1. The tyrosine motif (-YDEV-) in the C-terminus of kAE1 is the critical determinant for basolateral targeting of kAE1. However, the removal of the PDZ-interacting domain does appear to increase intracellular kAE1. The PDZ-binding domain (-AMPV) may therefore be necessary for efficient trafficking to, or retention at, the basolateral membrane. It is suggested that the PDZ-binding domain of kAE1 may be involved in a direct route of targeting to the plasma membrane, whereas the Tyr^{904} motif may function as a basolateral/endosomal recycling motif, which enables the return of internalized kAE1 protein back to the cell surface. TJ, tight junction.

Y^{904}DEV motif may function as a basolateral/endosomal recycling motif, which enables the return of internalized kAE1 protein back to the cell surface.

A potential role for tyrosine phosphorylation in kAE1 trafficking

Tyrosine phosphorylation can influence the localization of proteins by regulating their interactions with cellular targeting and/or internalization machinery [41,67–70]. Both Tyr^{359} in the N-terminal domain and Tyr^{904} in the C-terminus of AE1 are phosphorylated in red blood cells [71], and are present in kAE1. A potential role for phosphorylation in the basolateral localization of kAE1 was recently suggested by the observation that the substitution of phenylalanine for Tyr^{904} alters kAE1 targeting [30]. Further work is necessary to investigate the potential role of this Tyr^{904} phosphorylation in the polarized targeting of kAE1. However, it is tempting to draw parallels with other proteins where phosphorylation is known to regulate the levels of the protein at the cell surface. In particular, the Y^{904}DEV motif is very similar to Y^{515}SEV of CEACAM1-L, which, when phosphorylated, causes internalization of CEACAM1-L from the basolateral plasma membrane [41]. In addition, substitution of alanine or phenylalanine for Tyr^{515} in CEACAM1-L restricts this protein to the apical membrane. Furthermore, substitution of alanine for valine at the Tyr+3 position did not affect the basolateral localization of CEACAM1-L (as with kAE1), but did reduce the level of phosphorylation of the tyrosine residue (and therefore decreased internalization of this protein).

Currently, it is unknown what regulates the delivery of kAE1 to the basolateral plasma membrane of α-intercalated cells or its internalization from that membrane. However, if it is assumed that phosphorylation of the Tyr^{904} residue in kAE1 may occur (analogous to the phosphorylated Tyr^{515} in CEACAM1-L), then it is possible to suggest a scheme that explains why both the N- and C-termini are required for basolateral targeting (see Figure 6). Thus, if phosphorylation of Tyr^{904} enhances the internalization of kAE1 (as phosphorylation of Tyr^{515} does in the case of CEACAM1-L), then the Tyr^{904} residue must also be dephosphorylated in order for the Y^{904}DEV basolateral-targeting motif to be recognized and allow recycling of kAE1 back to the basolateral membrane. This is where I would speculate that the N-terminal Tyr^{359} residue (of AE1) may be necessary, as, in the red blood cell, phosphorylation of Tyr^{359} recruits the phosphatase SHP-2 (Src-homology-2-domain-containing tyrosine phosphatase 2), which in turn dephosphorylates the Tyr^{904} residue [72] (see Figure 6). If a similar process occurs in the α-intercalated cell, this would enable the recycling of kAE1 back to the basolateral plasma membrane. This hypothesis, if correct, also provides a mechanism for the intercalated cells to regulate the levels of kAE1 at the basolateral plasma membrane in response to cues such as metabolic acidosis.

Conclusion

Until relatively recently, very little was known about the properties of kAE1 targeting and regulation in the α-intercalated cell. The observation that specific AE1 gene mutations cause dominant dRTA in the kidney, but have little effect on the red blood cell AE1 or when kAE1 mutants were expressed in *Xenopus* oocytes, strongly suggested that these mutations affected regions of kAE1 protein important for its correct function in the kidney [13,15,19,24]. The recent development of transfected polarizable cell models of kAE1 has confirmed this hypothesis, suggesting that dominant dRTA mutations can cause the intracellular retention of the mutant protein, or result in the protein being mistargeted to the apical plasma membrane [16,30,31]. This aberrant targeting of kAE1 [16,24,28–31], alongside the retained ability of these mutant proteins to form hetero-oligomers with the wild-type kAE1 protein [28,29], explains the molecular mechanism behind dominant dRTA.

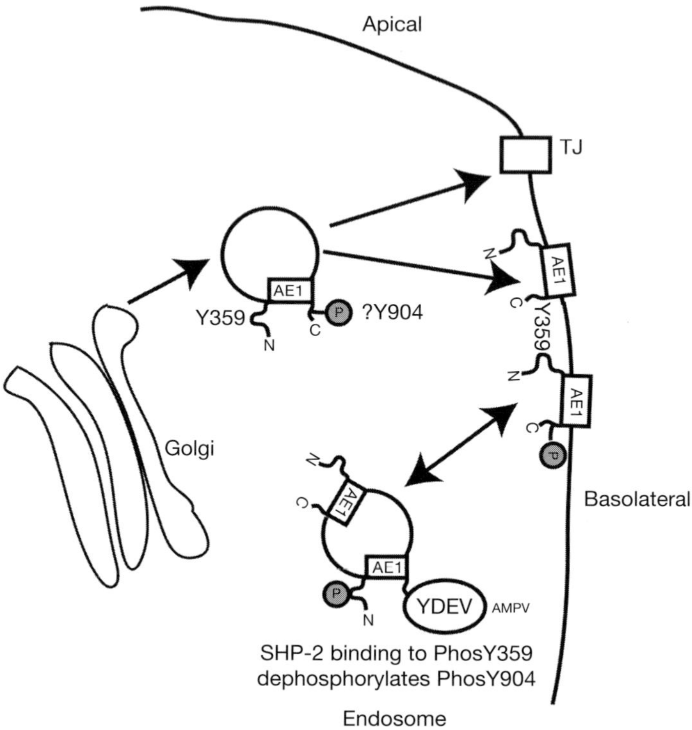

Figure 6 Potential role for tyrosine phosphorylation in kAE1 trafficking. A possible explanation as to why both the N-terminus and C-terminus are required for basolateral targeting, and why Tyr^{904}→Phe or Tyr^{904}→Ala mutations disrupt kAE1 trafficking, is that the phosphorylation of Tyr^{904} (PhosY904) is necessary for basolateral localization of the protein. The diagram illustrates how potentially, the phosphorylation of Tyr^{904} may result in the internalization of kAE1 from the basolateral plasma membrane to the endosomes. Tyr^{904} would then require the action of a phosphatase to dephosphorylate this residue, to then re-expose the -YDEV- basolateral sorting motif to enable recycling of kAE1 back to the basolateral membrane. This dephosphorylation is proposed to be carried out in the endosomes by the phosphatase SHP-2 (or an analogous phosphatase) that is recruited to kAE1 by the phosphorylation of residue Tyr^{359} (PhosY359). TJ, tight junction.

It is now clear that specific regions within kAE1 (the cytoplasmic N-terminus, residues between TM6 and TM7 and the cytoplasmic C-terminal tail) [30,31] are critical for correct localization of this basolateral protein. Therefore, by investigating the effects of dRTA mutations and also the regions of the kAE1 that they influence, we have opened up new avenues of research into kAE1 trafficking. It is hoped that the continuation of these studies will provide valuable insights into how mutations in oligomeric multispanning membrane proteins can cause human disease, and also provide valuable information regarding protein trafficking in polarized cells.

I thank Professor Mike Tanner and Professor George Banting for their support and encouragement during the studies described in this article. Many thanks to Professor Banting for critically reading this manuscript. The studies carried out by the author quoted in this article were funded by the Wellcome Trust and National Kidney Research Fund in grants to Professor Tanner and also Professor Banting.

References

1. Tanner, M.J. (2002) Curr. Opin. Hematol. **9**, 133–139
2. Zhang, D., Kiyatkin, A., Bolin, J.T. and Low, P.S. (2000) Blood **96**, 2925–2933
3. Vince, J.W. and Reithmeier, R.A. (1998) J. Biol. Chem. **273**, 28430–28437
4. Vince, J.W., Carlsson, U. and Reithmeier, R.A. (2000) Biochemistry **39**, 13344–13349
5. Reithmeier, R.A. (2001) Blood Cells Mol. Dis. **27**, 85–89
6. Kollert-Jons, A., Wagner, S., Hubner, S., Appelhans, H. and Drenckhahn, D. (1993) Am. J. Physiol. **265**, F813–F821
7. Ding, Y., Casey, J.R. and Kopito, R.R. (1994) J. Biol. Chem. **269**, 32201–32208
8. Chen, J., Vijayakumar, S., Li, X. and Al-Awqati, Q. (1998) J. Biol. Chem. **273**, 1038–1043
9. Hubner, S., Bahr, C., Gossmann, H., Efthymiadis, A. and Drenckhahn, D. (2003) Eur. J. Cell Biol. **82**, 240–252
10. Schuster, V.L., Bonsib, S.M. and Jennings, M.L. (1986) Am. J. Physiol. **251**, C347–C355
11. Drenckhahn, D., Schluter, K., Allen, D.P. and Bennett, V. (1985) Science **230**, 1287–1289
12. Alper, S.L., Natale, J., Gluck, S., Lodish, H.F. and Brown, D. (1989) Proc. Natl. Acad. Sci. U.S.A. **86**, 5429–5433
13. Bruce, L.J., Cope, D.L., Jones, G.K., Schofield, A.E., Burley, M., Povey, S., Unwin, R.J., Wrong, O. and Tanner, M.J. (1997) J. Clin. Invest. **100**, 1693–1707
14. Bruce, L.J., Wrong, O., Toye, A.M., Young, M.T., Ogle, G., Ismail, Z., Sinha, A.K., McMaster, P., Hwaihwanje, I., Nash, G.B. et al. (2000) Biochem. J. **350**, 41–51
15. Jarolim, P., Shayakul, C., Prabakaran, D., Jiang, L., Stuart-Tilley, A., Rubin, H.L., Simova, S., Zavadil, J., Herrin, J.T., Brouillette, J. et al. (1998) J. Biol. Chem. **273**, 6380–6388
16. Rungroj, N., Devonald, M.A., Cuthbert, A.W., Reimann, F., Akkarapatumwong, V., Yenchitsomanus, P.T., Bennett, W.M. and Karet, F.E. (2004) J. Biol. Chem. **279**, 13833–13838
17. Vasuvattakul, S., Yenchitsomanus, P.T., Vachuanichsanong, P., Thuwajit, P., Kaitwatcharachai, C., Laosombat, V., Malasit, P., Wilairat, P. and Nimmannit, S. (1999) Kidney Int. **56**, 1674–1682
18. Yenchitsomanus, P.T., Vasuvattakul, S., Kirdpon, S., Wasanawatana, S., Susaengrat, W., Sreethiphayawan, S., Chuawatana, D., Mingkum, S., Sawasdee, N., Thuwajit, P. et al. (2002) Am. J. Kidney Dis. **40**, 21–29
19. Karet, F.E., Gainza, F.J., Gyory, A.Z., Unwin, R.J., Wrong, O., Tanner, M.J., Nayir, A., Alpay, H., Santos, F., Hulton, S.A. et al. (1998) Proc. Natl. Acad. Sci. U.S.A. **95**, 6337–6342
20. Ribeiro, M.L., Alloisio, N., Almeida, H., Gomes, C., Texier, P., Lemos, C., Mimoso, G., Morle, L., Bey-Cabet, F., Rudigoz, R.C. et al. (2000) Blood **96**, 1602–1604
21. Ruf, R., Rensing, C., Topaloglu, R., Guay-Woodford, L., Klein, C., Vollmer, M., Otto, E., Beekmann, F., Haller, M., Wiedensohler, A. et al. (2003) Pediatr. Nephrol. **18**, 105–109
22. Smith, A.N., Skaug, J., Choate, K.A., Nayir, A., Bakkaloglu, A., Ozen, S., Hulton, S.A., Sanjad, S.A., Al-Sabban, E.A., Lifton, R.P. et al. (2000) Nat. Genet. **26**, 71–75
23. Karet, F.E., Finberg, K.E., Nelson, R.D., Nayir, A., Mocan, H., Sanjad, S.A., Rodriguez-Soriano, J., Santos, F., Cremers, C.W., Di Pietro, A. et al. (1999) Nat. Genet. **21**, 84–90
24. Toye, A.M., Bruce, L.J., Unwin, R.J., Wrong, O. and Tanner, M.J. (2002) Blood **99**, 342–347
25. Cheidde, L., Vieira, T.C., Lima, P.R., Saad, S.T. and Heilberg, I.P. (2003) Pediatrics **112**, 1361–1367

26. Sritippayawan, S., Sumboonnanonda, A., Vasuvattakul, S., Keskanokwong, T., Sawasdee, N., Paemanee, A., Thuwajit, P., Wilairat, P., Nimmannit, S., Malasit, P. and Yenchitsomanus, P.T. (2004) Am. J. Kidney Dis. **44**, 64–70
27. Kuma, H., Abe, Y., Askin, D., Bruce, L.J., Hamasaki, T., Tanner, M.J. and Hamasaki, N. (2002) Biochemistry **41**, 3311–3320
28. Quilty, J.A., Cordat, E. and Reithmeier, R.A. (2002) Biochem. J. **368**, 895–903
29. Quilty, J.A., Li, J. and Reithmeier, R.A. (2002) Am. J. Physiol. Renal Physiol. **282**, F810–F820
30. Toye, A.M., Banting, G. and Tanner, M.J. (2004) J. Cell Sci. **117**, 1399–1410
31. Devonald, M.A., Smith, A.N., Poon, J.P., Ihrke, G. and Karet, F.E. (2003) Nat. Genet. **33**, 125–127
32. Popov, M., Li, J. and Reithmeier, R.A. (1999) Biochem. J. **339**, 269–279
33. Tang, X.B., Fujinaga, J., Kopito, R. and Casey, J.R. (1998) J. Biol. Chem. **273**, 22545–22553
34. Tang, X.B., Kovacs, M., Sterling, D. and Casey, J.R. (1999) J. Biol. Chem. **274**, 3557–3564
35. Zhu, Q., Lee, D.W. and Casey, J.R. (2003) J. Biol. Chem. **278**, 3112–3120
36. Li, J., Quilty, J., Popov, M. and Reithmeier, R.A. (2000) Biochem. J. **349**, 51–57
37. Cheung, J.C. and Reithmeier, R.A. (2004) Biochem. J. **378**, 1015–1021
38. Beckmann, R., Smythe, J.S., Anstee, D.J. and Tanner, M.J. (1998) Blood **92**, 4428–4438
39. Beckmann, R., Toye, A.M., Smythe, J.S., Anstee, D.J. and Tanner, M.J. (2002) Mol. Membr. Biol. **19**, 187–200
40. Adair-Kirk, T.L., Cox, K.H. and Cox, J.V. (1999) J. Cell Biol. **147**, 1237–1248
41. Sundberg, U., Beauchemin, N. and Obrink, B. (2004) J. Cell Sci. **117**, 1091–1104
42. Denning, G.M., Anderson, M.P., Amara, J.F., Marshall, J., Smith, A.E. and Welsh, M.J. (1992) Nature (London) **358**, 761–764
43. Groves, J.D., Wang, L. and Tanner, M.J. (1998) FEBS Lett. **433**, 223–227
44. Zhu, Q. and Casey, J.R. (2004) J. Biol. Chem. **279**, 23565–23573
45. Cordat, E., Li, J. and Reithmeier, R.A. (2003) Traffic **4**, 642–651
46. Groves, J.D. and Tanner, M.J. (1992) J. Biol. Chem. **267**, 22163–22170
47. Groves, J.D. and Tanner, M.J. (1995) J. Biol. Chem. **270**, 9097–9105
48. Vince, J.W. and Reithmeier, R.A. (2000) Biochemistry **39**, 5527–5533
49. Asai, T., Kuwahara, M., Kurihara, H., Sakai, T., Terada, Y., Marumo, F. and Sasaki, S. (2003) Kidney Int. **64**, 2–10
50. Marr, N., Bichet, D.G., Lonergan, M., Arthus, M.F., Jeck, N., Seyberth, H.W., Rosenthal, W., van Os, C.H., Oksche, A. and Deen, P.M. (2002) Hum. Mol. Genet. **11**, 779–789
51. Koivisto, U.M., Hubbard, A.L. and Mellman, I. (2001) Cell **105**, 575–585
52. Shayakul, C., Jarolim, P., Zachlederova, M., Prabakaran, D., Cortez-Campeao, D., Kalabova, D., Stuart-Tilley, A.K., Ideguchi, H., Haller, C. and Alper, S.L. (2004) Nephrol. Dial. Transplant **19**, 371–379
53. Kamsteeg, E.J., Bichet, D.G., Konings, I.B., Nivet, H., Lonergan, M., Arthus, M.F., van Os, C.H. and Deen, P.M. (2003) J. Cell Biol. **163**, 1099–1109
54. Aguilar, R.C., Boehm, M., Gorshkova, I., Crouch, R.J., Tomita, K., Saito, T., Ohno, H. and Bonifacino, J.S. (2001) J. Biol. Chem. **276**, 13145–13152
55. Boehm, M. and Bonifacino, J.S. (2001) Mol. Biol. Cell **12**, 2907–2920
56. Dell'Angelica, E.C., Ohno, H., Ooi, C.E., Rabinovich, E., Roche, K.W. and Bonifacino, J.S. (1997) EMBO J. **16**, 917–928
57. Ohno, H., Fournier, M.C., Poy, G. and Bonifacino, J.S. (1996) J. Biol. Chem. **271**, 29009–29015
58. Stephens, D.J. and Banting, G. (1998) Biochem. J. **335**, 567–572
59. Cowan, C.A., Yokoyama, N., Bianchi, L.M., Henkemeyer, M. and Fritzsch, B. (2000) Neuron **26**, 417–430
60. Sheng, M. and Sala, C. (2001) Annu. Rev. Neurosci. **24**, 1–29
61. Moyer, B.D., Denton, J., Karlson, K.H., Reynolds, D., Wang, S., Mickle, J.E., Milewski, M., Cutting, G.R., Guggino, W.B., Li, M. and Stanton, B.A. (1999) J. Clin. Invest. **104**, 1353–1361

62. Olsen, O., Liu, H., Wade, J.B., Merot, J. and Welling, P.A. (2002) Am. J. Physiol. Cell Physiol. **282**, C183–C195
63. Folsch, H., Ohno, H., Bonifacino, J.S. and Mellman, I. (1999) Cell **99**, 189–198
64. Simmen, T., Honing, S., Icking, A., Tikkanen, R. and Hunziker, W. (2002) Nat. Cell Biol. **4**, 154–159
65. Sun, A.Q., Arrese, M.A., Zeng, L., Swaby, I., Zhou, M.M. and Suchy, F.J. (2001) J. Biol. Chem. **276**, 6825–6833
66. Sans, N., Prybylowski, K., Petralia, R.S., Chang, K., Wang, Y.X., Racca, C., Vicini, S. and Wenthold, R.J. (2003) Nat. Cell Biol. **5**, 520–530
67. Quick, M.W., Hu, J., Wang, D. and Zhang, H.Y. (2004) J. Biol. Chem. **279**, 15961–15967
68. Sterling, H., Lin, D.H., Gu, R.M., Dong, K., Hebert, S.C. and Wang, W.H. (2002) J. Biol. Chem. **277**, 4317–4323
69. Bradshaw, J.D., Lu, P., Leytze, G., Rodgers, J., Schieven, G.L., Bennett, K.L., Linsley, P.S. and Kurtz, S.E. (1997) Biochemistry **36**, 15975–15982
70. Stephens, D.J. and Banting, G. (1997) FEBS Lett. **416**, 27–29
71. Yannoukakos, D., Meyer, H.E., Vasseur, C., Driancourt, C., Wajcman, H. and Bursaux, E. (1991) Biochim. Biophys. Acta **1066**, 70–76
72. Bordin, L., Brunati, A.M., Donella-Deana, A., Baggio, B., Toninello, A. and Clari, G. (2002) Blood **100**, 276–282

Biochem. Soc. Symp. **72**, 65–70
(Printed in Great Britain)

6

Regulation of the clathrin-coated vesicle cycle by reversible phosphorylation

Alexander Flett*, Sophia Semerdjieva*, Antony P. Jackson† and Elizabeth Smythe*[1]

*Department of Biomedical Science, University of Sheffield, Western Bank, Sheffield S10 2TN, U.K., and †Department of Biochemistry, University of Cambridge, Tennis Court Road, Cambridge CB2 1QW, U.K.

Abstract

Reversible phosphorylation has long been an attractive mechanism to control cycles of coat assembly and disassembly during clathrin-mediated endocytosis. Many of the coat proteins are phosphorylated *in vivo* and *in vitro*. Our work has focused on the role of phosphorylation of the μ2 subunit of AP-2 (adaptor protein 2), which appears to be necessary for efficient cargo recruitment. Studies to probe the regulation of μ2 phosphorylation demonstrated that clathrin is a specific activator of the μ2 kinase, and, in permeabilized cells, cargo sequestration, driven by exogenously added clathrin, results in elevated levels of μ2 phosphorylation. Furthermore, phosphorylated μ2 is mainly associated with assembled clathrin *in vivo* and its steady-state level is strongly reduced in cells depleted of clathrin heavy chain. Our results imply a central role for clathrin in the regulation of cargo selection via modulation of phospho-μ2 levels. This is therefore a novel regulatory role for clathrin that is independent of its structural role and that provides elegant spatial control of AP-2 and cargo interactions, ensuring that AP-2 is only activated at the correct cellular location and in the correct functional context. Ongoing studies are exploring further the roles of reversible phosphorylation in the coated vesicle cycle.

Introduction

Most material internalized into eukaryotic cells follows the pathway of clathrin-mediated endocytosis. This includes essential nutrients, such as iron

[1]To whom correspondence should be addressed (email E.Smythe@sheffield.ac.uk).

and cholesterol (bound to transferrin and low-density lipoprotein respectively), signalling molecules, such as growth factors, and immune complexes, and the pathway is also hijacked by pathogens for entry into the cell [1]. Coated pits are morphologically distinct areas of the plasma membrane that are distinguished by their bristle-like appearance in transverse sections of cells viewed by electron microscopy [2]. In freeze–fracture deep-etch images, coated pits appear as regular lattices [3]. In both cases, the morphology results from the polymerization of the protein clathrin from the cytosol on to the membrane. Many other proteins are involved in the assembly of coated pits and contribute to their increase in curvature or invagination, recruitment of cargo and subsequent scission to form coated vesicles [4].

Coated vesicles can be isolated with relative ease from tissue such as brain and liver, and, after clathrin, the other major component of isolated coated vesicles is the adaptor protein, AP-2. AP-2 is a heterotetrameric complex consisting of two 100 kDa subunits (α and β2), a 50 kDa subunit (μ2) and an approx. 17–19 kDa subunit (σ2) [5]. As its name suggests, AP-2 acts as an adaptor that links cargo molecules that are to be internalized with the clathrin lattice. AP-2 also interacts with many other components that are implicated in coated vesicle formation, and, indeed, many of these components undergo a complex set of interactions with each other, suggesting that these interactions must be very tightly regulated in space and time during the vesicle cycle. Among these components are eps15 (epidermal growth factor receptor pathway substrate 15), epsin, endophilin, AP180 (assembly protein 180), auxilin, Hsc70 (heat-shock cognate 70) and Rab5, and the large-molecular-mass GTPase, dynamin, is required for the invagination and scission of coated pits [4].

μ2 phosphorylation is required for endocytosis

Reversible phosphorylation has always been suggested as a plausible regulatory mechanism to control the assembly of coated pits and the subsequent disassembly of the coat from the vesicle following scission. Our research has addressed the question as to whether and how phosphorylation of μ2 affects the activity of AP-2. Yeast two-hybrid analysis has shown that the μ2 subunit of AP-2 recognizes specific internalization motifs in cargo molecules, such as the transferrin receptor, which have the following general consensus sequence (YXXΦ, where Φ is a bulky hydrophobic residue) [6]. Solution of the crystal structure of μ2 complexed to peptides containing this motif indicated that tyrosine-based motifs bind to an extended groove on the μ2 subunit [7]. Phosphorylation of μ2 *in vitro* occurs on Thr^{156} [8], which is located within an exposed region of the molecule.

One of the approaches that we use extensively in our work to understand spatio-temporal interactions in the coated vesicle cycle is the reconstitution of different stages of the cycle in permeabilized cells [9–13]. For example, we can measure the sequestration of cargo (in this case, transferrin receptor) into deeply invaginated coated pits that form *de novo* in permeabilized cells [10] (Figure 1). This step is stimulated by the addition of AP-2 to the assay system. The availabil-

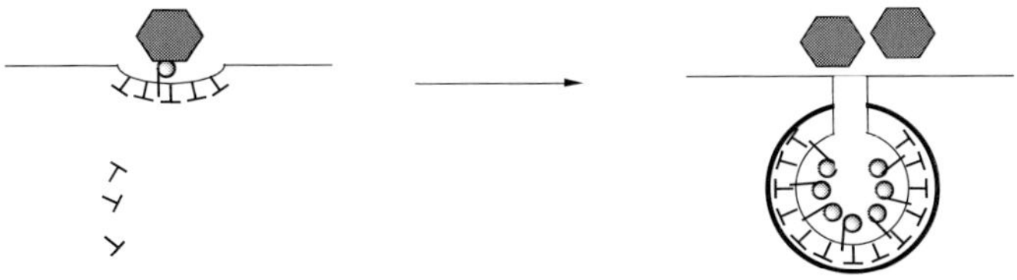

Figure 1 AP-2 (T-shapes) stimulates the sequestration of cargo into newly formed coated pits in permeabilized cells. These pits can sequester cargo molecules, such as biotinylated transferrin, that become inaccessible to avidin (indicated by the hexagons).

ity of an assay that is dependent on added AP-2 allowed us to investigate the regulation of AP-2 function by phosphorylation.

We found that if μ2 phosphorylation of the added AP-2 is blocked, then cargo sequestration is also blocked, suggesting that μ2 phosphorylation regulates cargo sequestration. The role of μ2 phosphorylation was confirmed *in vivo* by replacement of endogenous μ2 with a mutant form of μ2 where the single phosphorylation site, Thr^{156}, had been mutated to an alanine residue. Cells expressing this mutant form showed a reduced rate of transferrin internalization [12].

What does μ2 phosphorylation do?

Our data indicated that μ2 phosphorylation might have a role either in invagination or cargo recruitment, and several complementary studies indicated that μ2 phosphorylation was likely to promote cargo recruitment. Using BiaCore approaches, Honing and colleagues demonstrated that phosphorylated μ2 binds with a much higher affinity to peptides containing tyrosine-based internalization motifs compared with non-phosphorylated μ2 [14]. AP-1 is related to AP-2 in subunit composition and is involved in clathrin-coated vesicle formation at the *trans*-Golgi network. Phosphorylation of its medium subunit, μ1, similarly enhances its association with cargo [15].

This begs the question of how μ2 phosphorylation actually affects cargo sequestration. The structure of the AP-2 adaptor complex has been solved at atomic resolution, and, revealingly, the binding site for tyrosine-based peptides is partially blocked in the solved structure [16]. This suggests that one possible role for μ2 phosphorylation would be to induce a conformational change in AP-2, revealing the cargo-binding site [17]. Alternatively, it has been suggested that AP-2 might undergo many different conformations to enhance the flexibility with which it is able to bind sorting signals [18], and it is possible that μ2 phosphorylation stabilizes some of these conformations, thus enhancing cargo–adaptor interactions.

Regulation of μ2 phosphorylation

To pursue our interests in μ2 phosphorylation, we investigated the regulation of the μ2 kinase. The most likely candidate *in vivo* for the μ2 kinase is

AAK1 (adaptor-associated kinase 1), a kinase identified by a phage-display interaction screen using the C-terminal appendage or ear domain of α-adaptin as bait [19]. AAK1 is a serine/threonine kinase that belongs to the Ark/Prk family of protein kinases, which has been implicated in regulation of endocytosis and the actin cytoskeleton in yeast [20]. AAK1 is enriched at sites of endocytosis *in vivo*, and fractions from bovine brain that are enriched in μ2 kinase activity are also enriched in AAK1 [14,19]. Structurally, AAK1 may be divided into three main domains. It has an N-terminal kinase domain which is the domain of the protein with the strongest homology with the kinase domains of other members of the Ark/Prk family [20]. The middle domain has a high proportion of glutamine, proline and alanine residues. In general, sequences with this amino acid composition tend to be quite flexible and form so-called 'loopy' domains. The C-terminal domain of AAK1 contains motifs that allow interaction with other components of the endocytic machinery. These include DPF (Asp-Pro-Phe), NPF (Asn-Pro-Phe) and DLL (Asp-Leu-Leu) motifs that have been demonstrated to interact with α-adaptin, EH (eps15 homology)-domain-containing proteins, such as eps15, and clathrin respectively [19]. Interestingly, however, the interaction of AAK1 with the appendage domain of α-adaptin has been shown to be via a novel α-adaptin-interacting motif, WNPF (Trp-Asn-Pro-Phe) [21].

The presence of a DLL motif within the C-terminus of AAK1 suggested that clathrin, through binding of AAK1, might modulate its activity. In the first instance, we investigated this possibility by preparing a highly enriched fraction of the μ2 kinase from bovine brain, and examined the levels of μ2 phosphorylation over a range of AP-2 in the presence of increasing concentrations of clathrin. We found that clathrin acts as a competitive activator of μ2 kinase, acting to reduce the apparent K_m without having a significant effect on the apparent V_{max} [13].

Further studies in permeabilized cells allowed us to investigate whether clathrin also modulates phospho-μ2 levels under more physiological conditions. To pursue these studies, we generated antibodies that specifically recognize the phosphorylated form of μ2. Using this reagent, we were able to demonstrate that clathrin can also modulate phospho-μ2 levels in permeabilized cells, and this modulation correlates with enhanced cargo sequestration. These results were strengthened further by studies in a chicken DT-40 B-cell line where clathrin can be ablated inducibly [29]. Under conditions where clathrin is expressed, a large proportion of the clathrin is found in an assembled pool, whereas AP-2 is distributed more evenly between the assembled and unassembled pools. Strikingly, almost all of the phospho-μ2 is found associated with assembled clathrin. Under conditions where clathrin expression is knocked-out, there is no change in either the distribution or level of expression of AP-2, but the level of phospho-μ2 is reduced dramatically, indicating that clathrin can also modulate phospho-μ2 levels *in vivo* [13].

Mechanism of AAK1 activation by clathrin

Our investigations into clathrin activation of the μ2 kinase were initiated by the presence of a DLL motif within the C-terminus of AAK1. However, we have generated recombinant wild-type AAK1 and a mutant form where the three residues in the DLL motif have all been mutated to alanine. Although mutation of the DLL motif does reduce binding of AAK1 to the clathrin terminal domain by approx. 50%, there is no effect on the activation of the enzyme by clathrin (A. Flett and E. Smythe, unpublished work). There are no other defined clathrin-binding motifs present within the sequence of AAK1, suggesting that the activation of AAK1 by clathrin must occur via a novel motif. This is consistent with other studies that show that clathrin activation of AAK1 occurs via multiple interactions [22]. Our model for the activation of the μ2 kinase by clathrin is based on the idea that the kinase is closely associated with AP-2 in the cytosol and appears to be in a low-activity state. This could be because of an intramolecular interaction with a potential pseudosubstrate located at the C-terminus of AAK1. Many protein kinases are regulated in this manner [23]. Presumably activation by clathrin induces a conformational change such that this inhibition is relieved and the kinase becomes maximally active [13]. Our kinetic data are consistent with such a mechanism. If clathrin activates AAK1 in this manner, then the central loopy domain is sufficiently long and flexible to allow AAK1 to phosphorylate other μ2 subunits within the coated pit.

Clathrin acts to regulate cargo recruitment

These results suggest that clathrin, in addition to its well-established structural role in coated pit assembly, also plays a key regulatory role in the coated vesicle cycle by ensuring that productive adaptor–cargo interactions only occur in the context of a coated pit. This reappraisal of the role of clathrin is supported by two other lines of investigation. First, it has become apparent that many cargos use alternative adaptors to AP-2 to facilitate their internalization. These include β-arrestin for G-protein-coupled receptors, numb, Dab and ARH (autosomal recessive hypercholesterolaemia) for cargo bearing FXNPXY (Phe-Xaa-Asn-Pro-Xaa-Tyr) motifs and AP180 possibly for SNAREs (soluble *N*-ethylmaleimide-sensitive fusion protein attachment protein receptor) (for a review see [24]). Many of these other adaptors also interact with AP-2, but studies where AP-2 was ablated with RNAi (RNA interference) indicate that many cargos can be efficiently internalized in the absence of AP-2, and, importantly, coated pits can still form [25,26]. This supports further the idea that, although AP-2 clearly has a major role in pit formation, it may be more cargo-specific than previously thought. This brings clathrin back centre-stage as a regulator, in addition to a structural component, of coated pits. Intriguingly, the concept of clathrin as a regulator is supported by studies from other research groups. For example, inducible expression of antisense RNA against clathrin heavy chain resulted in a reduction of clathrin heavy chain expression of only 10%, but caused an almost complete inhibition of transferrin endocytosis [27]. This appeared to be due to the generation of coated pits with long necks around which spirals of dynamin were wound. This indicated cross-talk between clathrin and dynamin. Clathrin also binds the Class II phosphoinositide 3-kinase C2α and stimulates its activity [28]. Since this enzyme localizes to coated pits, it appears to be involved in local generation of phosphoinositides necessary for coated pit formation.

We propose that clathrin may act as a general regulator of adaptor–cargo interactions, and it will be interesting to explore in the future whether clathrin also modulates cargo recruitment during other clathrin-mediated sorting events within the cell.

Work in our laboratories is supported by the Medical Research Council (MRC) and Biotechnology and Biological Sciences Research Council (BBSRC) (to E.S.) and the Wellcome Trust (to A.P.J.).

References

1. Conner, S.D. and Schmid, S.L. (2003) Nature (London) **422**, 37–44
2. Marsh, M. and McMahon, H.T. (1999) Science **285**, 215–220
3. Heuser, J.E. and Evans, L. (1980) J. Cell Biol. **84**, 560–583
4. Brodsky, F.M., Chen, C.Y., Knuehl, C., Towler, M.C. and Wakeham, D.E. (2001) Annu. Rev. Cell Dev. Biol. **17**, 517–568
5. Robinson, M.S. (2004) Trends Cell Biol. **14**, 167–174
6. Ohno, H., Stewart, J., Fournier, M.C., Bosshart, H., Rhee, I., Miyatake, S., Saito, T., Gallusser, A., Kirchhausen, T. and Bonifacino, J.S. (1995) Science **269**, 1872–1875
7. Owen, D.J. and Evans, P.R. (1998) Science **282**, 1327–1332
8. Pauloin, A. and Thurieau, C. (1993) Biochem. J. **296**, 409–415
9. Smythe, E., Carter, L.L. and Schmid, S.L. (1992) J. Cell Biol. **119**, 1163–1171
10. McLauchlan, H., Newell, J., Morrice, N., Osborne, A., West, M. and Smythe, E. (1998) Curr. Biol. **8**, 34–45
11. Hill, E., van der Kaay, J., Downes, C.P. and Smythe, E. (2001) J. Cell Biol. **152**, 1–16
12. Olusanya, O., Andrews, P.D., Swedlow, J.R. and Smythe, E. (2001) Curr. Biol. **11**, 896–900
13. Jackson, A.P., Flett, A., Smythe, C., Hufton, L., Wettey, F.R. and Smythe, E. (2003) J. Cell Biol. **163**, 231–236
14. Ricotta, D., Conner, S.D., Schmid, S.L., von Figura, K. and Honing, S. (2002) J. Cell Biol. **156**, 791–795
15. Ghosh, P. and Kornfeld, S. (2003) J. Cell Biol. **160**, 699–708
16. Collins, B.M., McCoy, A.J., Kent, H.M., Evans, P.R. and Owen, D.J. (2002) Cell **109**, 523–535
17. Smythe, E. (2002) Trends Cell Biol. **12**, 352–354
18. Kirchhausen, T. (2002) Cell **109**, 413–416
19. Conner, S.D. and Schmid, S.L. (2002) J. Cell Biol. **156**, 921–929
20. Smythe, E. and Ayscough, K. (2003) EMBO Rep. **4**, 246–251
21. Jha, A., Agostinelli, N.R., Mishra, S.K., Keyel, P.A., Hawryluk, M.J. and Traub, L.M. (2004) J. Biol. Chem. **279**, 2281–2290
22. Conner, S.D., Schroter, T. and Schmid, S.L. (2003) Traffic **4**, 885–890
23. Kemp, B.E. and Pearson, R.B. (1991) Biochim. Biophys. Acta **1094**, 67–76
24. Traub, L.M. (2003) J. Cell Biol. **163**, 203–208
25. Motley, A., Bright, N.A., Seaman, M.N. and Robinson, M.S. (2003) J. Cell Biol. **162**, 909–918
26. Hinrichsen, L., Harborth, J., Andrees, L., Weber, K. and Ungewickell, E.J. (2003) J. Biol. Chem. **278**, 45160–45170
27. Iversen, T.G., Skretting, G., van Deurs, B. and Sandvig, K. (2003) Proc. Natl. Acad. Sci. U.S.A. **100**, 5175–5180
28. Gaidarov, I., Smith, M.E., Domin, J. and Keen, J.H. (2001) Mol. Cell **7**, 443–449
29. Wettey, F.R., Hawkins, S.F., Stewart, A., Luzio, J.P., Howard, J.C. and Jackson, A.P. (2002) Science **297**, 1521–1525

Biochem. Soc. Symp. **72**, 71–76
(Printed in Great Britain)

7

Single-molecule live-cell imaging of clathrin-based endocytosis

Tomas Kirchhausen[*1], Werner Boll[*], Antoine van Oijen[†] and Marcelo Ehrlich[*]

*Department of Cell Biology, Harvard Medical School and Center for Blood Research, 200 Longwood Avenue, Boston, MA 02115, U.S.A., and †Department of Biological Chemistry and Molecular Pharmacology, Harvard Medical School, 240 Longwood Avenue, Boston, MA 02115, U.S.A.

Abstract

Clathrin-coated vesicles carry traffic from the plasma membrane to endosomes. We report here the first real-time visualization of cargo sorting and endocytosis by clathrin-coated pits in living cells. We have visualized the formation of coats by monitoring the incorporation of fluorescently tagged clathrin or its adaptor AP-2 (adaptor protein 2), and have followed clathrin-mediated uptake of transferrin, single LDL (low-density lipoprotein) and single reovirus particles. The intensity of a cargo-loaded clathrin cluster grows steadily during its lifetime, and the time required to complete assembly is proportional to the size of the cargo particle. These results are consistent with a nucleation-growth mechanism and an approximately constant growth rate. There are no preferred nucleation sites. A proportion of the nucleation events appear to be abortive. Cargo incorporation occurs primarily or exclusively in a newly formed coated pit, and loading appears to commit that pit to finish assembly. Our data led to a model in which coated pits initiate randomly, but collapse with high likelihood unless stabilized, presumably by cargo capture.

Background

Thirty years ago, while still a graduate student, I would browse cell physiology textbooks, and be fascinated by the number of ideas and models dwelling on the transport of water, ions and small molecules across cell membranes. What was notoriously lacking then, however, were molecular and mechanistic thoughts on how much larger molecules, such as protein hormones, lipopro-

[1]To whom correspondence should be addressed
(email Kirchhausen@crystal.harvard.edu)

teins and viruses in the cell milieu and membrane proteins bound to the cell periphery, the cargo, entered cells. The situation is vastly different today, as we now realize that this type of entry requires the regulated formation of membrane invaginations that, after sealing, form vesicular or tubular carriers that deliver the cargo typically to endosomes.

The clathrin-dependent pathway was the first system of membrane traffic to be recognized and studied in detail. This was due primarily to the distinctive morphological characteristics of budding coated pits, to the ease with which coated vesicles could be purified and to the widespread importance of clathrin-coated structures for receptor-mediated endocytosis. The clathrin pathway has thus acquired special status for analysing molecular mechanisms in membrane traffic.

The 'life-cycle' of a clathrin-coated vesicle, from coat assembly and cargo loading to coat disassembly and cargo delivery, involves a number of highly regulated events occurring on a rapid time scale (seconds). New tools of live-cell imaging, which provide the time resolution required to separate key steps, are starting to provide important new data of temporal sequences of protein associations and translocations. These results can now be integrated with static information, such as the identification of molecular partners and participating structures, derived largely from genetics and biochemistry, and the determination of three-dimensional structures of components, from X-ray crystallography and cryo-EM (cryo-electron microscopy). The goal now is to generate a mechanistic picture of clathrin-based membrane vesicular traffic — a 'molecular movie' of the molecular endocytic machinery.

High-resolution structure of clathrin and molecular anatomy of a coat

Clathrin was discovered 25 years ago [1]. It forms the striking lattices that surround coated pits and vesicles. It is a spider-like molecule, with three legs radiating from a central hub (a 'triskelion') [2]. Each leg contains a 190 kDa 'heavy chain' (the species discovered by Pearse [1]) and an associated ~25 kDa 'light chain' [2,3]. Antiparallel interactions of the legs from triskelions centred on adjacent vertices of the lattice allow assembly of a coat [3,4].

High-resolution structural studies of clathrin fragments by X-ray diffraction [5–7] and lower-resolution analyses of clathrin coats by cryo-EM to resolutions of 21 Å (1 Å=0.1 nm) [8,9] and 7.9 Å [10] have led to a relatively detailed molecular picture of the framework of a coat and to a view of how clathrin assembly can be coupled to cargo incorporation.

Clathrin interacts directly with adaptor proteins, such as the heterotetrameric AP (adaptor protein) complexes [11–14] and other, specialized, adaptors, such as β-arrestins [15] and GGAs (Golgi-localized, γ-ear-containing, ADP-ribosylation-factor-binding proteins) [16]. These adaptors in turn link to the cytoplasmic segments of membrane-bound protein receptors (the cargo). The β-propeller structure of the clathrin N-terminal domain binds peptides such as the 'clathrin-box' motifs (LΦ{D/E}Φ{D/E}, where Φ indicates a large hydrophobic residue) [5] and the W box motif (Pro-Trp-Xaa-Xaa-Trp)

[17] present in disordered loops of various adaptor proteins and in other regulatory proteins. Combining the X-ray structure of the 55 kDa N-terminal fragment of clathrin with the 21 Å resolution cryo-EM image reconstruction of a coat, we were able to show that the clathrin terminal domain projects inward beneath a vertex, close to the presumed radial position of an adaptor complex [8]. The same work compared the conformation of the clathrin triskelion legs in each of the nine different packing environments within the coat, and suggested that the α-zigzag characteristic of most of the clathrin heavy chain provided a stiff, but slightly flexible, structure that could adapt to lattices of different size and design.

Coat assembly and disassembly

Despite this structural progress, we still do not fully understand the process of coat assembly and disassembly. Opinions are divided regarding the formation of coated pits. We believe that a curved lattice forms by sequential addition of individual cytosolic clathrin trimers to a growing shell [8]. Others have suggested that a flat hexagonal array of clathrin can restructure into the mixture of hexagons and pentagons in a curved assembly [18]. Such a rearrangement is physically impossible without substantial local dissociation, and we have proposed that flat arrays (observed in many electron micrographs) function as reservoirs for rapid assembly of curved coated pits at the edges of the arrays [8].

The nature of a clathrin coat is such that triskelions can assemble and disassemble by rotations centred on any vertex (for a computer animation of our model for clathrin assembly, consult http://www.cbrinstitute.org/kirchhausen/research.html). In fact, it has been discovered that such exchange occurs in living cells [19]. A large fraction of the clathrin structures bound to membranes underwent constant exchange with the cytosolic pool. The *in vivo* exchange was a surprise, however, as the prevailing model for coat formation predicted a simple sequential recruitment of coat components culminating with the formation of a coated vesicle. It is possible that Hsc70 (heat-shock cognate 70) and auxilin, proteins that drive the ATP-dependent disassembly of the clathrin lattice [20], are involved in the exchange reaction as well, but, at present, this is only a speculation. Although many mechanistic aspects of the Hsc70/auxilin-uncoating reaction are known, including the domains of these proteins required in the uncoating reaction [21], essentially nothing is known about where these proteins bind on the lattice, what type of contacts they establish with each other and with clathrin, and whether a rigid rotation of the triskelion driven by a ratchet mechanism combined with slight changes in leg conformation, rather than a very large conformational change, explains the process.

Coat formation *in vivo*

We can distinguish the following steps in coated-vesicle formation: nucleation or initiation of a coated pit; propagation of the clathrin lattice, accompanied by bilayer invagination and cargo recruitment; completion of the clathrin lattice and pinching off of the vesicle (budding); transport or diffusion of the vesicle away from the membrane; removal of the clathrin coat (uncoating); fusion of the uncoated vesicle with a target membrane [22,23]. Each of the steps is very fast, and it is estimated that the whole cycle occurs in 1 min or less [19,24]. Many of these steps can now be resolved with the aid of live-cell imaging tools [24–27].

Visualization of clathrin-coated pits and vesicles

Keen's group was the first to tag clathrin coats with an EGFP (enhanced green fluorescent protein)–LCa (light chain A) fusion protein and visualize clathrin clusters at the plasma membrane of living cells [24]. They found that clathrin tends to assemble at defined sites or 'hot spots', and that they exhibit a characteristic 'blinking' in time-lapse imaging. They suggested that each such on–off cycle might represent clathrin recruitment (increase in fluorescence intensity), budding and uncoating (abrupt disappearance of signal). A closer look at their results, at those of other groups [19], and at new data we have obtained [28] suggests a more complex and richer set of phenomena.

To visualize clathrin structures in living cells, we created a number of cell lines derived from monkey BSC1 epithelial cells stably expressing chimaeras of EGFP (and variants) attached to clathrin light chains as originally described by Keen's group [24] and of σ2-adaptin of the clathrin adaptor AP-2. We also detected the incorporation of EGFP-tagged dynamin to the clathrin structures and followed the incorporation to the forming coats of fluorescently tagged transferrin, single LDL (low-density lipoprotein) or reovirus particles. We examined the formation of coats with the aid of a spinning-disk confocal microscope, and found conditions that allow long-term imaging for more than 150 images before noticeable photo-quenching and undertook an unbiased analysis of clathrin dynamics.

One can detect four classes of fluorescent clathrin clusters: (i) those that do not fluctuate in intensity; (ii) those that 'blink' at a fixed position, tend to be short-lived and do not reach the fluorescence intensity associated with a fully formed coated vesicle; (iii) those that appear and do reach the fluorescence intensity associated with a coated vesicle, and typically show abrupt changes in lateral position before disappearing; and (iv) those that tend to be larger, brighter and highly mobile. In general, new spots do not appear at the location at which these last sorts of clusters formed. We suggest that many of the stationary spots (class i) might represent reservoirs for coat components. We propose that the clusters that disappear before becoming a coated vesicle (class ii) correspond to abortive events that fail to endocytose. We believe that the clusters that acquired rapid lateral mobility before disappearing (class iii) represent true endocytic budding events. The highly mobile clusters (class iv) correspond to clathrin associated with endosomes.

We noticed that cargo incorporation occurs primarily in a newly formed coated pit, and, in general, we could not detect incorporation of cargo into the abortive coats. The fluorescence intensity of a cargo-loaded clathrin cluster grows steadily during its lifetime, and the time required to complete coat assembly is proportional to the size of the cargo particle. These characteristics are consistent with the requirements of a system that must entrap cargo of highly variable size and shape.

Correlated dynamics of clathrin and AP-2 assemblies

By automatically tracking the fluorescent signals of clathrin and AP-2 along the time series, we discovered that every spot contains signals for both, and, regardless of their age, their fluorescent signals always appeared and disappeared together with similar rates during the life of a cluster. Thus clathrin and adaptor complexes co-associate in coated pits, and clathrin uncoating and adaptor dissociation also occur together.

Cargo capture and commitment to completion

To investigate the relationship between cargo capture and formation of a coated pit, we used fluorescently tagged transferrin, LDL and reovirus, and monitored their fate. It became apparent that these cargo molecules failed to associate with abortive clathrin clusters (class ii), and that the only successful interactions occurred with young clathrin clusters, which then proceed to grow, bud and uncoat (class iv). Budding followed by uncoating (e.g. endocytosis) was established with certainty for these events, because both fluorescent signals (cargo and clathrin or AP-2) were monitored as part of the same cluster before uncoating. Our results indicate a casual relationship between cargo capture and completion of the coat. It shows that assembly aborts unless cargo is captured. The molecular basis for the abortive step remains to be determined, however, and, at present, we cannot distinguish between an active dissolution process and a passive reaction where a molecular clamp fails to stabilize the nascent coat.

Exploratory model

Our observations suggest that the clathrin-based endocytic machinery has an exploratory character. The coated pits appear to dissolve abruptly unless some intervening event, perhaps cargo loading, determines that they continue. Our model separates the mechanism of initiation from the process of cargo recognition. It allows a variety of different cargo adaptors to participate equally, and it avoids the need to postulate differentiated budding sites on the membrane. Important tests of this description of endocytosis as an exploratory process will be to establish the mechanism by which abortive coated pits collapse and finding the signalling process that allows assembly to proceed.

References

1. Pearse, B.M. (1976) Proc. Natl. Acad. Sci. U.S.A. **73**, 1255–1259
2. Ungewickell, E. and Branton, D. (1981) Nature (London) **289**, 420–422
3. Kirchhausen, T. and Harrison, S.C. (1981) Cell **23**, 755–761

4. Crowther, R.A. and Pearse, B.M. (1981) J. Cell Biol. **91**, 790–797
5. Ter Haar, E., Harrison, S.C. and Kirchhausen, T. (2000) Proc. Natl. Acad. Sci. U.S.A. **97**, 1096–1100
6. Ter Haar, E., Musacchio, A., Harrison, S.C. and Kirchhausen, T. (1998) Cell **95**, 563–573
7. Ybe, J.A., Brodsky, F.M., Hofmann, K., Lin, K., Liu, S.H., Chen, L., Earnest, T.N., Fletterick, R.J. and Hwang, P.K. (1999) Nature (London) **399**, 371–375
8. Musacchio, A., Smith, C.J., Roseman, A.M., Harrison, S.C., Kirchhausen, T. and Pearse, B.M.F. (1999) Mol. Cell **3**, 761–770
9. Smith, C.J., Grigorieff, N. and Pearse, B.M. (1998) EMBO J. **17**, 4943–4953
10. Fotin, A., Cheng, Y., Sliz, P., Grigorieff, N., Harrison, S.C., Kirchhausen, T. and Walz, T. (2004) Nature (London), in the press
11. Ahle, S. and Ungewickell, E. (1989) J. Biol. Chem. **264**, 20089–20093
12. Gallusser, A. and Kirchhausen, T. (1993) EMBO J. **12**, 5237–5244
13. Pearse, B.M. and Robinson, M.S. (1984) EMBO J. **3**, 1951–1957
14. Zaremba, S. and Keen, J.H. (1983) J. Cell Biol. **97**, 1339–1347
15. Goodman, Jr, O.B., Krupnick, J.G., Santini, F., Gurevich, V.V., Penn, R.B., Gagnon, A.W., Keen, J.H. and Benovic, J.L. (1996) Nature (London) **383**, 447–450
16. Puertollano, R., Randazzo, P.A., Presley, J.F., Hartnell, L.M. and Bonifacino, J.S. (2001) Cell **105**, 93–102
17. Miele, A.E., Watson, P.J., Evans, P.R., Traub, L.M. and Owen, D.J. (2004) Nat. Struct. Mol. Biol. **11**, 242–248
18. Jin, A.J. and Nossal, R. (1993) Biophys. J. **65**, 1523–1527
19. Wu, X.F., Zhao, X.H., Baylor, L., Kaushal, S., Eisenberg, E. and Greene, L.E. (2001) J. Cell Biol. **155**, 291–300
20. Ungewickell, E., Ungewickell, H., Holstein, S.E.H., Lindner, R., Prasad, K., Barouch, W., Martin, B., Greene, L.E. and Eisenberg, E. (1995) Nature (London) **378**, 632–635
21. Scheele, U., Kalthoff, C. and Ungewickell, E. (2001) J. Biol. Chem. **276**, 36131–36138
22. Kirchhausen, T. (2000) Annu. Rev. Biochem. **69**, 699–727
23. Lamaze, C., Fujimoto, L.M., Yin, H.L. and Schmid, S.L. (1997) J. Biol. Chem. **272**, 20332–20335
24. Gaidarov, I., Santini, F., Warren, R.A. and Keen, J.H. (1999) Nat. Cell Biol. **1**, 1–7
25. Merrifield, C.J., Feldman, M.E., Wan, L. and Almers, W. (2002) Nat. Cell Biol. **4**, 691–698
26. Merrifield, C.J., Qualmann, B., Kessels, M.M. and Almers, W. (2004) Eur. J. Cell Biol. **83**, 13–18
27. Rappoport, J.Z. and Simon, S.M. (2003) J. Cell Sci. **116**, 847–855
28. Ehrlich, M., Boll, W., van Oijen, A., Hariharan, R., Chandran, K., Nibert, M.L. and Kirchhausen, T. (2004) Cell **118**, 591–605

Biochem. Soc. Symp. **72**, 77–86
(Printed in Great Britain)

8

Membrane traffic to and from lysosomes

J. Paul Luzio*†[1], Paul R. Pryor*†, Sally R. Gray*†, Matthew J. Gratian*, Robert C. Piper‡ and Nicholas A. Bright*†

*Cambridge Institute for Medical Research, University of Cambridge, Wellcome Trust/MRC Building, Addenbrooke's Hospital, Hills Road, Cambridge CB2 2XY, U.K., †Department of Clinical Biochemistry, University of Cambridge, Wellcome Trust/MRC Building, Addenbrooke's Hospital, Hills Road, Cambridge CB2 2XY, U.K., and ‡Department of Physiology and Biophysics, University of Iowa, Iowa City, IA 52242, U.S.A.

Abstract

In the late endocytic pathway, it has been proposed that endocytosed macromolecules are delivered to a proteolytic environment by 'kiss-and-run' events or direct fusion between late endosomes and lysosomes. To test whether the fusion hypothesis accounts for delivery to lysosomes in living cells, we have used confocal microscopy to examine content mixing between lysosomes loaded with rhodamine–dextran and endosomes subsequently loaded with Oregon-Green–dextran. Both kissing and explosive fusion events were recorded. Data from cell-free content-mixing assays have suggested that fusion is initiated by tethering, which leads to formation of a *trans*-SNARE (soluble *N*-ethylmaleimide-sensitive fusion protein attachment protein receptor) protein complex and then release of lumenal Ca^{2+}, followed by membrane bilayer fusion. We have shown that the R-SNARE (arginine-containing SNARE) protein VAMP (vesicle-associated membrane protein) 7 is necessary for heterotypic fusion between late endosomes and lysosomes, whereas a different R-SNARE, VAMP 8 is required for homotypic fusion of late endosomes. After fusion of lysosomes with late endosomes, lysosomes are re-formed from the resultant hybrid organelles, a process requiring condensation of content and the removal/recycling of some membrane proteins.

[1]To whom correspondence should be addressed (email jpl10@cam.ac.uk).

Introduction

Lysosomes are intracellular membrane-bound organelles containing many hydrolytic enzymes which are optimally active at an acidic pH. They may be distinguished from late endosomes because they lack MPRs (mannose 6-phosphate receptors) and recycling cell-surface receptors. They are usually regarded as the terminal degradation compartment of the endocytic pathway in which endocytosed macromolecules are first delivered to early endosomes, then to late endosomes and finally to lysosomes [1]. In mammalian cells, lysosomes characteristically appear as organelles of ~0.5 μm diameter with heterogeneous morphology, often containing electron-dense deposits and membrane whorls (reviewed in [2]). They can make up 0.5–5% of the cell volume, and are concentrated near the microtubule-organizing centre [3]. In yeast, the functional equivalent of the lysosome is the vacuole [4]. Both are acidic compartments containing a variety of hydrolytic enzymes that are involved in the degradation of macromolecules. Recently, there has been much interest in the idea that lysosomes are dynamic organelles, rather than simply a terminal compartment of the endocytic pathway. In some cell types, notably cells in the haematopoietic lineage, there is evidence that the lysosomal compartment is modified to form secretory lysosomes or lysosome-related organelles which contain cell-type-specific proteins that are subject to regulated secretion [5]. A regulated secretory function for conventional lysosomes has also been recognized because of evidence, in many cell types, that both conventional lysosomes and lysosome-related organelles can repair plasma membrane damage occurring in response to experimental, physiological or pathological trauma [6,7].

Biosynthetic delivery to lysosomes

In both yeast and mammalian cells, there is now a plethora of information available concerning how newly synthesized hydrolases and membrane proteins are delivered from the TGN (*trans*-Golgi network) to the lysosome/vacuole (for reviews see [8–10]). In mammalian cells, most newly synthesized acid hydrolases are delivered to lysosomes after being tagged with mannose 6-phosphate in the *cis*-Golgi and binding to MPRs in the TGN. The bound hydrolases are first delivered to endosomes, where they dissociate from the receptors as a result of the acidic lumenal pH, allowing the receptors to recycle to the TGN and onward traffic of the hydrolases to lysosomes. Clathrin, together with its heterotetrameric adaptor AP-1 (activator protein 1) and the monomeric adaptors known as GGAs (Golgi-localized, γ-ear-containing, ADP-ribosylation-factor-binding proteins) are required for MPR traffic from the TGN to endosomes [11]. Trafficking of newly synthesized lysosomal membrane proteins from the TGN does not require binding to MPRs, and occurs by either an indirect route via the plasma membrane or a direct intracellular route in which the heterotetrameric adaptor AP-3 (activator protein 3) is implicated. In yeast, there is evidence that the AP-3-mediated route from the TGN to the vacuole avoids endosomal compartments [12]. In mammalian cells, the site at which AP-3 functions has been ascribed to both the TGN and endosomes [13–15].

Dynamic equilibrium in the late endocytic pathway

Although it has been relatively easy to identify and characterize late endosomes and lysosomes by subcellular fractionation and morphological techniques (late endosomes often have the appearance of multivesicular bodies, or MVBs), establishing cell-free assays to examine interactions between them has been more difficult. This is because many endocytosed macromolecules are digested rapidly once they reach the degradative environment in these compartments. Nevertheless, content mixing and/or exchange of membrane proteins has been observed between late endosomes, between lysosomes, and between late endosomes and lysosomes (reviewed in [10]), showing that these organelles interact with each other and are in dynamic equilibrium both *in vivo* and *in vitro*. Not only are they interacting with each other and receiving proteins from the biosynthetic delivery route, but they also receive membrane proteins and lumenal content from early endosomes in the endocytic pathway and from autophagosomes in the autophagocytic pathway. In each of these pathways, proteins required at specific vesicle formation and fusion steps have been characterized. Many have been identified first from genetic screens in yeast, with the role of mammalian orthologues being established subsequently in experiments using transfected mammalian cells or cell-free assays.

Delivery from endosomes to lysosomes

Our own studies have focused on delivery from late endosomes to lysosomes. This delivery step is essential to deliver macromolecules destined for degradation by lysosomal hydrolases. Such macromolecules include proteins and lipoproteins endocytosed after binding to cell-surface receptors and released into the endosome lumen as a result of dissociation at the acidic pH found there. They also include membrane proteins tagged by ubiquitination and delivered to lumenal vesicles in MVBs as a result of the activity of ESCRT (endosomal sorting complexes required for transport) proteins [16,17].

Four hypotheses have been proposed to explain transfer of lumenal content from endosomes to lysosomes, although they are not mutually exclusive [8,10,18] (Figure 1). The first proposes endosomal maturation. Although this is a popular explanation well described in textbooks (e.g. [19]), the evidence is weak. Thus changes in organelle density observed after incubating late endosomes with cytosol and ATP, are insufficient to be consistent with the formation of lysosomes [20,21]. Moreover, the maturation hypothesis cannot explain content mixing observed in cell-free and living-cell assays (see below). The second hypothesis is vesicular transport, for which the best evidence comes from the results of a yeast cell-free assay in which pre-incubation of a donor pre-vacuolar fraction with cytosol and ATP gave rise to a separable membrane fraction, containing a transportable content marker, that could be used as a functional intermediate in a second incubation with acceptor membranes [22]. No specific identification of the putative transfer vesicles was made in these experiments, and the authors

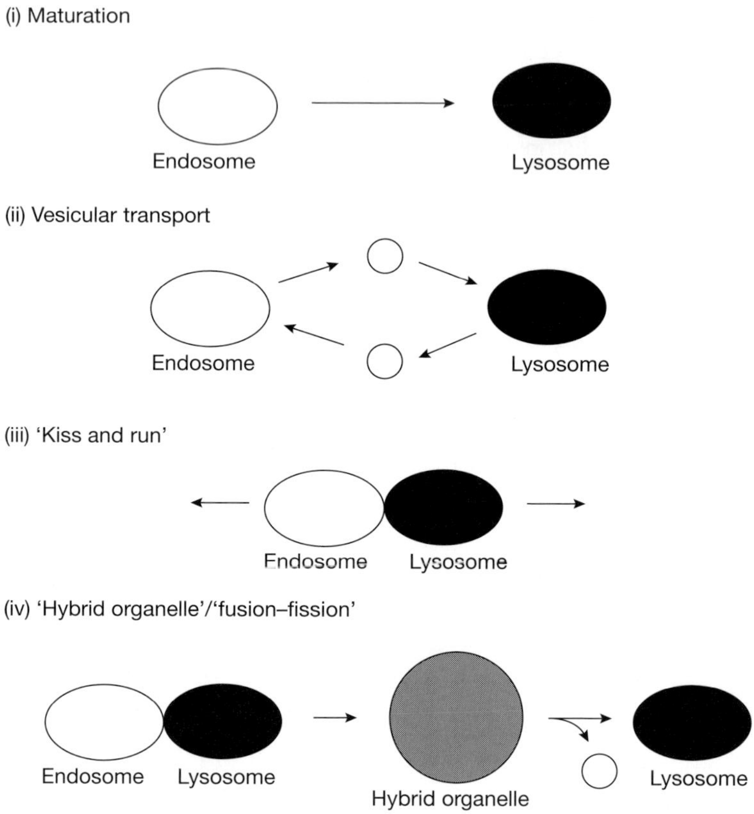

Figure 1 Diagrammatic representation of hypotheses to explain delivery of endocytosed macromolecules from late endosomes to lysosomes

did not address the question of whether their separable membrane fraction could have been a 'matured' endosome fraction itself capable of fusing with vacuoles. The third hypothesis is termed 'kiss and run', and proposes that late endosomes and lysosomes undergo continuous repeated transient fusion and fission events [23]. This hypothesis explains data from experiments in which small soluble markers appeared to move more readily between lysosomes and endosomes than larger soluble markers. The fourth hypothesis is the 'hybrid organelle' or 'fusion–fission' hypothesis, in which direct fusion between late endosomes and lysosomes is proposed with subsequent re-formation of lysosomes for re-use [18]. Experimental evidence leading to this hypothesis came initially from EM (electron microscopy) studies [24,25] and from cell-free assays in which content mixing between late endosomes and lysosomes was studied using membrane fractions derived from rat liver [26]. In this system, the immunoprecipitable product of content mixing between late endosomes and lysosomes was recovered in membrane-bound organelles intermediate in density between late endosomes and lysosomes. Using EM, these organelles, which were less electron-dense than lysosomes, contained markers derived from both lysosomes and late endosomes, consistent with them being late endosome–lysosome hybrid organelles. There was no evidence for vesicles being involved in content transfer between late endosomes and lysosomes in the rat liver cell-free system.

Delivery from endosomes to lysosomes in living cells

To address the question of how endocytosed macromolecules are delivered to lysosomes in living cells, we have recently established a content-mixing assay using different fluorochrome-conjugated dextrans in living cells. Our assay is based on previous experiments where we showed using EM that when BSA–gold conjugates were endocytosed by fluid-phase uptake in NRK (normal rat kidney) fibroblastic cells, they accumulated in dense core lysosomes [25]. We loaded lysosomes in NRK cells with dextran conjugated to rhodamine and then incubated these cells with dextran conjugated to Oregon Green 488 for short periods of time, such that late endosomes were the main organelles containing the Oregon Green. The cells were transferred to the heated stage of an inverted microscope equipped with a confocal laser-scanning attachment, and time-lapse images were recorded to determine how content mixing of the two fluorochrome-laden organelle populations was achieved. We have performed experiments that clearly show that the content mixing we observed between the two organelle populations had the following characteristics: (i) content mixing was only seen between organelles in physical contact; (ii) no vesicles trafficking between the organelles were observed; (iii) both kissing events and explosive fusion events were seen, with kissing sometimes, but not always, preceding explosive fusion; (iv) tubulation sometimes occurred from either type of organelle, and tubules were able to exchange contents with other organelles by both kissing and explosive fusion events; and (v) tubulation could be prevented by nocodazole, which causes microtubule depolymerization, but this did not inhibit content mixing completely (N.A. Bright, M.J. Gratian and J.P. Luzio, unpublished work). These data are in agreement with a role for microtubules similar to that reported for tubulation-dependent fusion of phagosomes [27]. Overall, our data from experiments on living cells are consistent with both ‘kiss and run’ and direct fusion occurring between late endosomes and lysosomes, but not with vesicular transfer. In some experiments, we have determined the ultrastructure of the compartments that exchange content using CLEM (correlative live cell and EM). By adding fixative during the kissing phase of some endosome–lysosome interactions, we have captured images showing that the point of explosive fusion had been reached with electron-dense lysosomal content diffusing into the endosome lumen. In the future, we hope to use CLEM to identify the location of the components of the molecular machinery for fusion during kissing and explosive fusion events.

Molecular machinery for endosome–lysosome fusion

Formation of hybrid organelles in the rat liver cell-free system is ATP-, cytosol- and temperature-dependent [26] and requires the presence of NSF (*N*-ethylmaleimide-sensitive factor), SNAPs (soluble NSF-attachment proteins), a small GTPase of the Rab family, probably Rab7, and Ca^{2+}/calmodulin [26,28]. Together, these characteristics are consistent with fusion being mediated, according to the SNARE (SNAP receptor) hypothesis, by specific integral membrane protein SNAREs together with the common cytosolic fusion machinery used at many other sites on the secretory and endocytic pathways [29]. The Rab and Ca^{2+} requirements were identified because fusion was inhibited respectively by Rab-GDI (GDP-dissociation inhibitor) and the fast-acting calcium chelator, BAPTA (1,2-bis[2-aminophenoxy]ethane-N,N,N',N'-tetra-acetic acid). The availability of these two inhibitors, Rab-GDI and BAPTA, has allowed the experimental ordering of the processes that they inhibit [28]. Thus Rab-GDI acts very early in the fusion process and BAPTA very late. By analogy with what is known about homotypic yeast vacuole fusion, where the Rab7 orthologue Ypt7p is involved in recruiting putative tethering proteins and there is a late calcium/calmodulin requirement for membrane fusion, it has been possible to propose a model for the mechanistic steps involved in late endosome–lysosome fusion [10,18] (Figure 2). In this model, the initial Rab7-dependent tethering of the organelles is followed by *trans*-SNARE complex formation and the release of lumenal Ca^{2+}, leading to membrane fusion.

Consistent with this model are data from experiments in which Rab7 was overexpressed in mammalian cells and caused lysosome aggregation [30]. Overexpression of dominant-negative mutants caused dispersion of lysosomes, and they became inaccessible to endocytosed molecules. The best candidates for proteins involved in the tethering complexes functioning late in the endocytic pathway of mammalian cells are the mammalian homologues of the yeast Vps (vacuolar protein sorting) proteins Vps11p, Vps16p, Vps18p, Vps33p, Vps39p and Vps41p. Studies of overexpression and knockdown of mammalian Vps18p and Vps39p have shown that they are involved in clustering of late endosomes and lysosomes [31,32]. Although they function early in the overall fusion process, these tether proteins are not specific for heterotypic late endosome–lysosome fusion. They also appear to play a role in homotypic fusion events in the endocytic pathway, including homotypic early endosome fusion [33].

For homotypic fusion of late endosomes, antibody-inhibition data have shown that the partners of syntaxin 7 in the *trans*-SNARE complex are syntaxin 8, Vti1b, and VAMP 8 [34]. These four SNAREs provide respectively the necessary Qa, Qb, Qc [glutamine-containing t-SNAREs (target SNAREs)] and R [arginine-containing v-SNAREs (vesicle transport SNAREs)] SNAREs for a functional complex [35]. The core of this complex has been crystallized and its structure has been solved [36]. In contrast, a different combinatorial *trans*-SNARE complex is required for heterotypic late endosome–lysosome fusion consisting of the same Q-SNAREs (syntaxin 7, syntaxin 8 and Vti1b) and a different R-SNARE, VAMP

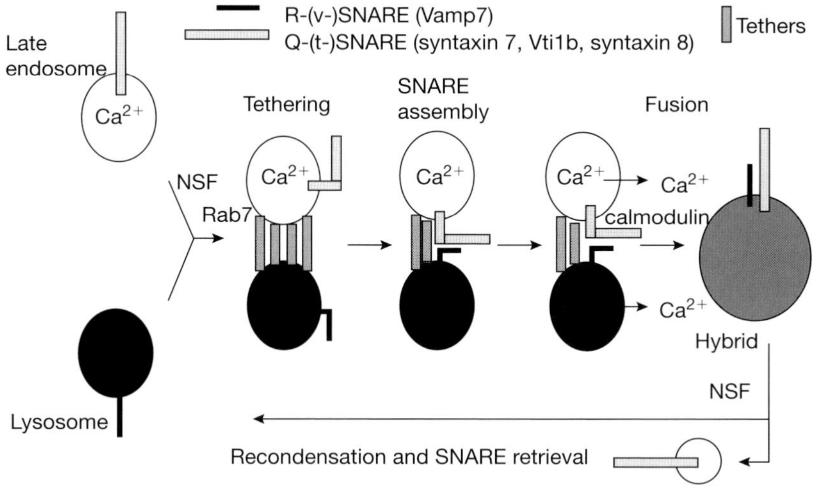

Figure 2 The mechanism of late endosome–lysosome fusion. The diagram shows the possible stages of fusion of late endosomes with lysosomes and reformation of lysosomes from the subsequent hybrid organelles. There is evidence that the R-SNARE VAMP 7 and the Q-SNARE complex are present on both late endosomes and lysosomes. In this representation, VAMP 7 is shown on the lysosome and the Q-SNARE complex on the late endosome. Tethering is shown preceding *trans*-SNARE assembly and Ca^{2+} release from the organelles is shown preceding fusion. Reformation of lysosomes from the hybrid organelles requires a process to recycle SNAREs for subsequent rounds of fusion and the condensation of lumenal content to produce dense core lysosomes.

7 [37]. VAMP 7 [also known as TI-VAMP (toxin-insensitive VAMP) or VAMP SYBL1 (synaptobrevin-like 1)] is an unusual R-SNARE in having a relatively long (~110-amino-acid) regulatory domain at its N-terminus which is required for delivery of this VAMP 7 to late endocytic compartments as a result of binding to AP-3 [38]. Intriguingly, VAMP 7 is able to form combinatorial *trans*-SNARE complexes with Q-SNAREs other than those used for heterotypic fusion between late endosomes and lysosomes. Thus lysosomal VAMP 7 forming a *trans*-SNARE complex with the plasma membrane Q-SNAREs syntaxin 4 and SNAP-23 is responsible for Ca^{2+}-regulated exocytosis of lysosomes [39]. VAMP 7 has also been implicated as an R- (v-) SNARE in vesicle traffic, in polarized epithelial MDCK (Madin–Darby canine kidney) cells, from the TGN to the apical membrane where it interacts with syntaxin 3 and SNAP-23 [40], as well as in vesicle delivery for neurite outgrowth, where it interacts with syntaxin 1 and SNAP-25 [38,41]. Release of lumenal Ca^{2+}, which is necessary for membrane fusion between late endosomes and lysosomes, is probably promoted by *trans*-SNARE complex formation, as has been described for yeast vacuole fusion [42].

Reformation of lysosomes from hybrids: membrane traffic out of lysosomes

The process of fusion of late endosomes with lysosomes would consume both organelles if no recovery process occurred. We have proposed that lysosome reformation from hybrid organelles occurs by content condensation and a membrane-retrieval process which removes endosomal membrane proteins and recycles SNAREs [18]. Lysosomes can be re-formed from hybrid organelles in a cell-free system, and we found that content condensation in this system required both a proton-pumping ATPase and lumenal Ca^{2+} [28]. We have also observed small vesicular tubular structures budding off hybrid organelles, after endosome–lysosome fusion, in experiments on living cells (N.A. Bright, M.J. Gratian and J.P. Luzio, unpublished work). One set of candidate machinery that might mediate the membrane-retrieval process is the retromer complex first described in yeast as a complex of Vps5p, Vps17p, Vps26p, Vps29p and Vps35p. Loss of any of these proteins results in the accumulation of endosomal and recycling Golgi proteins at the vacuole, resulting in a type of endosome–vacuole hybrid. Depletion of the Vps26p orthologue in mammalian cells leads to a similar phenotype, including some swelling and vacuolarization of lysosomal compartments [43,44].

It has been suggested that the absence of the Ca^{2+}-transporter mucolipin 1 in cells from patients with the autosomal recessive disease mucolipidosis type IV, and its orthologue CUP-5 in *Caenorhabditis elegans*, results in failure to re-form lysosomes from hybrids and thereby causes disease [45]. This may be due to the requirement for intraorganellar Ca^{2+} for the condensation of content in the reformation of lysosomes from hybrid organelles. However, loss of mucolipin 1 or CUP-5 may also cause defects in the retrieval of components from endosome–lysosome hybrids, possibly by inhibiting formation of specific transport vesicles or other intermediates.

Conclusions

An overall description of membrane traffic in the late endocytic pathway has now emerged with both 'kiss and run' and explosive fusion events being required for delivery of lumenal content from late endosomes to lysosomes. The *trans*-SNARE complex for heterotypic fusion of late endosomes and lysosomes has been described with a key role for the R-SNARE VAMP 7. A challenge now is to understand how the regulatory domain of VAMP 7 regulates heterotypic fusion and whether any interacting proteins are important for this regulation. Similarly, the mechanistic details are scant to explain the process whereby lysosomes are re-formed from endosome–lysosome hybrid organelles.

Unpublished experimental work referred to in this article was funded by a Programme Grant (G9310915) from the Medical Research Council to J.P.L. The Cambridge Institute for Medical Research (CIMR) is supported by a Strategic Initiative Grant (066438) from the Wellcome Trust.

References

1. Kornfeld, S. and Mellman, I. (1989) Annu. Rev. Cell Biol. **5**, 483–525
2. Holtzmann, E. (1989) Lysosomes, Plenum Press, New York
3. Matteoni, R. and Kreis, T.E. (1987) J. Cell Biol. **105**, 1253–1265
4. Klionsky, D.J., Herman, P.K. and Emr, S.D. (1990) Microbiol. Rev. **54**, 266–292
5. Blott, E.J. and Griffiths, G.M. (2002) Nat. Rev. Mol. Cell Biol. **3**, 122–131
6. Andrews, N.W. (2000) Trends Cell Biol. **10**, 316–321
7. McNeil, P.L. (2002) J. Cell Sci. **115**, 873–879
8. Mullins, C. and Bonifacino, J.S. (2001) BioEssays **23**, 333–343
9. Ghosh, P., Dahms, N.M. and Kornfeld, S. (2003) Nat. Rev. Mol. Cell Biol. **4**, 202–212
10. Luzio, J.P., Poupon, V., Lindsay, M.R., Mullock, B.M., Piper, R.C. and Pryor, P.R. (2003) Mol. Membr. Biol. **20**, 141–154
11. Doray, B., Ghosh, P., Griffith, J., Geuze, H.J. and Kornfeld, S. (2002) Science **297**, 1700–1703
12. Piper, R.C., Bryant, N.J. and Stevens, T.H. (1997) J. Cell Biol. **138**, 531–545
13. Rous, B.A., Reaves, B.J., Ihrke, G., Briggs, J.A.G., Gray, S.R., Stephens, D.J., Banting, G. and Luzio, J.P. (2002) Mol. Biol. Cell **13**, 1071–1082
14. Peden, A.A., Oorschot, V., Hesser, B.A., Austin, C.D., Scheller, R.H. and Klumperman, J. (2004) J. Cell Biol. **164**, 1065–1076
15. Ihrke, G., Kyttälä, A, Russell, M.R.G., Rous, B.A. and Luzio, J.P. (2004) Traffic, in the press
16. Katzmann, D.J., Odorizzi, G. and Emr, S.D. (2002) Nat. Rev. Mol. Cell Biol. **3**, 893–905
17. Raiborg, C., Rusten, T.E. and Stenmark, H. (2003) Curr. Opin. Cell Biol. **15**, 446–455
18. Luzio, J.P., Rous, B.A., Bright, N.A., Pryor, P.R., Mullock, B.M. and Piper, R.C. (2000) J. Cell Sci. **113**, 1515–1524
19. Alberts, B., Johnson, A., Lewis, J., Raff, M., Roberts, K. and Walter, P. (2002) Molecular Biology of the Cell, 4th edn., p. 742, Garland Science, New York
20. Roederer, M., Barry, J.R., Wilson, R.B. and Murphy, R.F. (1990) Eur. J. Cell Biol. **51**, 229–234
21. Mullock, B.M., Perez, J.H., Kuwana, T., Gray, S.R. and Luzio, J.P. (1994) J. Cell Biol. **126**, 1173–1182
22. Vida, T. and Gerhardt, B. (1999) J. Cell Biol. **146**, 85–97
23. Storrie, B. and Desjardins, M. (1996) BioEssays **18**, 895–903
24. Futter, C.E., Pearse, A., Hewlett, L.J. and Hopkins, C.R. (1996) J. Cell Biol. **132**, 1011–1023
25. Bright, N.A., Reaves, B.J., Mullock, B.M. and Luzio, J.P. (1997) J. Cell Sci. **110**, 2027–2040
26. Mullock, B.M., Bright, N.A., Fearon, C.W., Gray, S.R. and Luzio, J.P. (1998) J. Cell Biol. **140**, 591–601
27. Harrison, R.E., Bucci, C., Vieira, O.V., Schroer, T.A. and Grinstein, S. (2003) Mol. Cell. Biol. **23**, 6494–6506
28. Pryor, P.R., Mullock, B.M., Bright, N.A., Gray, S.R. and Luzio, J.P. (2000) J. Cell Biol. **149**, 1053–1062
29. Rothman, J.E. (1994) Nature (London) **372**, 55–63
30. Bucci, C., Thomsen, P., Nicoziani, P., McCarthy, J. and van Deurs, B. (2000) Mol. Biol. Cell **11**, 467–480
31. Caplan, S., Hartnell, L.M., Aguilar, R.C., Naslavsky, N. and Bonifacino, J.S. (2001) J. Cell Biol. **154**, 109–122
32. Poupon, V., Stewart, A., Gray, S.R., Piper, R.C. and Luzio, J.P. (2003) Mol. Biol. Cell **14**, 4015–4027
33. Richardson, S.C.W., Winistorfer, S.C., Poupon, V., Luzio, J.P. and Piper, R.C. (2004) Mol. Biol. Cell **15**, 1197–1210
34. Antonin, W., Holroyd, C., Fasshauer, D., Pabst, S., Von Mollard, G.F. and Jahn, R. (2000) EMBO J. **19**, 6453–6464
35. Bock, J.B., Matern, H.T., Peden, A.A. and Scheller, R.H. (2001) Nature (London) **409**, 839–841
36. Antonin, W., Fasshauer, D., Becker, S., Jahn, R. and Schneider, T.R. (2002) Nat. Struct. Biol. **9**, 107–111

37. Pryor, P.R., Mullock, B.M., Bright, N.A., Lindsay, M.R., Gray, S.R., Richardson, S.C.W., Stewart, A., James, D.E., Piper, R.C. and Luzio, J.P. (2004) EMBO Rep. **5**, 590–595
38. Martinez-Arca, S., Rudge, R., Vacca, M., Raposo, G., Camonis, J., Proux-Gillardeaux, V., Daviet, L., Formstecher, E., Hamburger, A., Filippini, F. et al. (2003) Proc. Natl. Acad. Sci. U.S.A. **100**, 9011–9016
39. Rao, S.K., Huynh, C., Proux-Gillardeaux, V., Galli, T. and Andrews, N.W. (2004) J. Biol. Chem. **279**, 20471–20479
40. Lafont, F., Verkade, P., Galli, T., Wimmer, C., Louvard, D. and Simons, K. (1999) Proc. Natl. Acad. Sci. U.S.A. **96**, 3734–3738
41. Martinez-Arca, S., Alberts, P., Zahraoui, A., Louvard, D. and Galli, T. (2000) J. Cell Biol. **149**, 889–900
42. Merz, A.J. and Wickner, W.T. (2004) J. Cell Biol. **164**, 195–206
43. Arighi, C.N., Hartnell, L.M., Aguilar, R.C., Haft, C.R. and Bonifacino, J.S. (2004) J. Cell Biol. **165**, 123–133
44. Seaman, M.N. (2004) J. Cell Biol. **165**, 111–122
45. Treusch, S., Knuth, S., Slaugenhaupt, S.A., Goldin, E., Grant, B.D. and Fares, H. (2004) Proc. Natl. Acad. Sci. U.S.A. **101**, 4483–4488

Biochem. Soc. Symp. **72**, 87–97
(Printed in Great Britain)

9

Dynamin I phosphorylation and the control of synaptic vesicle endocytosis

Karen J. Smillie and Michael A. Cousin[1]

Membrane Biology Group, Division of Biomedical Sciences, University of Edinburgh, George Square, Edinburgh EH8 9XD, U.K.

Abstract

The GTPase dynamin I is essential for synaptic vesicle endocytosis in nerve terminals. It is a nerve terminal phosphoprotein that is dephosphorylated on nerve terminal stimulation by the calcium-dependent protein phosphatase calcineurin and then rephosphorylated by cyclin-dependent kinase 5 on termination of the stimulus. Because of its unusual phosphorylation profile, the phosphorylation status of dynamin I was assumed to be inexorably linked to synaptic vesicle endocytosis; however, direct proof of this link has been elusive until very recently. This review will describe current knowledge regarding dynamin I phosphorylation in nerve terminals and how this regulates its biological function with respect to synaptic vesicle endocytosis.

Introduction

SV (synaptic vesicle) recycling in nerve terminals is essential for the maintenance of neurotransmission. On neuronal depolarization, Ca^{2+} influx through voltage-gated Ca^{2+} channels stimulates SV fusion with the plasma membrane (exocytosis), and thus neurotransmitter release. Following exocytosis, SVs are reformed from the plasma membrane in a process called endocytosis. SVE (SV endocytosis) is tightly coupled to SV exocytosis, both temporally and spatially, since SVE is activated by the same Ca^{2+} influx that stimulates exocytosis [1]. After SVE, SVs are recycled locally within the nerve terminal where they are filled with neurotransmitter and prepared for the next round of exocytosis.

[1]To whom correspondence should be addressed (email m.cousin@ed.ac.uk).

Dynamin I

Many proteins are essential for SVE, reflecting the complex nature of *de novo* generation of SVs from the plasma membrane. One of the most studied and best characterized is the pre-synaptic GTPase dynamin I. Dynamin I was initially discovered as a microtubule-binding protein [2]; however, its essential role in SVE was highlighted when it was identified as the homologue of the *shibire* gene product in *Drosophila* [3]. *Shibire* mutants display a temperature-sensitive paralysis originating from a block at a late stage of SVE [4]. Electron micrographs of *shibire* nerve terminals show large infoldings of membrane or deeply invaginated pits, both originating from the plasma membrane with electron-dense collars. These collars were identified as dynamin I, suggesting that dynamin I was required for SV fission from the plasma membrane.

Dynamin I is a member of a larger family of dynamin-like proteins [5,6] of which three mammalian isoforms exist. Dynamin I is expressed solely in the brain and is enriched in nerve terminals where SVE occurs. Dynamin II is expressed ubiquitously and is essential for receptor-mediated endocytosis [6]. Dynamin III expression was thought originally to be restricted to testis; however, it is also expressed in the dendritic spines of neurons, suggesting a role in post-synaptic receptor internalization [7].

Dynamin I structure

Dynamin I is a modular protein, with four identified domains (Figure 1): an N-terminal GTPase domain, a PH (pleckstrin homology) domain, a GED (GTPase effector domain) and a C-terminal PRD (proline-rich domain). The GTPase activity of dynamin is essential for endocytosis. This was initially revealed by the *shibire* mutant, which has a defect in GTP binding as a result of a point mutation within the GTPase domain [3]. Subsequent studies using overexpression of similar dynamin point mutants abolish both receptor-mediated endocytosis and SVE [7–9]. The most likely role for dynamin GTPase activity in endocytosis is to produce a mechanical force to drive SV fission, either by constriction or expansion of the dynamin collar surrounding the neck of the invaginated vesicle [10–12].

The PH domain is principally for binding to phosphoinositides, specifically PtdIns(4,5)P_2. PtdIns(4,5)P_2 binding to the PH domain greatly increases dynamin GTPase activity, and mutations that inhibit PtdIns(4,5)P_2 binding abolish receptor-mediated endocytosis [13].

When dynamin self-assembles into collars at the neck of an invaginating vesicle, its GTPase activity is activated. The GED controls this process, since it regulates both dynamin I assembly and functions as a GTPase-activating domain. The role of these two functions in receptor-mediated endocytosis has been investigated independently, with overexpression of either assembly-deficient or GTPase-inactive mutants increasing the amount of endocytosis [14]. This has lead to an alternative hypothesis for the role of dynamin GTPase activity in endocytosis, where activity is required to terminate interactions with downstream effectors [14,15].

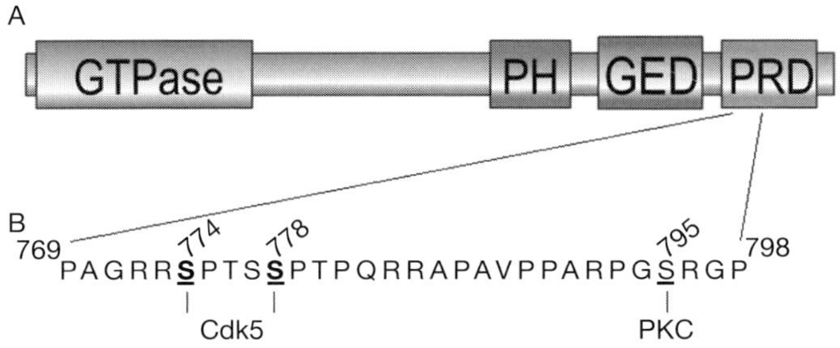

Figure 1 Domain structure of dynamin I and its phosphorylation sites. (**A**) Dynamin is a modular protein. The GTPase domain is essential for SVE. The PH domain interacts with PtdIns(4,5)P_2 and is also essential for SVE. The GED controls both dynamin I assembly and activation of its GTPase domain. The C-terminal PRD interacts with many proteins, including the endocytosis proteins amphiphysin I and 2, endophilin and syndapin, and signalling/cytoskeletal proteins, such as p85 (regulatory subunit of phosphoinositide 3-kinase), Grb-2, phospholipase C γ, actin-filament-binding protein, profilin, cortactin, intersectin and calcineurin. (**B**) Location of the *in vitro* and *in vivo* phosphorylation sites on the dynamin I PRD. The amino acid sequence corresponds to residues 769 to 798 of the Dynamin I PRD. The bold underlined serine residues (774 and 778) indicate the endogenous Cdk5 phosphorylation sites. The underlined serine (795) indicates the *in vitro* phosphorylation site for PKC; however, this site is not phosphorylated *in vivo*. Amino acid sequences are denoted using the single letter code.

The C-terminal PRD is the major protein-interaction region of dynamin, encompassing multiple binding sites for proteins containing SH3 (Src-homology 3) domains such as amphiphysin and endophilin [16–18]. Binding of several of these SH3-domain-containing proteins stimulate dynamin I GTPase activity, albeit to much lower levels than PtdIns(4,5)P_2 [19].

Dynamin I is a phosphoprotein that is dephosphorylated on nerve terminal stimulation

Dynamin I is a phosphoprotein and was originally discovered in nerve terminals due to its ability to undergo a rapid (<5 s) depolarization-dependent dephosphorylation [20]. On termination of nerve terminal stimulation, dynamin I is slowly ($t_{1/2}$ approx. 40 s) rephosphorylated by its endogenous kinase. Dynamin I is present in both the cytosol and membrane of nerve terminals; however, phosphorylated dynamin I is restricted to the cytosol [21]. Thus the phosphorylation status of dynamin I would appear to regulate its localization within the nerve terminal. It is tempting to speculate that the dephosphorylation of dynamin I facilitates its translocation to the plasma membrane for SVE and its rephosphorylation directs its return to the nerve terminal cytosol.

These initial observations on the physiology of dynamin I phosphorylation in nerve terminals were made approx. 10 years ago [20,21]. In the intervening time, many questions related to dynamin I phosphorylation and SVE have been answered, while many have still not been addressed or are contentious. The remainder of the review will summarize the current understanding of dynamin I phosphorylation with respect to its biological activity and SVE.

Calcineurin is the *in vivo* dynamin I phosphatase

Dynamin I is dephosphorylated in a Ca^{2+}-dependent manner on nerve terminal depolarization. Thus the obvious candidate for the dynamin I phosphatase was calcineurin (PP2B), a Ca^{2+}-dependent protein phosphatase that is enriched in nerve terminals. Calcineurin dephosphorylates purified phospho-dynamin I *in vitro* [22] and more importantly *in vivo*, since pharmacological calcineurin antagonists abolish KCl-evoked dynamin I dephosphorylation in nerve terminals [23,24]. Calcineurin co-ordinately dephosphorylates at least seven other nerve terminal proteins on nerve terminal stimulation, in addition to dynamin I. These proteins are termed the dephosphins, and every member of this group of proteins [PIPK1γ (phosphoinositide phosphate kinase 1γ), AP180 (assembly protein 180), epsin, eps15, amphiphysin 1, amphiphysin 2, synaptojanin and dynamin I] is essential for SVE in their own right [25].

Dynamin I binds calcineurin with high affinity within its C-terminal PRD [22]. Interestingly, calcineurin interacts with the PRD in a Ca^{2+}-dependent manner, with binding stimulated by Ca^{2+} changes in the physiological range (EC_{50}, 100–400 nM) [26]. This adds another potential level of control to the dephosphorylation of dynamin I, with Ca^{2+}-dependent recruitment of calcineurin preceding dynamin I dephosphorylation.

Cdk5 (cyclin-dependent kinase 5) is the *in vivo* dynamin I kinase

On termination of nerve terminal depolarization, dynamin I is slowly rephosphorylated. The identification of the endogenous dynamin I kinase was more problematic than its phosphatase, with a number of potential protein kinases identified within the past 10 years. Protein kinases that phosphorylate dynamin I *in vitro* include microtubule-associated protein kinase, casein kinase II, Cdk2, minibrain kinase and PKC (protein kinase C) [27–30]. PKC was assumed to be the dynamin I kinase for many years. This was because dynamin I was an excellent PKC substrate *in vitro* [21], and pharmacological PKC antagonists abolished dynamin I rephosphorylation in intact nerve terminals [31,32]. However, there were always several inconsistencies with this assumption. For example, PKC substrates, such as MARCKS (myristoylated alanine-rich C kinase substrate) and GAP-43 (growth-associated protein 43) were phosphorylated on nerve terminal stimulation, whereas dynamin I was dephosphorylated [33]. Also, dynamin I rephosphorylation only occurred after intracellular Ca^{2+} levels had returned to basal levels, thus after the removal of the stimulus for the Ca^{2+}-dependent PKCs [20].

Recent evidence now proves that the endogenous dynamin I kinase is not PKC, but Cdk5 [34]. Cdk5 is a nerve terminal enriched serine/threonine kinase that is active at rest, and phosphorylates multiple proteins implicated in SV recycling *in vitro* [35–40]. Cdk5 phosphorylates either purified dynamin I or a GST (glutathione S-transferase) fusion protein of the dynamin I PRD *in vitro* [34,39]. The *in vitro* phosphorylation sites for cdk5 on dynamin I were determined by MS to be Ser^{774} and Ser^{778} [34]. Both of these residues are located in the dynamin I PRD (Figure 1) and were distinct from the previously determined *in vitro* PKC phosphorylation site (Ser^{795}, [41]). Importantly, endogenous dynamin I purified from intact nerve terminals was phosphorylated on identical sites to that observed with *in vitro* Cdk5-dependent phosphorylation [34]. Cdk5 also phosphorylates dynamin I *in vivo*, since dynamin I rephosphorylation is abolished by either Cdk5 antagonists in intact nerve terminals or overexpression of dominant-negative Cdk5 in neuronal cultures [34]. The identified Cdk5 phosphorylation sites are physiologically relevant, since blotting with phosphospecific antibodies raised against them showed decreases and increases in signal in depolarized and repolarized nerve terminal lysates respectively [34].

A recent *in vitro* study proposed that Cdk5-dependent phosphorylation of dynamin I occurred on Thr^{780} and not Ser^{774}/Ser^{778} [39]. However, when this MS data is re-evaluated it would seem that dynamin I is phosphorylated on Ser^{778} and not Thr^{780}. This was confirmed when the phosphorylation sites on endogenous dynamin I were re-examined using a refined protocol employing graphite powder microcolumns [42]. There was no detectable phosphate on Thr^{780}, with all the phosphate on Ser^{774} and Ser^{778}. It was confirmed that dynamin I can be doubly phosphorylated, but, in singly phosphorylated peptides, phosphate was predominantly found on Ser^{774} [42]. This suggests that Ser^{774} may potentially be the most important site for the control of SVE.

Cdk5 rephosphorylates some, but not all, of the dephosphin group of proteins, in addition to dynamin I. The lipid phosphatase synaptojanin is rephosphorylated by Cdk5 in intact nerve terminals, whereas amphiphysin 1 and 2 and AP180 are not [32,34,40]. This highlights the difficulties in extrapolating *in vitro* phosphorylation data to *in vivo*, since amphiphysin is an *in vitro* Cdk5 substrate [38,39].

The identification of Cdk5 as the dynamin I kinase has resolved some of the paradoxes raised by previous studies. For example, the inhibition of dynamin I rephosphorylation by PKC antagonists in nerve terminals was explained by the fact that the same antagonists potently inhibit Cdk5 activity [34]. Likewise, the reciprocal relationship between the Ca^{2+}-dependence of the dynamin I phosphorylation cycle and PKC activation is now resolved. Since Cdk5 is constitutively active in nerve terminals, there is no requirement for the activation of dynamin I phosphorylation after termination of stimulation. Thus the amount of phosphorylated or dephosphorylated dynamin I in nerve terminals is determined by the stimulation-dependent activity of calcineurin, and not by Cdk5.

How does dynamin I phosphorylation regulate its biological activity?

The identification of calcineurin and Cdk5 as the *in vivo* dynamin I phosphatase and kinase allows the biological role of dynamin I phosphorylation to be investigated. Previous *in vitro* studies have shown that changes in dynamin I phosphorylation affect its GTPase activity and its interactions with both proteins and lipids.

Dynamin I phosphorylation and GTPase activity

In vitro phosphorylation of dynamin I by PKC stimulates its GTPase activity approx. 12-fold [27], with calcineurin reversing this activation [22]. However, these studies now have to be re-interpreted (supplementary data in [34]). The conditions used to phosphorylate dynamin I with PKC in previous studies were found to stimulate dynamin self-assembly, resulting in an increase in GTPase activity that was independent of phosphorylation. When dynamin I self-assembly was disrupted after phosphorylation by PKC, no stimulation of GTPase activity was observed [34]. Using this modified protocol, the GTPase activity of dynamin I phosphorylated by Cdk5 showed a modest 3-fold stimulation. Thus phosphorylation of dynamin I by Cdk5 may regulate its GTPase activity in nerve terminals. However, phosphorylation cannot be the major activator, since nanotubules that incorporate PtdIns(4,5)P_2 can increase GTPase activity by approx. 1000-fold [11].

Dynamin I phosphorylation and protein–protein interactions

Dynamin I shares many interactions with other essential endocytosis proteins; however, the understanding of how dynamin I phosphorylation regulates these interactions is still embryonic. Studies using rat brain detergent extracts suggested that phosphorylation of different endocytosis proteins may control their interactions [43]. The multiple protein kinase activities present in these extracts reduced the interaction of dynamin I with both amphiphysin and AP-2, an effect that was reversed by calcineurin antagonists. It is difficult to extrapolate these results to nerve terminals, since many of the protein kinases present in the lysates would not normally be able to phosphorylate these proteins, because they would be in separate subcellular compartments.

Some *in vitro* studies have been performed to establish whether or not dynamin I phosphorylation regulates its protein interactions. For example, *in vitro* phosphorylation of purified dynamin I with either minibrain kinase or Cdk5 inhibits its binding to the SH3 domain of amphiphysin [30,39]. However, *in vitro* phosphorylation of only the dynamin I PRD by Cdk5 did not inhibit this interaction. Pre-incubation of the dynamin I PRD with the amphiphysin SH3 domain prevents its phosphorylation by Cdk5 [34]. These studies suggest that other dynamin I regions may influence its putative phosphorylation-dependent interaction with amphiphysin and also indicate that dynamin I may have to be released from amphiphysin before it can be phosphorylated by Cdk5, possibly on termination of SVE.

Dynamin I phosphorylation and protein–lipid interactions

Phosphorylated dynamin I is restricted to the cytosol of nerve terminals, suggesting a possible phosphorylation-dependent interaction with plasma membrane lipids [21]. In support, *in vitro* phosphorylation of dynamin I by either PKC or Cdk5 inhibits its ability to bind to acidic phospholipids [34,41]. Dynamin I interacts with PtdIns(4,5)P_2 via its PH domain. This interaction is essential for dynamin I translocation to the plasma membrane, since point mutations within the PH domain that abolish PtdIns(4,5)P_2 binding restrict dynamin I to the cytosol [13]. If the PtdIns(4,5)P_2–dynamin I interaction is regulated by phosphorylation, it should provide a mechanism for dynamin I movement between different subcellular compartments.

The physiological role of dynamin I phosphorylation in SVE

The phosphorylation of dynamin I regulates its biological activity, but does its phosphorylation status play any essential role in SVE? Studies in isolated nerve terminals and primary cell cultures have shown that the enzyme activities of both calcineurin and Cdk5 are essential for SVE. Calcineurin is essential, since the internalization of the amphiphilic membrane marker FM2-10 was inhibited by calcineurin antagonists in isolated nerve terminals (synaptosomes) [24,32]. Cdk5 activity is also essential for SVE, with one important distinction: inhibition of Cdk5 activity does not affect SVE following nerve terminal stimulation, unlike calcineurin antagonists [32,34]; however, if nerve terminals continue to be incubated with Cdk5 antagonists, the next cycle of SVE is abolished. This is consistent with Cdk5 being required to rephosphorylate dynamin I after SVE and thus reset the endocytic machinery for the next round of SV recycling. Morphological studies using electron microscopy also highlight the essential role for calcineurin and Cdk5 in SVE. Synaptosomes pre-incubated with calcineurin antagonists display a severe SV depletion (indicative of a block in SVE), with no obvious changes in membrane appearance, showing SVE was arrested at an early stage (M.A. Cousin, unpublished work). Synaptosomes pre-incubated with Cdk5 blockers had a normal appearance after the first stimulation of SVE, but displayed extensive SV depletion and multiple membrane invaginations after the second stimulation, indicative of a block in SVE at a specific step [34]. In addition, overexpression of dominant-negative Cdk5 in primary neuronal cultures greatly retarded the internalization of the membrane marker FM4-64, highlighting its essential role in SVE [34].

The activities of calcineurin and Cdk5 are essential for the activation and maintenance of SVE. However, this does not reveal whether the phosphorylation status of dynamin I itself is essential, since both calcineurin and Cdk5 have multiple substrates in nerve terminals. The specific role of dynamin I phosphorylation can now be addressed, since the physiological phosphorylation sites on dynamin I have been identified. We have conclusive evidence that a cycle of dynamin I phosphorylation at Ser^{774} and Ser^{778} is essential for SVE (K.J. Smillie, unpublished work). Thus both the initial dephosphorylation and subsequent rephosphorylation of dynamin I are required for SVE to proceed. The model in Figure 2 proposes a possible mechanism of action for this process.

Perspectives

Since the identification of dynamin I as a phosphoprotein that is dephosphorylated on nerve terminal stimulation, it has been hypothesized that its phosphorylation status may regulate its essential role in SVE. However, it has taken many years for direct proof to emerge. This delay was mainly a result of identifying the physiological dynamin I kinase. However, the proof of its essential role in SVE has rapidly followed the identification of its physiological phosphorylation sites. Now that calcineurin activity, Cdk5 activity and the phosphorylation status of dynamin I itself have been demonstrated to be essential for SVE, this would seem to bring this story to a conclusion. However, some questions regarding the phosphorylation status of dynamin I in SVE still remain unanswered.

For example, can dynamin I phosphorylation be modulated within the nerve terminal and does this control SVE? Cdk5 is activated by the p39 or p35 protein. p35 itself is a Cdk5 substrate and is an unstable protein, with a half-life of 20 min. Inhibition of Cdk5 activity greatly increases p35 stability, as does mutagenesis of its Cdk5-dependent phosphorylation sites, suggesting a possible negative-feedback mechanism [44]. Positive regulation of Cdk5 activity could originate from stimulation of growth factor cascades. For example, nerve growth factor induces p35 expression in PC12 cells and brain-derived growth factor activates Cdk5 activity in cultured primary neurons [45]. Also, the extent of dynamin I dephosphorylation may be controlled by the endogenous calcineurin inhibitor, cain [46]. Cain is expressed in nerve terminals, binds amphiphysin, and its overexpression abolishes endocytosis in non-neuronal cells. Thus cain should be in the correct localization to regulate dynamin I phosphorylation and possibly SVE. Further studies will determine whether this is the case.

Another question is how does dynamin I phosphorylation control SVE? *In vitro* studies have hinted at three possible mechanisms, with the regulation of either or both protein or lipid interactions being most likely. However, this question will not be definitively addressed until *in vivo* interaction studies are performed in real time, to establish whether the temporal and spatial properties of candidate interactions match their proposed role in SVE. The development of FRET (fluorescence resonance energy transfer)-based interaction assays should facilitate these experiments.

We have shown that dual Ser^{774} and Ser^{778} mutants abolish SVE (K.J. Smillie, unpublished work); however, do these sites control SVE independently of each other? Ser^{774} may be the essential residue for SVE, since it is the predominately phosphorylated site on singly phosphorylated dynamin I [42]. This can be tested by overexpressing dynamin I mutants in neurons that encompass single substitutions at either Ser^{774} or Ser^{778} and examining the effect on SVE.

Now that the essential role for the phosphorylation status of dynamin I in SVE is established, the remaining immediate questions will become more and more targeted. One final wider question is do the phosphorylation status of all

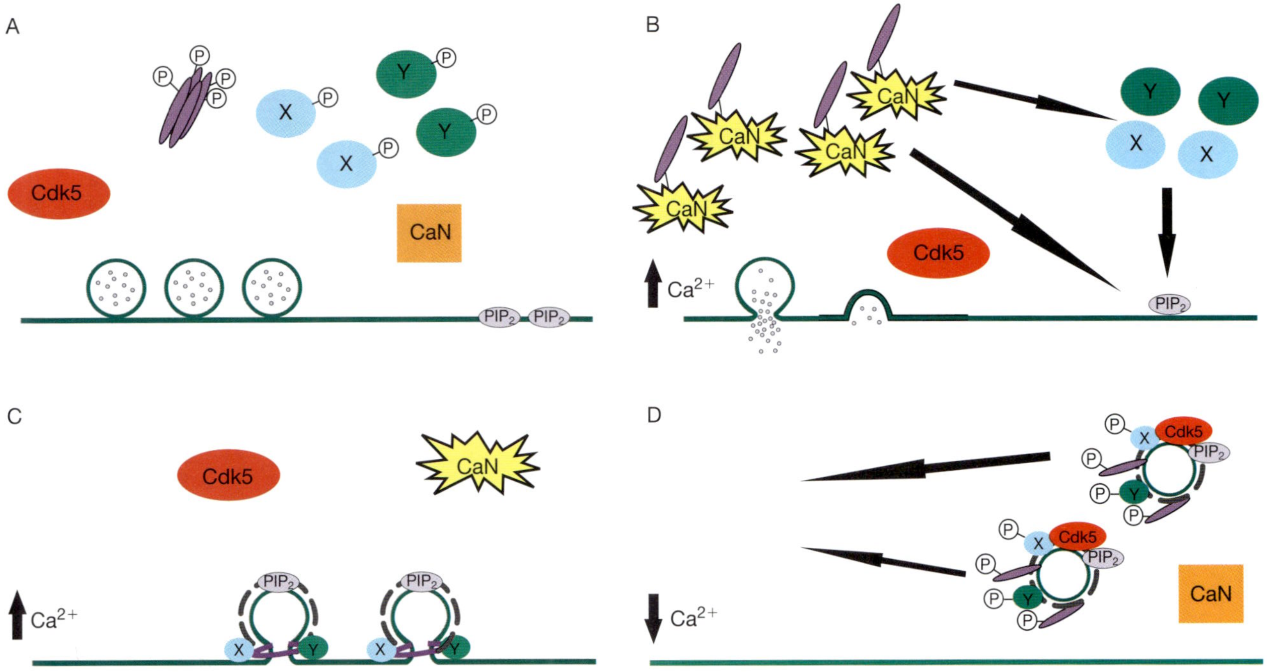

Figure 2 Model for the control of SVE by dynamin I phosphorylation. (**A**) In a resting nerve terminal, all phosphorylated dynamin I (purple bars) and its phosphorylation-dependent binding partner(s) (X and Y) are in the cytosol. Calcineurin (CaN) is inactive. (**B**) On nerve terminal depolarization, Ca^{2+} influx activates calcineurin, which dephosphorylates dynamin I, and proteins X and Y. Dephosphorylation allows dynamin I to translocate to the plasma membrane via interactions with either or both PtdIns(4,5)P_2 (PIP_2) or its phosphorylation-dependent binding partners (X and Y). (**C**) During depolarization and SVE, dynamin I forms a collar round the neck of the retrieving SV, and causes fission from the plasma membrane. (**D**) On termination of nerve terminal stimulation, intracellular Ca^{2+} levels fall, inactivating CaN. This allows Cdk5 to rephosphorylate dynamin I, and X and Y. This rephosphorylation reduces its affinity for PtdIns(4,5)P_2, and proteins X and Y, facilitating the return of dynamin I to the cytosol for the next cycle of SVE.

the other dephosphins also regulate SVE? This question will have to be addressed in a similar manner to dynamin I, via the identification of the physiologically relevant *in vivo* phosphorylation sites of the other dephosphins. When these questions are answered, then we should be a long way to discovering the mechanism by which global dephosphorylation of the dephosphins controls SVE.

This work was supported by grants from the Wellcome Trust (Ref: GR070569) and a University of Edinburgh Medical Faculty Scholarship (to K.J.S.).

References

1. Cousin, M.A. (2000) Mol. Neurobiol. **22**, 115–128
2. Shpetner, H.S. and Vallee, R.B. (1989) Cell **59**, 421–432
3. van der Bliek, A.M. and Meyerowitz, E.M. (1991) Nature (London) **351**, 411–414
4. Koenig, J.H. and Ikeda, K. (1989) J. Neurosci. **9**, 3844–3860
5. Liu, J.P. and Robinson, P.J. (1995) Endocr. Rev. **16**, 590–607
6. Urrutia, R., Henley, J.R., Cook, T. and McNiven, M.A. (1997) Proc. Natl. Acad. Sci. U.S.A. **94**, 377–384
7. Gray, N.W., Fourgeaud, L., Huang, B., Chen, J., Cao, H., Oswald, B.J., Hemar, A. and McNiven, M.A. (2003) Curr. Biol. **13**, 510–515
8. Herskovits, J.S., Burgess, C.C., Obar, R.A. and Vallee, R.B. (1993) J. Cell Biol. **122**, 565–578
9. van der Bliek, A.M., Redelmeier, T.E., Damke, H., Tisdale, E.J., Meyerowitz, E.M. and Schmid, S.L. (1993) J. Cell Biol. **122**, 553–563
10. Sweitzer, S.M. and Hinshaw, J.E. (1998) Cell **93**, 1021–1029
11. Stowell, M.H., Marks, B., Wigge, P. and McMahon, H.T. (1999) Nat. Cell Biol. **1**, 27–32
12. Marks, B., Stowell, M.H., Vallis, Y., Mills, I.G., Gibson, A., Hopkins, C.R. and McMahon, H.T. (2001) Nature (London) **410**, 231–235
13. Vallis, Y., Wigge, P., Marks, B., Evans, P.R. and McMahon, H.T. (1999) Curr. Biol. **9**, 257–260
14. Sever, S., Muhlberg, A.B. and Schmid, S.L. (1999) Nature (London) **398**, 481–486
15. Song, B.D. and Schmid, S.L. (2003) Biochemistry **42**, 1369–1376
16. David, C., McPherson, P.S., Mundigl, O. and De Camilli, P. (1996) Proc. Natl. Acad. Sci. U.S.A. **93**, 331–335
17. Ringstad, N., Nemoto, Y. and De Camilli, P. (1997) Proc. Natl. Acad. Sci. U.S.A. **94**, 8569–8574
18. De Heuvel, E., Bell, A.W., Ramjaun, A.R., Wong, K., Sossin, W.S. and McPherson, P.S. (1997) J. Biol. Chem. **272**, 8710–8716
19. Barylko, B., Binns, D.D. and Albanesi, J.P. (2001) Methods Enzymol. **329**, 486–496
20. Robinson, P.J., Liu, J.P., Powell, K.A., Fykse, E.M. and Südhof, T.C. (1994) Trends Neurosci. **17**, 348–353
21. Liu, J.P., Powell, K.A., Südhof, T.C. and Robinson, P.J. (1994) J. Biol. Chem. **269**, 21043–21050
22. Liu, J.P., Sim, A.T.R. and Robinson, P.J. (1994) Science **265**, 970–973
23. Bauerfeind, R., Takei, K. and De Camilli, P. (1997) J. Biol. Chem. **272**, 30984–30992
24. Marks, B. and McMahon, H.T. (1998) Curr. Biol. **8**, 740–749
25. Cousin, M.A. and Robinson, P.J. (2001) Trends Neurosci. **24**, 659–665
26. Lai, M.M., Hong, J.J., Ruggiero, A.M., Burnett, P.E., Slepnev, V.I., De Camilli, P. and Snyder, S.H. (1999) J. Biol. Chem. **274**, 25963–25966
27. Robinson, P.J., Sontag, J.-M., Liu, J.P., Fykse, E.M., Slaughter, C., McMahon, H.T. and Südhof, T.C. (1993) Nature (London) **365**, 163–166

28. Hosoya, H., Komatsu, S., Shimizu, T., Inagaki, M., Ikegami, M. and Yazaki, K. (1994) Biochem. Biophys. Res. Commun. **202**, 1127–1133
29. Earnest, S., Khokhlatchev, A., Albanesi, J.P. and Barylko, B. (1996) FEBS Lett. **396**, 62–66
30. Chen-Hwang, M.-C., Chen, H.-R., Elzinga, M. and Hwang, Y.-W. (2002) J. Biol. Chem. **277**, 17597–17604
31. Robinson, P.J. (1992) J. Biol. Chem. **267**, 21637–21644
32. Cousin, M.A., Tan, T.C. and Robinson, P.J. (2001) J. Neurochem. **76**, 105–116
33. Robinson, P.J. (1992) Mol. Neurobiol. **5**, 87–142
34. Tan, T.C., Valova, V.A., Malladi, C.S., Graham, M.E., Berven, L.A., Jupp, O.J., Hansra, G., McClure, S.J., Sarcevic, B., Boadle, R.A. et al. (2003) Nat. Cell Biol. **5**, 701–710
35. Matsubara, M., Kusubata, M., Ishiguro, K., Uchida, T., Titani, K. and Taniguchi, H. (1996) J. Biol. Chem. **271**, 21108–21113
36. Shuang, R.Q., Zhang, L., Fletcher, A., Groblewski, G.E., Pevsner, J. and Stuenkel, E.L. (1998) J. Biol. Chem. **273**, 4957–4966
37. Rosales, J.L., Nodwell, M.J., Johnston, R.N. and Lee, K.Y. (2000) J. Cell. Biochem. **78**, 151–159
38. Floyd, S.R., Porro, E.B., Slepnev, V.I., Ochoa, G.C., Tsai, L.H. and De Camilli, P. (2001) J. Biol. Chem. **276**, 8104–8110
39. Tomizawa, K., Sunada, S., Lu, Y.F., Oda, Y., Kinuta, M., Ohshima, T., Saito, T., Wei, F.Y., Matsushita, M., Li, S.T. et al. (2003) J. Cell Biol. **163**, 813–824
40. Lee, S.Y., Wenk, M.R., Kim, Y., Nairn, A.C. and De Camilli, P. (2004) Proc. Natl. Acad. Sci. U.S.A. **101**, 546–551
41. Powell, K.A., Valova, V.A., Malladi, C.S., Jensen, O.N., Larsen, M.R. and Robinson, P.J. (2000) J. Biol. Chem. **275**, 11610–11617
42. Larsen, M.R., Graham, M.E., Robinson, P.J. and Roepstorff, P. (2004) Mol. Cell Proteomics **3**, 456–465
43. Slepnev, V.I., Ochoa, G.C., Butler, M.H., Grabs, D. and De Camilli, P. (1998) Science **281**, 821–824
44. Patrick, G.N., Zhou, P., Kwon, Y.T., Howley, P.M. and Tsai, L.H. (1998) J. Biol. Chem. **273**, 24057–24064
45. Dhavan, R. and Tsai, L.H. (2001) Nat. Rev. Mol. Cell Biol. **2**, 749–759
46. Lai, M.M., Luo, H.R., Burnett, P.E., Hong, J.J. and Snyder, S.H. (2000) J. Biol. Chem. **275**, 34017–34020

Biochem. Soc. Symp. **72**, 99–108
(Printed in Great Britain)

10

Early endocytic Rabs: functional prediction to functional characterization

Jeremy C. Simpson* and Arwyn T. Jones†[1]

*Cell Biology and Biophysics Programme, European Molecular Biology Laboratory (EMBL), Meyerhofstrasse 1, D69117, Heidelberg, Germany, and †Welsh School of Pharmacy, Cardiff University, Redwood Building, Cardiff CF10 3XF, Wales, U.K.

Abstract

Endocytic pathways are highly dynamic gateways for molecules to enter cells. Functionality and specificity is in part controlled by a number of small GTPases called Rabs. In defined cellular locations, Rabs mediate multiple functions in membrane trafficking via their specific interaction with organelle membranes and a host of affector and effector molecules. On endocytic pathways, Rabs have been shown to control the formation of vesicles on the plasma membrane and the downstream delivery of internalized molecules to a number of cellular locations. As numerous Rabs are located to endocytic pathways, an internalized molecule may traverse a number of Rab specific substations or subdomains en route to its final destination. Rabs 5, 21 and 22 have all been localized to the early endocytic pathway and have been shown to share a number of characteristics to merit their segregation into a single functional endocytic group. In this review, we compare experiments that describe similarities and differences in endosome morphology and function that is mediated by their expression in cells.

Endocytosis and Rab GTPases

Endocytic pathways serve as cellular entry ports for traffic of molecules from the plasma membrane to a number of subcellular stations [1–3]. These pathways are important for an array of cellular functions, including nutrient uptake, down-regulation of signalling receptors and antigen processing in the immune response. Uptake may be mediated by either ligand binding to plasma membrane receptors which are then internalized in clathrin-coated vesicles or,

[1]To whom correspondence should be addressed (email jonesat@cardiff.ac.uk).

alternatively, less well characterized mechanisms, such as macropinocytosis, clathrin-independent endocytosis and caveoleae [2]. Once internalized, these vesicles are thought to fuse with the first sorting endocytic station, which we define here as the early endosome, and this structure segregates molecules to distinct transport pathways.

The endocytic pathway along which a molecule traverses from entry to final destination will be mediated in part by one or several members of a family of small GTPases called Rabs [4–7]. Rab proteins belong to the Ras superfamily of small GTPases that shuttle between GDP- and GTP-bound states to mediate their cellular effects. The GDP/GTP cycle is modulated by factors that catalyse nucleotide exchange and hydrolysis, and most Rabs alternate between membrane (active) and cytosolic (inactive) states. Shuttling is aided by the cytosolic Rab chaperones GDI (GDP-dissociation inhibitor) and REP (Rab-escort protein). Rabs control numerous cellular trafficking steps, and functional identity is characterized by their specific binding to distinct organelles on, for example, the secretory and endocytic pathways. Membrane attachment and further specificity is mediated by their post-translational modification via C-terminal geranylgeranylation [8].

Today, the full sequences of hundreds of Rabs from different species are known, including the 60 or so variants that are found in the human genome. A number of these have been studied, although few extensively so, and a general emerging theme is that, via interaction with a multitude of affector and effector molecules, they control several cellular functions, including some outside the classical remit of organization of membrane traffic and dynamics.

In distinct membrane locations, Rabs orchestrate organelle biogenesis and fusion, and directed transport of vesicles along cytoskeletal frameworks, and there is increasing evidence of a defined link with cell signalling pathways [9,10].

Rab definition and specificity

Rabs, as members of the Ras superfamily of GTPases have highly conserved motifs that mediate binding to the nucleotide. Studies in Ras have identified specific point mutations that generated proteins that are either predominantly bound to GDP or that have very low GAP (GTPase-activating protein)-stimulated GTPase activity [11,12]. Based on their effects in cells, these are often described as dominant-negative and constitutively active, and also GDP-binding and GTP-hydrolysis mutants. For Rab proteins, mutation of a conserved Ser/Thr→Asn in the phosphate/magnesium-binding PM1 motif (e.g. Rab5a residue 34) or Asn/Thr→Ile mutation in a separate nucleotide-binding G2 motif (Rab5 residue 133) leads to a GDP-binding mutant. Conversely, mutation of a Gln/Thr→Leu in another distinct PM3 motif (e.g. Rab5a residue 79) generates a GTP-hydrolysis mutant [13]. The ability to express such Rab mutants in cells or purify them prior to their addition to membrane transport assays has been the main tool for studying the function of these proteins. Silencing Rab expression via RNAi (RNA interference) technology is also now gaining widespread importance.

Endocytic Rab proteins

As secretory, endocytic and recycling pathways are inexorably linked, it is sometimes difficult to assign proteins such as Rabs to just one of these. However, at least 13 Rab family members are now known to be involved in the dynamics of the endocytic pathway, and information on the majority of these was described by Somsel Rodman and Wandinger-Ness in 2000 [14]. Roles are described for Rab5 in early endosome dynamics, Rab7 as being a mediator of late endocytic traffic, and functions for Rabs 4 and 11 in recycling to the plasma membrane. Several endocytic Rabs have now been pinpointed to early endosomes, and an intriguing question is how so many Rabs, some of which have been postulated to co-inhabit single endosomes, relate to each other both temporally and functionally. Phylogenic scrutiny of Rabs suggests they can in fact be segregated into distinct functional groups, and one of these, group V, was specific for Rabs 5, 21 and 22 [5]. Rab5 has been studied extensively, but comparatively little information is available for Rab21 and 22. The present review will focus on these three variants and recent data that suggest common traits and unique features between them.

Extensive phylogenic analysis of Rabs from different species have identified Rab-specific motifs which were numbered RabF1–5 [5,15,16]. Further scrutiny of Rab sequences have also allowed for classification of Rabs into subfamilies that are homologous in sequences called specificity determining regions (RabSF). RabSF domains 1–4 are conserved across species, and are predicted to mediate binding of Rabs to specific effector proteins. These domains are often associated within the more commonly identified switch I and switch II domains, defined initially from Ras structure–activity studies. Thus Rabs belonging to the same subfamily would be predicted to share common interacting proteins. The emerging picture was that general Rab-binding proteins, such as GDI and REP, would bind to RabF regions and Rab specificity would be mediated via interactions with RabSF domains. Thus new criteria for Rab isoforms should include conservation of sequence and RabF domains, but also of RabSF domains. One RabSF motif is in the hypervariable C-terminal region, which is classically the largest differentiating denominator between individual Rabs.

When sequences from Rabs from different species were compared using these parameters with the functional data on Rabs from different species, there appeared to be functional segregation–phylogeny of function. This approach segregated some Rabs into functional groups that contained members previously shown to be dissimilar in localization and function. For example, group II contains Rab2, Rab4 and Rab11 and thus it remains to be seen what common functional characteristic is actually shared by these members. As mentioned above, functional group V was exclusive for human Rab5a, b and c, Rab22a, b and c, and Rab21, together with their isoforms in other species.

Comparison of endosome morphologies in cells expressing Rabs5, 21 and 22

Isoforms have been identified for Rabs 5 and 22, but to date only one for Rab21 (Table 1). Rab5 isoforms are highly homologous, but the individual variants are only 40–50% similar (Table 1).

Rab5 is one of the most studied Rab proteins and is localized to the plasma membrane, clathrin-coated vesicles, sorting endosomes and phagosomes [7]. The majority of work on Rab5 has been performed with the 5a isoform and, unless specified, Rab5 in the present review refers to Rab5a. The three Rab5 variants co-localize on early endosomes, and their overexpression in transfected cells leads to the formation of slightly enlarged transferrin-receptor-positive structures [17,18]. These organelles are even larger when the corresponding GTPase-defective mutants are expressed [18,19]. Thus, in unstimulated cells, there does not appear to be any obvious morphological difference in endosomes of cells overexpressing Rab5a, b or c. However, when cells are stimulated with EGF (epidermal growth factor), then only cells expressing wild-type Rab5a have swollen vacuoles akin to those observed in Rab5a Gln^{79}→Leu-expressing cells [20]. The functional significance of these findings with respect to the three isoforms is discussed below.

GFP (green fluorescent protein)–Rab22a labels the plasma membrane and early endosomes, and overexpression of the wild-type variant leads to the formation of very swollen vacuole-like structures that also harbour EEA1 (early endosome antigen 1), Rab5 and internalized probes [21,22]. In one study, the distribution pattern was cell-line-dependent, and, in HeLa cells, a significant fraction of the Rab22a protein also co-localized with Golgi markers [22]. Earlier reports suggested that Rab22a distribution was not cell-type-dependent [21], and, in the present study, a fraction of Rab22a was localized to late endosomes. Both of these reports indicate that the morphological effects of Rab22a expression are more pronounced than those of Rab5, and the Rab22a phenotype was more akin to that observed in cells expressing Rab5 Gln^{79}→Leu. More recent data challenges these findings, showing specifically that endogenous Rab22a localizes to tubules that emanate from a perinuclear area and are involved in recycling of MHCI (major histocompatibility complex class I) from endosomes to the plasma membrane [23]. As an anti-Rab22 antibody was not utilized in the former two studies, it was suggested that the observed swelling phenotypes were due to gross overexpression [23]. Rat Rab22b (rRab22b) shares 94% iden-

Table 1 Homology matrix for Rabs5, 21 and 22. We were unable to obtain the Rab22c sequence from the database.

Rab5a	Rab5b	Rab5c	Rab21	Rab22a	Rab22b
Rab5a	100%				
Rab5b	83%	100%			
Rab5c	89%	85%	100%		
Rab21	40%	39%	41%	100%	
Rab22a	47%	47%	48%	42%	100%
Rab22b	45%	44%	46%	44%	74%

tity with human Rab22b, and, in HeLa cells, the YFP (yellow fluorescent protein)-tagged protein associates with large swollen asymmetrical structures close to the nucleus that seemed to bud tubular Rab22b-positive processes that extended towards the cell periphery [24]. Human Rab22a $Gln^{64}{\rightarrow}Leu$ was also found associated with lysosomal markers, suggesting that the mutation results in enhanced endosome fusion similar to those observed in cells expressing Rab5 $Gln^{79}{\rightarrow}Leu$, and that these structures may be akin to the previously described Rab5 $Gln^{79}{\rightarrow}Leu$ early–late hybrid organelles [25,26].

In the case of Rab21, an anti-Rab21 antibody revealed that the subcellular distribution of this protein was dependent on cell polarity with endoplasmic-reticulum-like staining in non-polarized cells and apical (possibly late endosomal) distribution in polarized cells [27]. Since this report, Rab21 has been thought to be specific for polarized cells, despite the fact that it was shown to be expressed ubiquitously [27]. Using fluorescence and electron microscopy, we have shown that GFP–Rab21 localizes to the perinuclear region, on distinct peripheral vesicles and occasionally on the plasma membrane (Figure 1, and [28]). Co-localization with markers for subcellular organelles identified the perinuclear

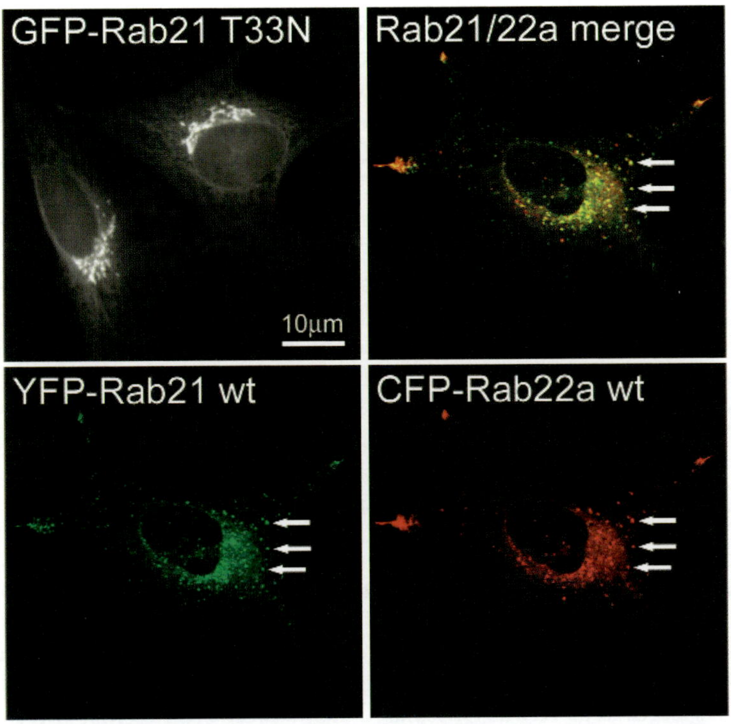

Figure 1 Localization of GFP–Rab21 $Thr^{33}{\rightarrow}Asn$ and YFP–Rab21 wild-type in comparison with CFP (cyan fluorescent protein)–Rab22a in HeLa cells. Upper left panel shows localization of GFP–Rab21 $Thr^{33}{\rightarrow}Asn$. Remaining panels show localization of YFP–Rab21 wild-type (pseudocoloured green) compared with Rab22a wild-type [tagged with CFP (pseudocoloured red)]. Arrows denote Rab21 and Rab22a co-localization.

labelling as Golgi and the vesicular structures as early endosomes. Comparison with the distribution of the early endosomal marker and Rab5 effector, EEA1, demonstrated that, in overexpressing cells, Rab21 localized to enlarged endosmes. The findings were confirmed using a previously characterized Rab21 antibody [27] and were supported further by co-localization of Rab21 with probes internalized to specifically label early endosomes (transferrin and EGF). Surprisingly, in Rab21 Gln^{78}→Leu-expressing cells, EEA1-labelled endosomes have normal morphologies, but gain the additional presence of an extensive reticular tubular network that has yet to be fully characterized.

The expression of the GDP-binding-defective mutants again revealed major differences between the three variants. Rab5a Ser^{34}→Asn- and Asn^{133}→Ile-expressing cells typically have small endosomes; vesiculotubular structures have also been described [17]. As expected, a much larger fraction of the Rab GDP-binding mutants is cytosolic. The phenotypes of cells expressing the GDP-binding- and GTP-hydrolysis-defective mutants parallel their effects in *in vitro* assays of early endosome fusion, which is stimulated and inhibited by Rab5 Gln^{79}→Leu and Rab5 Asn^{133}→Ile respectively [17,29]. There were no major morphological differences between cells expressing mutants of all three Rab5 isoforms and they behaved similarly in stimulating and inhibiting endosome-fusion assays [18]. Rab22a Ser^{19}→Asn labelling is also cytosolic [21–23], but the protein is additionally found associated with the Golgi apparatus, and this phenomenon was also been seen when the equivalent rRab22b mutation was expressed [24]. Surprisingly, and in contrast with the effects of the wild-type and GTP-hydrolysis mutants, Rab22a Ser^{19}→Asn expression did not significantly affect the morphologies of EEA1-labelled endosomes [22]. Rab21 Thr^{33}→Asn labels intensely in a juxta-nuclear region (Figure 1), co-localizing with TGN (*trans*-Golgi network) markers [28]. The significance of this Golgi localization for GDP-binding mutants of Rabs 21 and 22 is unknown, and although their expression does not significantly impair TGN morphology, a role for rRab22b in TGN–endosome traffic was postulated [24]. Addition of Rab21 and 22 variants to endosome fusion and Golgi–endosome transport assays will undoubtedly shed further insight into their cellular roles.

Co-localization of endocytic Rab proteins

Co-transfection of cells with Rabs 4, 5 and 11 was performed to compare the number of endocytic structures that contained more than one Rab and to analyse the temporal nature of the transferrin cycle with respect to co-localization of the ligand with these endocytic Rabs [30]. Transferrin was initially located in Rab5-positive compartments and then traversed Rab4 and Rab11 vesicles, as predicted. It was postulated from these studies that the transit of transferrin was not necessarily through Rab-specific endosomes, but rather through Rab domains on the same endosome [7]. To analyse further the distribution of Rabs on endosomes, Rab21 was co-expressed with Rabs previously localized to the secretory pathway and to early, late and recycling endosomes. Extensive (>75%) overlap of Rab21 was observed with a specific subset of Rabs, including Rab4, Rab5a and Rab22a (Figure 1), and also Rab17, an endocytic variant that is predominantly expressed in epithelial cells [31]. Interestingly, Rab17 is closely homologous with Rabs 5, 21 and 22 in the functional prediction model [5]. It therefore seems likely that a single endosome could contain several Rab variants, and that each is involved with mediating specific transport steps from and to this organelle. As most Rabs are ubiquitously expressed, a major challenge remains in defining their relationship to each other, their host organelle and the cargo they help deliver.

Endocytic function in cells expressing mutants of Rab5, Rab21 and Rab22

The initial characterization of Rab5 was based on its subcellular localization and the effects of the mutants on *in vitro* assays of early endosome fusion and on functional uptake experiments in cells [18,19,29]. In general, expression of the wild-type and Gln^{79}→Leu Rab5a protein had stimulatory effects, but the expression of the GTP-binding mutant was inhibitory to these assays. Endocytic pathways are extensively linked to signalling events at the plasma membrane [9,10,32–34], and insights on differential roles for Rab5 isoforms have come from experiments investigating the mechanism by which the dynamics of endocytic pathways are modified after cell stimulation with EGF [20]. Expression of the three Rab5 isoforms enhanced fluid-phase uptake in unstimulated cells, but only the Rab5a isoform promoted endocytosis further in response to EGF. Expression of Rab5a also had much more pronounced effects on EGF uptake compared with Rab5b and Rab5c. In contrast, earlier studies showed comparable abilities of the three isoforms to enhance transferrin uptake from the plasma membrane [18]. As the three Rab5 isoforms share >80% homology (Table 1), it should not be difficult to identify the domain(s) responsible for these EGF-specific effects. It also remains to be seen whether this is a phenomenon observed with other growth factors and whether expression of the GTP-binding-defective mutants of the three isoforms have different effects on EGF-stimulated endocytosis.

Several of these studies on the effects of Rab5 expression on EGF internalization and, to a certain extent, a role for Rab5 in the formation of clathrin-coated vesicles [35] at the plasma membrane were challenged recently [36]. HeLa and HepG2 cells overexpressing wild-type Rab5 and Rab5 mutants were found to have control levels of internalized EGF, and the effects of Rab5 Ser^{34}→Asn on EGF traffic were determined to be downstream of the plasma membrane. Additionally, Rab5 expression had no effect on uptake of a fluid-phase marker. There are other reports on the lack of Rab5 Ser^{34}→Asn effect or a very minor effect on uptake of receptors predicted to be internalized via clathrin-coated vesicles, including EGF [37,38]. In *Saccharomyces cerevisiae*, loss of function of one or a combination of the Rab5 homologues Ypt51, 52 and 53 (Ypt is yeast protein transport) was without effect on internalization of α-factor, but loss of Ypt51 led to significant reduction in α-factor degradation; trafficking abnormalities away from the plasma membrane were also reported for the other variants [39,40]. These data are supported by RNAi analysis showing no effect on uptake of transferrin or EGF after silencing expression, individually, of Rab5a, b and c [41]. It remains to be seen how these findings relate to the extensive literature showing that expression of Rab5 Ser^{34}→Asn significantly inhibits the uptake of several unrelated molecules and pathogens. Cell type and expression levels are undoubtedly important parameters.

Numerous endocytic defects have been attributed to overexpression or loss of function of Rab22a. These range from an ability of Rab22a and Rab22a Gln^{64}→Leu to partially inhibit degradation of EGF to a reduction of fluid-phase endocytosis in cells expressing Rab22a and Rab22a Ser^{19}→Asn [21,22]. Expression of Rab22a and the nucleotide-binding mutants inhibited the recycling of MHCI to varying degrees, whereas only the Ser^{19}→Asn mutant significantly reduced formation of MHCI tubules [23]. These data were supported by silencing Rab22a expression, leading to a 75% reduction in recycling of MHCI and a minor inhibition in recycling of transferrin. It will be interesting to see whether silencing Rab22 also inhibits EGF degradation and fluid-phase uptake, as this would suggest that Rab22 is required for the general integrity of different endocytic pathways.

No defects in endocytosis were observed in cells expressing Rab21 or Rab21 Gln^{78}→Leu; however, both uptake of EGF and transferrin are reduced in cells expressing Rab21 Thr^{33}→Asn and there is also a traffic defect in EGF delivery to late endosomes and lysosomes [28]. EGF degradation has therefore been shown to be inhibited to various extents in cells overexpressing, predominantly, the GDP-binding mutant of Rabs 4, 5, 7, 21 and 22 [22,42]. Since internalized ligands, such as EGF, move through different Rab stations or domains, all of these studies highlight the fact that common Rab-interacting partners may be sequestered via overexpression of one variant. Moreover, endocytic pathways and organelles for transferrin, EGF and MHCI traffic are linked, such that loss of function of one pathway may lead to the partial or complete failure of another. Thus an observed inhibition in a given trafficking step due to overexpression of a Rab does not necessarily relate to an absolute requirement for the protein in that pathway. If Rabs are indeed membrane organizers, then the same may be true when one Rab is silenced using RNAi.

Mice knockouts of Rabs5, 21 or 22 genes have not been described, but, interestingly, RNAi feeding experiments in *Caenorhabditis elegans* revealed that Rab21 inhibition was without a phenotype, whereas the Rab5 was larval lethal ([43] and http://www.wormbase.org/); Rab22 is not expressed in this organism.

Rab-interacting proteins

Functional specificity and activity of Rabs are mediated by their affectors and effectors located both in the cytosol and on membranes. New Rab5 affectors and effectors appear in the literature with striking consistency, and each new partner expands the complexity of the involvement of this protein with endocytic and, more recently, signalling and cytoskeletal processes [7,10,44,45]. Indeed, over 20 Rab5–GTP interacting partners were initially described from an affinity screen, and many of these await characterization [46]. These are soluble proteins and there is comparatively very little information regarding the interaction of endocytic Rabs, or indeed other Rabs, with integral membrane components.

Should each Rab have 20 soluble interacting partners, then approx. 5% of the human genome would be preoccupied with Rab protein function! But, as described above, the RabF and RabSF domains add two levels of similarity that ensure that proteins, such as GDI and REP, probably bind to all Rabs; Rab effectors themselves, though to a lesser degree, are now known to be promiscuous. Rabaptin-5 and Rabenosyn-5 bind to Rab4 [47,48], and, interestingly, the affinity-column approach used to identify Rab5-binding partners also showed that Rab33b had a direct or indirect interaction with these two proteins [49]. Rab33b was initially localized and proposed to function in the medial Golgi [49,50].

It is predicted that Rabs5, 21 and 22 will share many binding partners, but as research on the latter two variants is in its infancy, this remains to be tested. Rab22a has, however, been shown to interact with EEA1, but the interaction seems to be slightly different to that of Rab5 with this protein [22]. As Rab21 is also capable of enlarging early endosomal membranes, it is postulated that Rab21 and EEA1 interaction mediates this phenomenon, suggesting that all three Rab variants promote endosomal swelling via a common fusion mechanism. The phenotypes of the GDP-binding and GTP-hydrolysis mutants of Rab21, however, suggest major differences between their capacity to control endosomal traffic. The challenge now is to identify unique interacting proteins and trafficking pathways.

We thank Professor Marianne Wessling-Resnick, Harvard School of Public Health, Boston, MA, U.S.A., for critically reading this manuscript.

References

1. Gruenberg, J. (2001) Nat. Rev. Mol. Cell Biol. **2**, 721–730
2. Conner, S.D. and Schmid, S.L. (2003) Nature (London) **422**, 37–44
3. Maxfield, F.R. and McGraw, T.E. (2004) Nat. Rev. Mol. Cell Biol. **5**, 121–132
4. Bock, J.B., Matern, H.T., Peden, A.A. and Scheller, R.H. (2001) Nature (London) **409**, 839–841
5. Pereira-Leal, J.B. and Seabra, M.C. (2001) J. Mol. Biol. **313**, 889–901
6. Stenmark, H. and Olkkonen, V.M. (2001) Genome Biol. **2**, REVIEWS3007
7. Zerial, M. and McBride, H. (2001) Nat. Rev. Mol. Cell Biol. **2**, 107–117
8. Gomes, A.Q., Ali, B.R., Ramalho, J.S., Godfrey, R.F., Barral, D.C., Hume, A.N. and Seabra, M.C. (2003) Mol. Biol. Cell **14**, 1882–1899
9. Benmerah, A. (2004) Curr. Biol. **14**, R314–R316
10. Horazdovsky, B. (2004) Dev. Cell **6**, 161–162
11. Der, C.J., Finkel, T. and Cooper, G.M. (1986) Cell **44**, 167–176
12. Farnsworth, C.L. and Feig, L.A. (1991) Mol. Cell. Biol. **11**, 4822–4829
13. Zerial, M. and Huber, L.A. (1995) Guidebook to the Small GTPases, Oxford University Press, Oxford
14. Somsel Rodman, J. and Wandinger-Ness, A. (2000) J. Cell Sci. **113**, 183–192
15. Moore, I., Schell, J. and Palme, K. (1995) Trends Biochem. Sci. **20**, 10–12
16. Pereira-Leal, J.B. and Seabra, M.C. (2000) J. Mol. Biol. **301**, 1077–1087
17. Bucci, C., Parton, R.G., Mather, I.H., Stunnenberg, H., Simons, K., Hoflack, B. and Zerial, M. (1992) Cell **70**, 715–728
18. Bucci, C., Lutcke, A., Steele-Mortimer, O., Olkkonen, V.M., Dupree, P., Chiariello, M., Bruni, C.B., Simons, K. and Zerial, M. (1995) FEBS Lett. **366**, 65–71
19. Stenmark, H., Parton, R.G., Steele-Mortimer, O., Lutcke, A., Gruenberg, J. and Zerial, M. (1994) EMBO J. **13**, 1287–1296

20. Barbieri, M.A., Roberts, R.L., Gumusboga, A., Highfield, H., Alvarez-Dominguez, C., Wells, A. and Stahl, P.D. (2000) J. Cell Biol. **151**, 539–550
21. Mesa, R., Salomon, C., Roggero, M., Stahl, P.D. and Mayorga, L.S. (2001) J. Cell Sci. **114**, 4041–4049
22. Kauppi, M., Simonsen, A., Bremnes, B., Vieira, A., Callaghan, J., Stenmark, H. and Olkkonen, V.M. (2002) J. Cell Sci. **115**, 899–911
23. Weigert, R., Yeung, A.C., Li, J. and Donaldson, J.G. (2004) Mol. Biol. Cell **15**, 3758–3770
24. Rodriguez-Gabin, A.G., Cammer, M., Almazan, G., Charron, M. and Larocca, J.N. (2001) J. Neurosci. Res. **66**, 1149–1160
25. Rosenfeld, J.L., Moore, R.H., Zimmer, K.P., Alpizar-Foster, E., Dai, W., Zarka, M.N. and Knoll, B.J. (2001) J. Cell Sci. **114**, 4499–4508
26. Duclos, S., Corsini, R. and Desjardins, M. (2003) J. Cell Sci. **116**, 907–918
27. Opdam, F.J., Kamps, G., Croes, H., van Bokhoven, H., Ginsel, L.A. and Fransen, J.A. (2000) Eur. J. Cell Biol. **79**, 308–316
28. Simpson, J.C., Griffiths, G., Wessling-Resnick, M., Fransen, J.A.M., Bennett, H. and Jones, A.T. (2204) J. Cell Sci., in the press
29. Gorvel, J.P., Chavrier, P., Zerial, M. and Gruenberg, J. (1991) Cell **64**, 915–925
30. Sonnichsen, B., De Renzis, S., Nielsen, E., Rietdorf, J. and Zerial, M. (2000) J. Cell Biol. **149**, 901–914
31. Lutcke, A., Jansson, S., Parton, R.G., Chavrier, P., Valencia, A., Huber, L.A., Lehtonen, E. and Zerial, M. (1993) J. Cell Biol. **121**, 553–564
32. von Zastrow, M. (2001) Biochem. Soc. Trans. **29**, 500–504
33. Sorkin, A. (2001) Biochem. Soc. Trans. **29**, 480–484
34. Clague, M.J. and Urbe, S. (2001) J. Cell Sci. **114**, 3075–3081
35. McLauchlan, H., Newell, J., Morrice, N., Osborne, A., West, M. and Smythe, E. (1998) Curr. Biol. **8**, 34–45
36. Dinneen, J.L. and Ceresa, B.P. (2004) Exp. Cell Res. **294**, 509–522
37. Papini, E., Satin, B., Bucci, C., de Bernard, M., Telford, J.L., Manetti, R., Rappuoli, R., Zerial, M. and Montecucco, C. (1997) EMBO J. **16**, 15–24
38. Houle, S. and Marceau, F. (2003) Biochem. J. **375**, 151–158
39. Singer-Kruger, B., Stenmark, H., Dusterhoft, A., Philippsen, P., Yoo, J.S., Gallwitz, D. and Zerial, M. (1994) J. Cell Biol. **125**, 283–298
40. Horazdovsky, B.F., Busch, G.R. and Emr, S.D. (1994) EMBO J. **13**, 1297–1309
41. Huang, A., Khvorova, A., Marshall, W. and Sorlin, A. (2004) J. Biol. Chem. **279**, 16657–16661
42. McCaffrey, M.W., Bielli, A., Cantalupo, G., Mora, S., Roberti, V., Santillo, M., Drummond, F. and Bucci, C. (2001) FEBS Lett. **495**, 21–30
43. Tijsterman, M., May, R.C., Simmer, F., Okihara, K.L. and Plasterk, R.H. (2004) Curr. Biol. **14**, 111–116
44. Lanzetti, L., Palamidessi, A., Areces, L., Scita, G. and Di Fiore, P.P. (2004) Nature (London) **429**, 309–314
45. Miaczynska, M., Christoforidis, S., Giner, A., Shevchenko, A., Uttenweiler-Joseph, S., Habermann, B., Wilm, M., Parton, R.G. and Zerial, M. (2004) Cell **116**, 445–456
46. Christoforidis, S., McBride, H.M., Burgoyne, R.D. and Zerial, M. (1999) Nature (London) **397**, 621–625
47. de Renzis, S., Sonnichsen, B. and Zerial, M. (2002) Nat. Cell Biol. **4**, 124–133
48. Vitale, G., Rybin, V., Christoforidis, S., Thornqvist, P., McCaffrey, M., Stenmark, H. and Zerial, M. (1998) EMBO J. **17**, 1941–1951
49. Valsdottir, R., Hashimoto, H., Ashman, K., Koda, T., Storrie, B. and Nilsson, T. (2001) FEBS Lett. **508**, 201–209
50. Zheng, J.Y., Koda, T., Fujiwara, T., Kishi, M., Ikehara, Y. and Kakinuma, M. (1998) J. Cell Sci. **111**, 1061–1069

Biochem. Soc. Symp. **72**, 109–118
(Printed in Great Britain)

11

The 'lipid raft' microdomain proteins reggie-1 and reggie-2 (flotillins) are scaffolds for protein interaction and signalling

Claudia A.O. Stuermer[1] and Helmut Plattner

Department of Biology, University of Konstanz, 78467 Konstanz, Germany

Abstract

Reggie-1 and reggie-2 are two evolutionarily highly conserved proteins which are up-regulated in retinal ganglion cells during regeneration of lesioned axons in the goldfish optic nerve. They are located at the cytoplasmic face of the plasma membrane and are considered to be 'lipid raft' constituents due to their insolubility in Triton X-100 and presence in the 'floating fractions'; hence they were independently named flotillins. According to our current view, the reggies subserve functions as protein scaffolds which form microdomains in neurons, lymphocytes and many other cell types across species as distant as flies and humans. These microdomains are of a surprisingly constant size of $\leq$0.1 μm in all cell types, whereas the distance between them is variable. The microdomains co-ordinate signal transduction of specific cell-surface proteins and especially of GPI (glycosylphosphatidylinositol)-anchored proteins into the cell, as is demonstrated for PrP^c (cellular prion protein) in T-lymphocytes. These cells possess a pre-formed reggie cap scaffold consisting of densely packed reggie microdomains. PrP^c is targeted to the lymphocyte reggie cap when activated by antibody cross-linking, and induces a distinct Ca^{2+} signal. In developing zebrafish, reggies become concentrated in neurons and axon tracts, and their absence, after morpholino antisense RNA-knockdown, results in deformed embryos with reduced brains. Likewise, defects in *Drosophila* eye morphogenesis occur upon reggie overexpression in mutant flies. The defects observed in the organism, as well as in single cells in culture, indicate a morphogenetic function of the reggies, with emphasis on the nervous system. This complies with their role as scaffolds for the formation of multiprotein complexes involved in signalling across the plasma membrane.

[1]To whom correspondence should be addressed (claudia.stuermer@uni-konstanz.de).

Introduction

Reggie-1 and reggie-2 were discovered as two 48 kDa proteins that are up-regulated during axon regeneration of retinal ganglion cells following optic nerve lesion in goldfish and rats [1,2], suggesting an important role in axon growth and neuronal differentiation. On the other hand, their biochemical properties, such as insolubility in cold non-ionic detergents and presence in the 'floating' fraction after sucrose-density centrifugation, identifies reggies as 'lipid raft' proteins. Hence they were considered to be constituents of caveolae and were named flotillin [3]. Our analyses with LM and EM (light and electron microscopy respectively), however, defined reggies as non-caveolar microdomains of a surprisingly constant average size of $\leq$0.1 μm [1,4–7]. This is reflected by their typical punctate distribution along the plasma membrane, which extends into the finest cellular extensions in all cultured cells, including neurons, axons and filopodia (Figures 1A–1F) [1,4,5]. Reggies are associated with the cytoplasmic face of the plasma membrane via acylation [7,8], where they most likely exist as oligomers. Oligomerization of reggie-1 has been shown by velocity gradient centrifugation [7], and, since reggie-2 closely resembles reggie-1 in the region that is involved in oligomerization, we expect that reggie-1 and -2 associate to form hetero-oligomers, which in turn represent the $\leq$0.1 μm microdomains of our microscopic analysis [1,5].

Reggies are evolutionarily highly conserved from flies to fish and humans [9,10]. This predicts an evolutionarily important and possibly conserved role of the reggies, but their exact function is not yet known. The fact that reggies are implicated in functions such as process formation in epithelial, glial and neuronal cell lines [5,7,11–13], cell-membrane-directed transport of GluT (glucose transporter) in adipocytes [14], correct positioning of neuronal cell-adhesion proteins [15], lymphocyte signalling and cell division [5,16] suggests a very basic cellular function. Indeed, pathways involving flotillin-1 (reggie-2) and leading to actin cytoskeleton rearrangement have been proposed [14,17]. There is accumulating evidence that reggies have a developmentally early and a later, spatio-temporally restricted expression in the organisms [15,18], becoming concentrated mainly in neurons and axon tracts, and in addition in lymphocytes [4,5,16] and other blood cells [19]. This is consistent with their original isolation from brain [2,3]. In the present review, we suggest a unifying view for the role of reggies in light of these diverse functional aspects.

We have analysed the reggies in various cell types, most carefully in human lymphocytes, neuronal cells, zebrafish embryos and *Drosophila* [1,15,18]. Based on this work and together with suggestions concerning the role of podocin [20–23] and stomatin [19] (which are, like reggies, members of a superfamily of structurally related proteins [24]), we propose that reggie-1 and -2 form hetero-oligomeric co-clusters and scaffolds that promote the focal concentration and interaction of specific cell-surface and intracellular signal transduction proteins. The existence of such conspicuous microdomains in almost all cells, their dynamic rearrangement in specific situations and association with sets of cellular recognition and signalling proteins suggests that

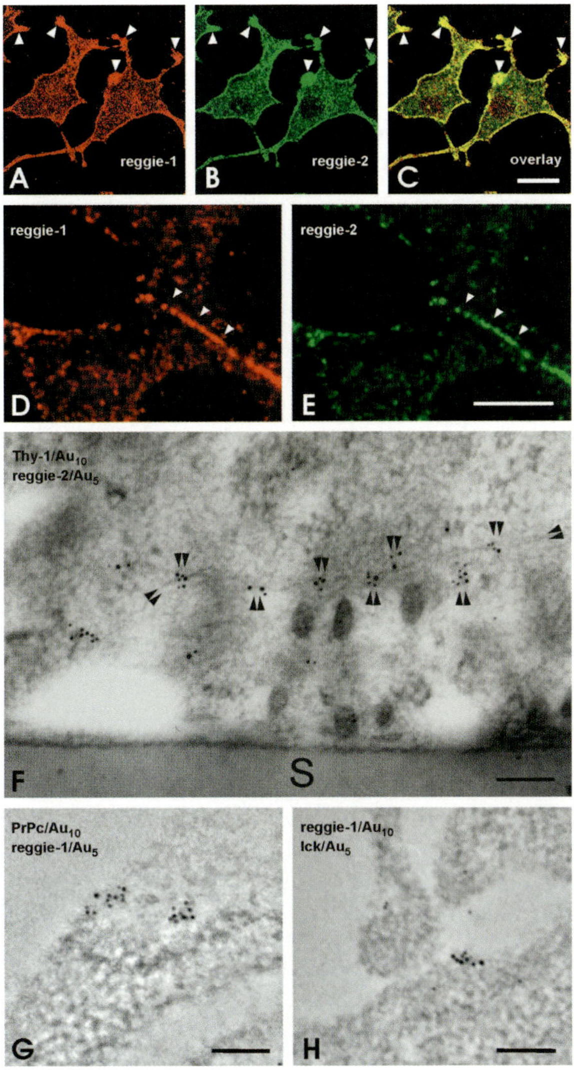

Figure 1 Reggie microdomains and associated proteins at the plasma membrane of PC12 and T-cells. Reggie-1 (**A**) and reggie-2 (**B**) are co-localized (**C**) in plasma membrane microdomains in PC12 cells and are enriched in growth-cone-like processes (arrowheads) and filopodia. Microdomains of reggie-1 (**D**) and reggie-2 (**E**) occur in increased density at cell–cell contact sites. Scale bars, 5 μm. (**F**) Double immuno-EM image showing co-clusters of Thy-1 and reggie-2 which are co-localized in $\leqslant$0.1 μm microdomains of increased density at cell–cell contact sites. (**D**, **E** and **F**) are reprinted from the Molecular Biology of the Cell (2001) **12**, 3031–3045, with permission by the American Society for Cell Biology. (**G**, **H**) In T-lymphocytes, stimulated by antibody cross-linking of PrP^c, aggregates of PrP^c are observed which co-localize with reggie-1 (**G**), and reggie-1 in turn co-clusters with the tyrosine kinase lck (**H**), demonstrating the formation of multiprotein signalling complexes in reggie microdomains. Scale bars, 0.1 μm.

reggies build up scaffolds for the assembly and signal transduction of multiprotein complexes at strategic sites (see below).

Reggie caps in T-lymphocytes and signal transduction by PrP^c (cellular prion protein)

Consistent with our view that reggies form microdomains and pre-formed signalling centres in neurons [5] is the pattern of reggie expression in T-lymphocytes [5,16]. Reggie-1 and reggie-2 are concentrated in one aspect of Jurkat T-cells and T-cells isolated from human blood, and represent pre-formed scaffolds, where proteins involved in T-cell activation and signalling assemble in a process known as capping. Importantly, the reggies occupy this cap position before any stimulation, and thus represent pre-formed cap scaffolds much like the signalling platforms composed of reggie-1 and -2 in neurons [5,16].

It has long been known that GPI (glycosylphosphatidylinositol)-anchored proteins, such as Thy-1, cluster and undergo capping in T-cells when activated by antibody cross-linking [25,26]. It has also been shown that GPI-anchored protein cross-linking induces signal transduction across the plasma membrane, involving the Src family tyrosine kinases fyn and lck, and leading to the rearrangement of the actin cytoskeleton, which in turn promotes the local accumulation of T-cell receptor components [4,26].

Jurkat T-cells also express the GPI-anchored PrP^c, and we examined if PrP^c cross-linking would induce PrP^c capping, co-capping with the pre-formed reggie cap scaffolds and signal transduction with reorganization of the cytoskeleton. Indeed, antibody cross-linking in Jurkat T-cells resulted in PrP^c aggregates that were concentrated in the cap region, and co-localized with reggie [4]. This assembly also led to an increase of fyn, lck (Figure 1H), LAT (linker for activation of T-cells), tyrosine phosphorylation of proteins, and F-actin (filamentous actin) in the cap region, thus implying that PrP^c in association with reggie induces signal transduction. This was demonstrated directly by recordings of the intracellular Ca^{2+} concentration during PrP^c cross-linking, which showed a brief distinct Ca^{2+} pulse [4], together with moderate phosphorylation of the MAPK (mitogen-activated protein kinase). When the Ca^{2+} signal was blocked, PrP^c aggregates formed, but were widely distributed over the cell's circumference instead of being concentrated in the cap. Notably, the pre-formed reggie cap remained [4].

This implies that signals evoked by PrP^c cross-linking promote the focal concentration of PrP^c aggregates in regions of the reggie cap. This involves actin cytoskeletal reorganization ([25], and C.A.O. Stuermer, A. Reuter and H. Plattner, unpublished work). Furthermore, this result shows the participation of PrP^c in signal transduction in T-lymphocytes and emphasizes the importance of scaffolding molecules such as the reggies. Reggie caps accumulate specific sets of proteins. They contribute to the transmission of signals by concentrating PrP^c, and thereby enhance protein interactions and communication with signal transduction molecules which are guided to these sites.

This result opens new perspectives for the analysis of PrP^c functions which may only become clear in the context of its interactions in cell-specific protein complexes. Studying PrP^c in association with reggie scaffolds and proteins that preferentially co-cluster with them is a promising approach. It is not yet known if the misfolded PrP scrapie conformer associates with reggie scaffolds in T-lymphocytes. This knowledge is relevant for two reasons: misfolded PrP scrapie may lose its capability of clustering in reggie microdomains and of undergoing appropriate protein interactions, leading to the well-known fatal disease [27,28]. Moreover, since blood cells reach every single organ, PrP scrapie in T-cells could promote scrapie propagation.

Whether such association of PrP^c and reggies exists in neurons is currently under investigation in PC12 and N2A cells. To this end, it is safe to say that a fraction of these cells indeed shows a close correlation in the distribution of reggie and PrP^c.

Neurons

Reggie-1 and reggie-2 exhibit a strikingly punctate distribution indicative of microdomains in neurons and neuronal cell lines (Figures 1A–1E), and these microdomains have, according to LM and EM analyses using double labelling with antibodies, a constant 'unit size' of approx. 0.1 μm (Figures 1D–1H), regardless of whether the cells were stimulated by cross-linking of GPI-anchored proteins (Figures 1F–1H) or not [4,5]. These microdomains reach into axons, growth cones and filopodia [4,29], and, while their unit size remains constant, the distance between them is often reduced [5]. At strategic sites, the microdomains become more closely spaced (Figures 1D–1F), i.e. at cell–cell contact sites and in growth-cone-like extensions of PC12 cells (Figures 1A–1C). This suggests that cells can regulate the density of reggie microdomains [30] and provide many reggie microdomain scaffolds to sites of cell communication, adhesion, motility and growth [5]. That microdomain density and the concentration of reggie-1 and -2 are regulated according to the actual 'needs' of the cell and its processes is consistent with the original discovery of the reggies: proteins highly up-regulated during retinal axon regeneration in fish [2].

The impression that reggie expression is adapted to actual requirements [30] is confirmed by the analysis of their expression pattern in developing zebrafish embryos [10,18]. Here, reggies turned out to be maternal transcripts, first present in the fertilized egg, then in all cells during cell division and gastrulation, now through expression by the embryo proper, suggesting an early function of the reggies. A few hours later, reggies become particularly relevant for neuronal development, as is suggested by their presence in all neurons and developing axon tracts in the 1.5–5-day embryo. In fact, reggie staining in the nervous system closely resembles labelling with pan-neuronal markers that identify the typical scaffold of tracts and commissures in the embryo [18]. This spatially and temporally highly regulated expression of reggies during nervous system development in the organism not only implies a function in neuronal differentiation, axon growth and growth cone pathfinding, but also emphasizes their very strict regulation, which was unexpected from their presence in basically every cell tested in culture.

Which functions do reggies have?

Aside from scaffolding and signalling functions associated with T-cell caps and neuronal microdomains, new experiments with reggie siRNA (small interfering RNA) in HeLa cells, with morpholino knockdown approaches (a modified anti-sense RNA strategy) in zebrafish and with *Drosophila* mutants suggest a morphogenetic role of reggies in cells and organisms. Reggie and the proteins associating with them seem to contribute to actin cytoskeletal rearrangements, and, thereby, to so apparently diverse aspects as cell-shape control, adhesion, growth cone and filopodia formation and motility, as well as intracellular transport and protein targeting/positioning events.

Morphogenetic functions of reggie-1 and reggie-2 revealed by siRNA-treated HeLa cells

To determine if and into which direction down-regulation of reggie expression affects cells, we initiated experiments with reggie-1- and reggie-2-specific siRNA constructs which block protein translation. The most consistent results were obtained with HeLa cells [11]. Blockade of reggie-1 and -2 synthesis in siRNA-treated cells caused a reduction in reggie-1 and -2 protein concentration according to semi-quantitative Western blots. Plated cells changed their morphology quite strikingly by 3–4 days. Whereas the cells possess a more or less continuous row of filopodia on one side, they begin to distribute these filopodia to the tips of finger-like processes, which, in some cells, reached considerable length and led to cells of unusual shape. Control siRNA-treated cells did not show such effects. Reggie expression is apparently crucial for the formation and/or maintenance of processes, the distribution of filopodia and cell-shape control. This is consistent with a potential link of the reggies to the cytoskeleton [4,17].

Morphogenetic functions of reggie-1 and reggie-2 in zebrafish

The expression analysis of reggies in the developing zebrafish nervous system led us to expect severe defects in result of the inhibition of reggie protein synthesis. We employed the morpholino knockdown approach in zebrafish embryos to block protein translation, and examined the emerging defects.

Zebrafish embryos developing from reggie-1 and -2 morpholino-injected eggs exhibited severe abnormalities [18]. The first defects were noted during gastrulation. From 1 day onwards, and persisting over the subsequent days, were brain defects, significantly reduced brain size, immobility, and embryos with severely malformed heads and tails [18], whereas control morpholino-injected embryos were normal. This is fully in line with the view that reggies are involved in the co-ordination of interacting molecular components promoting signal transduction, and thereby they contribute to various aspects of cell–cell interaction dependent events: cell division, cell-shape changes and neuronal differentiation, i.e. morphogenetic events.

Reggie in *Drosophila*

In *Drosophila*, reggies are also lipid-raft-associated proteins, and are of functional relevance during nervous system development [15]. Antibody staining revealed that reggie proteins are highly enriched in axon fascicles, e.g. in the chiasmata of the optic lobe, which would be consistent with a role in cell adhesion. However, the P-element mutant for reggie-1 that lacks reggie-1 and has lost most of the reggie-2 protein did not show any obvious phenotype [15]. Yet it indicated that the reggie-2 protein requires reggie-1 for its stabilization which was then confirmed directly. Moreover, overexpression studies showed a functional contribution of reggies to appropriate positioning of the cell-adhesion molecule irreC-rst (irregular chiasm C roughest) in the eye [15]. By overexpressing reggie-1 or reggie-1 together with reggie-2 in the eye imaginal disc, we were able to interfere with ommatidial sorting that involves the cell-adhesion molecule irreC-rst (and its relatives) [31–33], and its highly restricted expression in the apical cell-adhesion belt. Together with reports [6,34–36] that show an association of homologous proteins with the reggie-related scaffolding protein podocin, a role of the reggie proteins in the establishment or co-ordination of multimolecular complexes in microdomains and communication across the plasma membrane is being supported.

Modelling

Interesting multimolecular complexes around the lipid raft scaffolding protein podocin were reported in slit diaphragm formation in human kidney cells [21,22,34,37]. Podocin belongs to the same superfamily as reggie-1 and reggie-2 [24,34], and provides microdomains for two large cell-adhesion proteins of the immunoglobulin superfamily [IgCAMs (Ig superfamily cell-adhesion molecules)], i.e. nephrin and Neph which undergo heterophilic binding like their *Drosophila* relatives irreC-rst and SnS (sticks and stones)/hibris [31–33,38]. These IgCAMs in turn are distantly related to zebrafish and goldfish neurolin, which functions in eye morphogenesis and retinal growth cone guidance and regeneration [39,40]. Nephrin and Neph are strictly localized to the apical side of the podocyte's filopodia-like extensions which is marked by podocin [20,34]. It appears likely that podocin forms a scaffold much like reggies do in the many other cells discussed above. The *Drosophila* IgCAMs irreC-rst and SnS/hibris, as well as Nephrin and Neph1 possess large intracellular domains which undergo interactions with Src family tyrosine kinases and adaptor proteins, such as CIN (Cbl-interacting protein) in case of the podocyte [20,34–36], to regulate the actin cytoskeleton, directly or via a series of other proteins.

A multiprotein signalling complex that regulates the actin cytoskeleton and is centred on the reggie scaffolds is depicted in Figure 2. Our results are derived from experiments with Thy-1, PrP^{c} and other GPI-anchored proteins which signal across the plasma membrane through their association with reggie scaffolds, Src tyrosine kinases and adaptor proteins (Figure 2). It is very likely that such multiprotein complexes can harbour a mixed population of interacting GPI-anchored and transmembrane proteins (Figure 2), depending on the cell

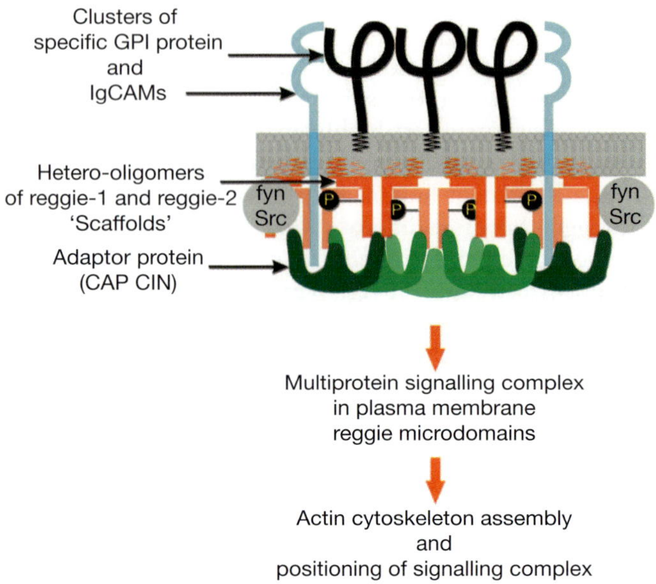

Figure 2 Reggie scaffolds and associated proteins. Our model proposes the existence of scaffolds consisting of hetero-oligomeric clusters of reggie-1 and reggie-2 (red T bars) in microdomains at the cytoplasmic face of the plasma membrane. The membrane association of reggies via acylation is shown by the red jagged features, and their (still unproven) phosphorylation by P. Specific GPI-anchored proteins, such as PrP^c and Thy-1 (black structures), co-cluster with the reggie scaffolds upon activation, and become capable of transmitting signals across the plasma membrane leading to a Ca^{2+} signal, Src tyrosine kinase activation, modified phosphorylation of the reggies and recruitment (or activation) of adaptor proteins (green) such as CAP (c-Cbl-associated protein) or CIN which effect actin cytoskeleton rearrangement via a series of intracellular proteins. It is possible that transmembrane proteins (light blue) also associate with the reggie scaffolds to contribute to or modify the multiprotein signal transduction cascade.

type and the affinity of the proteins to reggie scaffolds. It appears reasonable to assume that podocin recruits its specific partners for kidney morphogenesis and function, whereas reggies recruit their partners for nervous system morphogenesis and lymphocyte signalling. This implies evolutionary conservation of multiprotein signalling complexes and their scaffolds, from flies to fish and humans. Other members of the scaffolding protein family, such as stomatin [24], perform related functions in *Caenorhabditis elegans* morphogenesis and are important for the proper positioning of touch-receptor-associated ion channels [41]. We would predict similar functions for the remaining members of the family, such as prohibitin. Such a role of the reggie scaffolding proteins would account for their presence in different cell types, together with their role in differentiation, signalling and their evolutionary conservation.

This view is consistent with the original concept of signalling of GPI-anchored proteins via clustered lipid rafts although 'classical' rafts (distinct from caveolae) have not been considered as elements that are stabilized by scaffolding proteins [26] nor to form pre-formed signalling centres [5] or pre-formed caps [16]. Rather, new biophysical methods to estimate raft size and mobility suggest a quite different view: rafts being tiny regions with fewer than a dozen of inhabitants that cannot be resolved at the microscopic level [42,43]. More significantly, membrane rafts are thought to be found predominantly in the outer leaflet of the plasma membrane [26,43]. In contrast, our current view on reggies emphasizes their larger unit size (0.1 μm), presence on the cytoplasmic face of the plasma membrane and role as scaffolds for the formation of multiprotein complexes, involved in signalling across the plasma membrane upon ligand binding and communication with neighbouring cells.

We thank Dr Joachim Hentschel and Christopher Bleck for image composition, and Dr Alexander Reuter, Dr Edward Màlaga-Trillo and Matthias Langhorst for comments on the manuscript. Our work has been supported by the Ministerium Wissenschaft und Kunst (MWK) Baden-Württemberg, TSE programme, TR SFB 11 of the Deutsche Forschungsgemeinschaft (DFG) and Fonds der Chemischen Industrie (FCI).

References

1. Lang, D.M., Lommel, S., Jung, M., Ankerhold, R., Petrausch, B., Laessing, U., Wiechers, M.F., Plattner, H. and Stuermer, C.A.O. (1998) J. Neurobiol. **37**, 502–523
2. Schulte, T., Paschke, K.A., Laessing, U., Lottspeich, F. and Stuermer, C.A.O. (1997) Development **124**, 577–587
3. Bickel, P.E., Scherer, P.E., Schnitzer, J.E., Oh, P., Lisanti, M.P. and Lodish, H.F. (1997) J. Biol. Chem. **272**, 19793–19802
4. Stuermer, C.A.O., Langhorst, M., Wiechers, M., Legler, D., Hannbeck von Hanwehr, S., Guse, A. and Plattner, H. (2004) FASEB J., in the press
5. Stuermer, C.A.O., Lang, D.M., Kirsch, F., Deininger, S., Wiechers, M.F. and Plattner, H. (2001) Mol. Biol. Cell **12**, 3031–3045
6. Sumant, S.C., Kaw, B. and Kanwar, Y.S. (2003) Semin. Nephrol. **23**, 544–555
7. Neumann-Giesen, C., Falkenbach, B., Rajendran, L., Lüers, G., Illges, H., Stuermer, C.A.O., Herzog, V. and Tikkanen, R. (2004) Biochem. J. **378**, 509–518
8. Morrow, I.C., Rea, S., Martin, S., Prior , I.A., Prohaska, R., Hancock, J.F., James, D.E. and Parton, R.G. (2002) J. Biol. Chem. **277**, 48834–48841
9. Galbiati, F., Volonté, D., Goltz, J.S., Steele, Z., Sen, J., Jurcsak, J., Stein, D., Stevens, L. and Lisanti, M.P. (1998) Gene **210**, 229–237
10. Màlaga-Trillo, E., Laessing, U., Lang, D.M., Meyer, A. and Stuermer, C.A.O. (2002) J. Mol. Evol. **54**, 235–245
11. Haenisch, C. (2004) Ph.D. thesis, University of Konstanz
12. Hazarika, P., Dham, N., Patel, P., Cho, M., Weidner, D., Goldsmith, L. and Duvic, M. (1999) J. Cell. Biochem. **75**, 147–159
13. Schroeder, W., Stewart-Galetka, D.S., Parry, D.A., Goldsmith, D. and Duvic, M.L. (1994) J. Biol. Chem. **269**, 19983–19991
14. Baumann, C.A., Ribon,V., Kanzaki, M., Thurmond, D.C., Mora, S., Shigematsu,S., Bickel, P.E., Pessing, J.E. and Saltiel, A.R. (2000) Nature (London) **407**, 202–207
15. Höhne, M. (2003) Ph.D. thesis, University of Freiburg
16. Rajendran, L., Masilamani, M., Solomon, S., Tikkanen, R. and Stuermer, C.A.O. (2003) Proc. Natl. Acad. Sci. U.S.A. **100**, 8241–8246

17. Kioka, N., Ueda, K. and Amachi, T. (2002) Cell Struct. Funct. **27**, 1–7
18. Màlaga-Trillo, E., Rivera-Milla, E., Marx, M., Bastmeyer, M. and Stuermer, C.A.O. (2002) Society for Neuroscience Abstract A30
19. Salzer, U. and Prohaska, R. (2001) Blood **97**, 1141–1143
20. Chugh, S.H., Kaw, B. and Kanwar, Y. (2003) Semin. Nephrol. **23**, 544–555
21. Schwarz, K., Simons, M., Reiser, J., Saleem, M.A., Faul, C., Kriz, W., Shaw, A.S., Holzman, L.B. and Mundel, P. (2001) J. Clin. Invest. **108**, 1621–1629
22. Sellin, L., Huber, T.B., Gerke, P., Quack, I., Pavenstädt, H. and Walz, G. (2003) FASEB J. **17**, 115–117
23. Shih, N.-Y., Li, J., Karpitskii, V., Nguyen, A., Dustin, M.L., Kanagawa, O., Miner, J.H. and Shaw, A.S. (1999) Science **286**, 312–315
24. Tanaverakis, N., Driscoll, M. and Kyprides, N.C. (1999) Trends Biochem. Sci. **24**, 425–427
25. Harder, T., Scheiffele P., Verkade, P. and Simons, K. (1998) J. Cell Biol. **141**, 929–942
26. Simons, K. and Toomre, D. (2000) Nat. Mol. Cell Biol. **1**, 31–39
27. Prusiner, S.B. (1995) Proc. Natl. Acad. Sci. U.S.A. **95**, 13363–13383
28. Weissmann, C. (1999) J. Biol. Chem. **274**, 3–6
29. Deininger, S.O., Rajendran, L., Lottspeich, F., Przybylski, M., Illges, H., Stuermer, C.A.O. and Reuter, A. (2003) Mol. Cell. Neurosci. **22**, 544–554
30. López-Cases, P.P. and del Mazo, J. (2003) FEBS Lett. **555**, 223–228
31. Dworak, H.A. and Sink, H. (2002) BioEssays **24**, 591–601
32. Eroglu, C., Brügger, B., Wieland, F. and Sinning, I. (2003) Proc. Natl. Acad. Sci. U.S.A. **100**, 10219–10224
33. Taylor, M.V. (2004) Curr. Biol. **13**, R964–R966
34. Huber, T.B., Simons, M., Hartleben, B., Sernetz, L., Schmidts, M., Gundlach, E., Saleem, M.A., Walz, G. and Benzing, T. (2004) Hum. Mol. Genet. **12**, 3397–3405
35. Jalanko, H. (2003) Pediatr. Nephrol. **18**, 487–491
36. Li, C., Ruotsalainen, V., Tryggvason, K., Shaw, A.S. and Miner, J.H. (2000) Am. J. Physiol. Renal. Physiol. **279**, F785–F792
37. Putaala, H., Soininen, R., Kilpeläinen, P., Wartiovaara, J. and Tryggvason, K. (2001) Human Mol. Genet. **10**, 1–8
38. Strünkelnberg, M., De Couet, H.G., Hertenstein, A. and Fischbach, K.-F. (2003) J. Mol. Evol. **56**, 187–197
39. Leppert, C.A., Diekmann, H., Paul, C., Laessing, U., Marx, M., Bastmeyer, M. and Stuermer, C.A.O. (1999) J. Cell Biol. **144**, 339–349
40. Paschke, K.A., Lottspeich, F. and Stuermer, C.A.O. (1992) J. Cell Biol. **117**, 863–875
41. Sedensky, M.M., Siefker, J.M. and Morgan, P.G. (2001) Am. J. Physiol. Cell Physiol. **280**, C1340–C1348
42. Munro, S. (2003) Cell **115**, 377–388
43. van Meer, G. (2004) Traffic **5**, 211–212

Biochem. Soc. Symp. 72, 119–127
(Printed in Great Britain)

12

Spatial control of actin-based motility through plasmalemmal PtdIns(4,5)P_2-rich raft assemblies

Tamara Golub and Pico Caroni[1]

Friedrich Miescher Institut, Basel, Switzerland

Abstract

The interactions of cells with their environment involve regulated actin-based motility at defined positions along the cell surface. Sphingolipid- and cholesterol-dependent microdomains (rafts) order proteins at biological membranes, and have been implicated in most signalling processes at the cell surface. Many membrane-bound components that regulate actin cytoskeleton dynamics and cell-surface motility associate with PtdIns(4,5)P_2-rich lipid rafts. Although raft integrity is not required for substrate-directed cell spreading, or to initiate signalling for motility, it is a prerequisite for sustained and organized motility. Plasmalemmal rafts redistribute rapidly in response to signals, triggering motility. This process involves the removal of rafts from sites that are not interacting with the substrate, apparently through endocytosis, and a local accumulation at sites of integrin-mediated substrate interactions. PtdIns(4,5)P_2-rich lipid rafts can assemble into patches in a process depending on PtdIns(4,5)P_2, Cdc42 (cell-division control 42), N-WASP (neural Wiskott–Aldrich syndrome protein) and actin cytoskeleton dynamics. The raft patches are sites of signal-induced actin assembly, and their accumulation locally promotes sustained motility. The patches capture microtubules, which promote patch clustering through PKA (protein kinase A), to steer motility. Raft accumulation at the cell surface, and its coupling to motility are influenced greatly by the expression of intrinsic raft-associated components that associate with the cytosolic leaflet of lipid rafts. Among them, GAP43 (growth-associated protein 43)-like proteins interact with PtdIns(4,5)P_2 in a Ca^{2+}/calmodulin and PKC (protein kinase C)-regulated manner, and function as intrinsic determinants of motility and anatomical plasticity. Plasmalemmal PtdIns(4,5)P_2-rich raft assemblies thus provide powerful organizational principles for tight spatial and temporal control of signalling in motility.

[1]To whom correspondence should be addressed (email caroni@fmi.ch).

Introduction

Owing to their distinct biophysical properties and their relative abundance, sphingolipids and cholesterol play a predominant role in generating microdomains in biological membranes [1,2]. These sphingolipid- and cholesterol-dependent microdomains are also designated as lipid rafts. In the biosynthetic pathway, lipid rafts are first assembled at the Golgi, and play major roles in specific trafficking of proteins and lipids to and from cellular compartments [3]. Different types of cells at distinct developmental stages can differ substantially in their raft contents. Most importantly, association with lipid rafts influences signalling, trafficking and the assembly of cellular structures in specific ways [1].

Because they provide an ordering mechanism to accumulate defined molecular components at specific sites of membranes, and because they interact extensively with cellular signalling and trafficking processes, it has become clear that rafts could play major roles in spatial and temporal control of actin-based motility. Thus the mechanisms that engage specific forms of cell-surface motility and control their progression include mechanisms that link locally actin dynamics to the assembly and trafficking of specific membrane-associated components, orchestrating motility. Among them, the lipid second messenger PtdIns(4,5)P_2 is an attractive candidate to integrate signalling and co-ordinate actin and membrane dynamics in motility. PtdIns(4,5)P_2 accumulates at sites of cell-surface motility, and can modulate actin dynamics, endo- and exo-cytosis, and the assembly of membrane-associated protein coats, mediating morphogenesis [4–8]. Significantly, PtdIns(4,5)P_2 is concentrated at cholesterol-dependent lipid rafts [9]. These observations have thus raised the possibility that local PtdIns(4,5)P_2 levels and the local assembly and remodelling of raft domains enriched in PtdIns(4,5)P_2 may regulate cell motility [6,8,9].

The present review focuses on recent developments of how lipid rafts play critical roles in spatial and temporal control of actin-based motility, and on how actin cytoskeleton regulation involved in neurite outgrowth and the remodelling of synapses between neurons is controlled by intrinsic and extrinsic mechanisms, controlling raft local accumulation and trafficking.

Many of the components regulating the actin cytoskeleton target to plasmalemmal rafts

Lipid rafts are well suited to play major roles in regulating actin dynamics locally. Thus, among the molecular components that are involved in actin cytoskeleton regulation, those associated with rafts include receptor tyrosine kinases, activated integrins, the Rho-type GTPases RhoA, Rac and Cdc42 (cell-division control 42), their activated WASP (Wiskott–Aldrich syndrome protein)-family effectors, Src-like kinases, ERM (ezrin, radixin and moesin) proteins, PI5K (phosphoinositide 5-kinase), and the lipid second messenger PtdIns(4,5)P_2 (e.g. [6]). In addition, plasmalemmal raft-associated GAP43 (growth-associated protein 43)-like proteins modulate PtdIns(4,5)P_2-rich raft assembly and influence actin dynamics ([9,10]; see also below). Rafts of distinct compositions are concentrated at the leading and trailing edges of migrating neutrophils (see below), and Arp (actin-related protein) 2/3-directed actin assembly is particularly efficient at PtdIns(4,5)P_2-rich rafts [11]. Furthermore, Dynamin 2–cortactin–Arp2/3 complexes have emerged as versatile molecular scaffolds that are regulated by raft-associated components, such as Src kinases, Rac and PtdIns(4,5)P_2, to generate actin-based membrane protrusion [12].

The close association of actin-regulating components with plasmalemmal rafts provides for a rich repertoire of specific regulatory mechanisms to control actin-based motility, and couple it to cell organization and regulation. Thus lipid microdomains are highly dynamic entities, where interactions between specific components are thought to be short-lived and hence exquisitely sensitive to regulation. In addition, association with rafts influences both the signalling strength of individual components, as well as signal quality through differential recruitment of signalling components at and outside of rafts. For example, a fraction of the cellular pool of the cell-adhesion molecule NCAM (neural cell-adhesion molecule)-140 can associate with lipid rafts upon palmitoylation at juxtamembrane intracellular residues. For NCAM-140 to promote neurite outgrowth *in vitro*, activation of two parallel pathways appears to be required: mediation of FGF (fibroblast growth factor) signalling through non-raft NCAM, and activation of FAK (focal adhesion kinase) and ERK (extracellular-signal-regulated kinase) by an NCAM–Fyn complex at rafts [13]. In addition, by associating to GPI (glycosylphosphatidylinositol)-anchored receptor for GDNF (glial-derived neurotrophic factor) [GFRα1 (GDNF-family receptor α1)], NCAM-140 can bind GDNF with high affinity, and mediate signalling to Fyn and FAK [14]. Furthermore, NCAM associates with and signals through GAP43 and Fyn at distinct subpopulations of growth cone raft complexes [15].

Cell-surface motility involves a redistribution and local accumulation of plasmalemmal rafts

As they provide ordered membrane microdomains that can reorganize rapidly in the plane of the membrane, and can be trafficked through dedicated pathways, rafts are attractive candidates to efficiently set up defined sites of cell-surface signalling and motility in cells. Such polarization of rafts, and the ways in which rafts are involved in setting up cell polarity and polarized signalling are illustrated nicely in chemotactic cell migration. In an early report of raft polarization, Martinez-A and colleagues [16] showed that polarizing T-cells accumulate distinct raft components, including the gangliosides GM3 and GM1 at their leading and trailing edges respectively. A later study showed that, while the first signalling events in cell polarization are raft-independent, the presence of intact rafts is necessary for sustained Rac activation and actin polymerization [17]. Perhaps most significantly, in migrating leucocytes, CCR5 (chemokine receptor 5) and subsequently activated PI3K (phosphoinositide 3-kinase) redistribute to leading edge rafts in a raft-dependent manner [18]. Taken together, these studies show that in polarized cell migration, spatial signalling can be organized by concentration-gradient-sensing machinery through and at rafts. As discussed below, these principles probably play a major role in growth-cone guidance.

In a recent study to investigate the relation between cell-surface rafts and cell motility [19], we analysed the distribution of raft-associated components, such as cholesterol (filipin), GAP43 and PtdIns(4,5)P_2, in quiescent, replated or NGF (nerve growth factor)-treated cells. We found that quiescent cells, with little detectable cell-surface motility and no sites of intense F-actin (filamentous actin) labelling at the cell edge did not exhibit obvious sites of cell-surface raft concentrations. In contrast, raft components in replated cells were highly concentrated in patches at cell-surface ruffling lamellae, where they co-distributed with intense labelling for F-actin. Likewise, NGF-induced cell-surface motility coincided with the rapid accumulation of PtdIns(4,5)P_2-rich raft patches at prominent actin-assembling sites along the edge of PC12 cells [19]. Raft-patch accumulation at the cell edge (detectable from 30 s onwards) preceded the accumulation of prominent F-actin patches (from 45–60 s onwards) at these sites. As expected, the cholesterol-sequestering agent cyclodextrin suppressed assembly sites for raft-associated components. Significantly, cells attached and spread to a comparable extent in the absence or presence of cyclodextrin, indicating that raft disruption did not suppress all substrate-induced morphogenic interactions in a non-specific manner. However, lamellipod and ruffling motility at the leading edge were absent in the absence of intact rafts [19].

In experiments aimed at defining mechanisms through which rafts accumulate at the leading edge, we found that the accumulation of cell-surface raft patches involved in leading-edge motility depends on signals by Cdc42 and PtdIns(4,5)P_2, upstream of N-WASP (neuronal WASP) and actin assembly [19]. As Rac activity was required to generate PtdIns(4,5)P_2 locally at the cell edge, this led to a dual requirement for Cdc42 and Rac signalling in raft patching, which can be provided, e.g. through Vav2, downstream of integrin signalling [20]. Based on these results, and on those of previous studies, we proposed a mechanism for PtdIns(4,5)P_2-rich raft-patch accumulation at the cell surface, involving the recruitment and activation of N-WASP, through co-operative binding of PtdIns(4,5)P_2 and Cdc42, downstream of Rac and Cdc42 activation [19,21]. This mode of regulation could ensure tight spatial regulation of motility through Cdc42-recruiting complexes at the cell membrane, and tight dynamic regulation of motility through PtdIns(4,5)P_2 synthesis (PI5Ks) and hydrolysis (phospholipase C). These findings are consistent with those of other studies, which found dual requirements for Rac and Cdc42 signalling in the induction of lamellipodial-based cell migration [22], and in lamellipodial-based phagocytosis [23].

The local accumulation of raft assemblies at sites of motility can greatly enhance spatial specificity in Rho-GTPase signalling to the actin cytoskeleton. But how do signals that initiate and promote motility interact with raft trafficking to bring about local accumulation of rafts at the leading edge? An important recent study has shed light on how integrin signalling relates to raft distribution and Rac activity [24]. Del Pozo and colleagues showed that integrin engagement leads to the local accumulation of rafts at the cell surface [24]. In contrast, when cells are detached from their substrate, rafts are endocytosed and accumulate intracellularly. The study shows further that, in the absence of surface rafts, Rac-GTP fails to associate with the plasma membrane, and to activate the downstream effector of integrins, FAK. Together with those of previous studies [25], these results provide evidence that integrin signalling couples sensing of the local environment to the local accumulation of rafts required for Rac-GTP activity at the cell surface. The mechanism that mediates local delivery of rafts at defined sites on the cell surface are not well understood. However, for long-term polarization, they could involve dedicated targeting mechanisms, e.g. involving the exocyst, an octameric complex involved in targeting proteins and vesicles to specified plasma membrane domains, thereby acting as a polarization cue [26]. In addition, in processes likely to be particularly important to motility regulation, raft-associated components are internalized selectively through clathrin-independent/dynamin-dependent endocytosis under the control of Rho-GTPases [27]. A major specialized path involves association with caveolae, a stable plasmalemmal compartment where phosphorylation of caveolin-1 is required to induce endocytosis [28–30]. Specific raft-dependent processes of receptor internalization at ruffling lamellipods may provide powerful mechanisms to integrate local motility and signalling at the cell surface with cell regulation processes through sorting and signalling endosomes.

Actin assembly at PtdIns(4,5)P_2-rich plasmalemmal rafts

To investigate how the accumulation of PtdIns(4,5)P_2-rich rafts into patches might influence cell-surface motility, we have investigated their possible role in actin cytoskeleton regulation. We found a pronounced accumulation of F-actin at raft patches along the cell edge in the presence of NGF, which was greatly reduced in cells overexpressing dn (dominant-negative)-Cdc42 or dn-N-WASP [19]. Consistent with a role of PtdIns(4,5)P_2-rich raft patches in promoting actin assembly in motility, phase-contrast time-lapse recordings revealed that, in the presence of dn-Cdc42 or dn-N-WASP, both intensity and duration of NGF-induced lamellipodial motility were reduced greatly. These results suggested that a raft-patching process involving PtdIns(4,5)P_2, Cdc42 and N-WASP enhances signalling for actin assembly and motility in NGF-treated cells [19]. This is consistent with reports that rather than interfering with the initial triggering of motility, the disruption of rafts diminishes responses, such as membrane ruffling and pinocytosis, and prevents sustained activation of Rac and actin dynamics in neutrophils [17,26]. Since PtdIns(4,5)P_2-rich platforms assembled at regions of prospective motility, these results suggest that assembled rafts provide spatial domains of enhanced signalling for motility. This is probably achieved through the dynamic assembly of specific signalling scaffolds, and through positive-feedback loops that sustain the activation of key components such as PI3K, Rac, Cdc42 and PI5K [8,31–33].

A role for plasmalemmal rafts in steering actin-based motility

Since local clustering of rafts appears to play a critical role in promoting motility, processes of raft accumulation and assembly may be targets of regulation by external signals that guide motility. Recent studies illuminating the roles of lipid microdomains in growth-cone guidance suggest that this is indeed the case. An elegant study by Zheng and colleagues has demonstrated that raft integrity is required specifically for growth-cone-turning responses induced by BDNF (brain-derived neurotrophic factor), Netrin-1 and Sema3A, but not glutamate [34]. Treatments disrupting cholesterol-rich rafts affected BDNF-induced attraction specifically, whereas treatments affecting GM1-rich rafts interfered with BDNF-induced neurite extension. Importantly, signalling-competent rafts accumulated asymmetrically towards a BDNF gradient, suggesting that the turning response may be brought about by a spatial rearrangement of lipid microdomains on the surface of growth cones [34]. As in integrin-mediated adhesion, raft-associated Rac1 plays a central and complex role in growth-cone guidance, where it is implicated in both growth and collapse. In the absence of collapsing signals, Rac1 activity drives actin polymerization at the leading edge of growth cones, and raft-associated proteins, such as L1, can increase the levels of active Rac1 at the growth cone membrane. Upon collapsing signals (e.g. through ephrin A2), Rac1 activity is transiently lost, the growth cone switches to a 'proofreading' mode dominated by Rho-mediated contraction, and growth cone microdomains are redistributed.

Intrinsic determination of cell-surface PtdIns(4,5)P_2-rich raft accumulation and motility by GAP43-like proteins

While extracellular signals initiate raft-mediated signalling, the extent of this signalling can be influenced greatly by the expression levels of cell-intrinsic factors. Thus the controlled assembly of raft-dependent signalling domains is influenced by the expression of raft-associated components, such as flotillins, which affect both raft levels and processes of signal-induced raft clustering [1,3]. Furthermore, in the nervous system, process outgrowth and synaptic rearrangements coincide with the expression of specific sets of genes in neurons, including the GAP43-like proteins GAP43, CAP23 (cortical cytoskeleton-associated protein of 23 kDa), MARCKS (myristoylated alanine-rich C kinase substrate), and MacMARCKS (macrophage MARCKS) [9,35]. Functional studies *in vivo* have provided strong evidence that all GAP43-like proteins are intrinsic determinants of anatomical plasticity, conferring competence for morphogenetic processes to cells that express them [9,35,36]. Within cells, GAP43-like proteins accumulate at lipid rafts, where they co-distribute with the lipid second messenger PtdIns(4,5)P_2 [35]. GAP43-like proteins appear to function by modulating the accumulation and assembly of PtdIns(4,5)P_2-rich rafts at the cell membrane in a Ca^{2+}/calmodulin- and PKC-regulated manner [9]. For MARCKS and GAP43, this involves a direct interaction of the basic domains with PtdIns(4,5)P_2, which can promote clustering and sequestration of this important lipid second messenger [37]. PC12 cells stably expressing GAP43, CAP23 or MARCKS exhibit higher levels of raft markers at the cell surface, whereas cells expressing a

GAP43(ΔED) mutant lacking the basic activation domain exhibit particularly low labelling signals for cell-surface raft markers [35]. These differences in cell-surface raft accumulation are accompanied by corresponding differences in substrate-dependent spreading and adhesion in these cells, and in neurite outgrowth in transgenic mice [35]. In experiments that support further the notion that the accumulation of raft patches at the cell surface promotes motility, we found recently that GAP43(ΔED) interferes with both raft-patch accumulation at the surface and cell-surface motility in PC12 cells [19].

The mechanisms through which cell intrinsic components influence raft accumulation at the cell surface probably involve actin-mediated interactions with the cortical cytoskeleton. Thus the assembly of dynamic actin filaments at lipid rafts, and their loose association with the cortical cytoskeleton are thought to stabilize signalling domains, whereas actin disassembly mediates their dispersal. In addition, loose interactions in *trans* with other cells or extracellular components can serve to prolong and enhance signalling through rafts. Interestingly, and emphasizing the dynamic nature of raft-associated interactions in signalling, a stable association of individual components with the cytoskeleton can function to restrict their recruitment to signalling domains, and to restrict motility [9,38].

Conclusions and outlook

Recent studies have revealed how lipid rafts are rapidly redistributed and assembled locally in response to extracellular signals, and how components of raft-based signalling domains undergo rapid and regulated rearrangements influencing signal quality, duration and strength. In the present review, we have discussed how locally accumulating PtdIns(4,5)P_2-rich raft assemblies promote and steer motility. Briefly, the local accumulation of PtdIns(4,5)P_2-rich rafts enhances local signalling for motility, and their assembly state is regulated through captured microtubules and PKA, coupling local regulation of motility with cell organization. In addition, the expression of intrinsic components, such as GAP43-like proteins, that modulate PtdIns(4,5)P_2-rich raft accumulation and assembly at the cell surface influences the extent to which cells respond to external signals, to produce cell-surface motility and interact with their local environment. Major open issues for future studies include: (i) elucidating the precise mechanisms through which local PtdIns(4,5)P_2 pools mediate spatial and temporal regulation of motility at the leading edge, and the roles of raft assemblies in these processes; (ii) elucidating how raft-dependent membrane trafficking to and from the cell surface interacts with local regulation of motility, and couples this process to cell regulation; and (iii) elucidating the mechanisms through which intrinsic components such as GAP43-like proteins influence PtdIns(4,5)P_2-rich raft trafficking, assembly and function. The results of this research should provide insights into how cells interact and cope with their environment, in health and in disease.

The Friedrich Miescher Institut is a branch of the Novartis Research Foundation.

References

1. Simons, K. and Toomre, D. (2000) Nat. Rev. Mol. Cell Biol. **1**, 31–39
2. van Meer, G. (2002) Science **296**, 855–857
3. Lai, E.C. (2003) J. Cell Biol. **162**, 365–370
4. Tall, E.G., Spector, I., Pentyala, S.N., Bitter, I. and Rebecchi, M.J. (2000) Curr. Biol. **10**, 743–746
5. Botelho, R.J., Teruel, M., Dierckman, R., Anderson, R., Wells, A., York, J.D., Meyer, T. and Grinstein, S. (2000) J. Cell Biol. **151**, 1353–1367
6. Martin, T.F.J. (2001) Curr. Opin. Cell Biol. **13**, 493–499
7. Cremona, O. and De Camilli, P. (2001) J. Cell Sci. **114**, 1041–1052
8. Yin, H.L. and Janmey, P.A. (2003) Annu. Rev. Physiol. **65**, 761–789
9. Caroni, P. (2001) EMBO J. **20**, 4332–4336
10. McLaughlin, S., Wang, J., Gambhir, A. and Murray, D. (2002) Annu. Rev. Biophys. Biomol. Struct. **31**, 151–175
11. Rozelle, A.L., Machesky, L.M., Yamamoto, M., Driessens, M.H.E., Insall, R.H., Roth, M.G., Luby-Phelps, K., Marriott, K., Hall, A. and Yin, H.L. (2000) Curr. Biol. **10**, 311–320
12. Orth, J.D. and McNiven, M.A. (2003) Curr. Opin. Cell Biol. **15**, 31–39
13. Niethammer, P., Delling, M., Sytnyk, V., Dityatev, A., Fukami, K. and Schachner, M. (2002) J. Cell Biol. **157**, 521–532
14. Paratcha, G., Ledda, F. and Ibanez, C.F. (2003) Cell **113**, 867–879
15. He, Q. and Meiri, K.F. (2002) Mol. Cell. Neurosci. **19**, 18–31
16. Gomez-Mouton, C., Abad, J.L., Mira, E., Lacalle, R.A., Gallardo, E., Jimenez-Baranda, S., Illa, I., Bernad, A., Manes, S. and Martinez-A, C. (2001) Proc. Natl. Acad. Sci. U.S.A. **98**, 9642–9647
17. Pierini, L.M., Eddy, R.J., Fuortes, M., Seveau, S., Casulo, C. and Maxfield, F.R. (2003) J. Biol. Chem. **278**, 10831–10841
18. Gomez-Mouton, C., Lacalle, R.A., Mira, E., Jimenez-Baranda, S., Barber, D.F., Carrera, A.C., Martinez-A, C. and Manes, S. (2004) J. Cell Biol. **164**, 759–768
19. Golub, T. and Caroni, P. (2004) J. Cell Biol., in the press
20. DeMali, K.A., Wennerberg, K. and Burridge, K. (2003) Curr. Opin. Cell Biol. **15**, 572–582
21. Rohatgi, R., Nollau, P., Ho, H.H.-Y. and Kirschner, M.W. (2001) J. Biol. Chem. **276**, 26448–26452
22. Shahai, E. and Marshall, C.J. (2003) Nat. Cell Biol. **5**, 711–719
23. Caron, E. and Hall, A. (1998) Science **282**, 1717–1721
24. del Pozo, M.A., Alderson, N.B., Kiosses, W.B., Chiang, H.H., Anderson, R.G. and Schwartz, M.A. (2004) Science **303**, 839–842
25. Grimmer, S., van Deurs, B. and Sandvig, K. (2002) J. Cell Sci. **115**, 2953–2962
26. Inoue, M., Chang, L., Hwang, J., Chiang, S.H. and Saltiel, A.R. (2003) Nature (London) **422**, 629–633
27. Lamaze, C., Dujeancourt, A., Baba, T., Lo, C.G., Benmerah, A. and Dautry-Varsat, A. (2001) Mol. Cell **7**, 661–671
28. Thomsen, P., Roepstorff, K., Stahlhut, M. and van Deurs, B. (2002) Mol. Biol. Cell **13**, 238–250
29. Le, P.U., Guay, G., Altschuler, Y. and Nabi, I.R. (2002) J. Biol. Chem. **277**, 3371–3379
30. Upla, P., Marjomaki, V., Kankaanpaa, P., Ivaska, J., Hyypia, T., Van Der Goot, F.G. and Heino, J. (2004) Mol. Biol. Cell **15**, 625–636
31. Etienne-Manneville, S. and Hall, A. (2002) Nature (London) **420**, 629–635
32. Rodriguez, O.C., Schaefer, A.W., Mandato, C.A., Forscher, P., Bement, W.M. and Waterman-Storer, C. (2003) Nat. Cell Biol. **5**, 599–609
33. Srinivasan, S., Wang, F., Glavas, S., Ott, A., Hofmann, F. and Bourne, H.R. (2003) J. Cell Biol. **160**, 375–385
34. Guirland, C., Suzuki, S., Kojima, M., Lu, B. and Zheng, J.Q. (2004) Neuron **42**, 51–62
35. Laux, T., Fukami, K., Thelen, M., Golub, T., Frey, D. and Caroni, P. (2000) J. Cell Biol. **149**, 1455–1472

36. Bomze, H.M., Bulsara, K.R., Iskandar, B.J., Caroni, P. and Skene, J.H. (2001) Nat. Neurosci. **4**, 38–43
37. McLaughlin, S., Wang, J., Gambhir, A. and Murray, D. (2002) Annu. Rev. Biophys. Biomol. Struct. **31**, 151–75
38. Raucher, D., Stauffer, T., Chen, W., Shen, K., Guo, S., York, J.D., Sheetz, M.P. and Meyer, T. (2000) Cell **100**, 221–228

Biochem. Soc. Symp. **72**, 129–138
(Printed in Great Britain)

13

The conflicting role of brain cholesterol in Alzheimer's disease: lessons from the brain plasminogen system

Maria Dolores Ledesma[1] and Carlos G. Dotti[1]

Cavalieri Ottolenghi Scientific Institute, Universita degli Studi di Torino, A.O. San Luigi Gonzaga, Regione Gonzole 10, 10043 Orbassano (Torino), Italy

Abstract

Retrospective clinical studies indicate that individuals chronically treated with cholesterol synthesis inhibitors, statins, are at lower risk of developing AD (Alzheimer's disease). Moreover, treatment of guinea pigs with high doses of simvastatin or drastic reduction of cholesterol in cultured cells decrease Aβ (β-amyloid peptide) production. These data sustain the concept that high brain cholesterol is responsible for Aβ accumulation in AD, providing the scientific support for the proposed use of statins to prevent this disease. However, a number of unresolved issues raise doubts that high brain cholesterol is to blame. First, it has not been shown that higher neuronal cholesterol increases Aβ production. Secondly, it has not been demonstrated that neurons in AD have more cholesterol than control neurons. On the contrary, the brains of AD patients show a specific down-regulation of seladin-1, a protein involved in cholesterol synthesis, and low membrane cholesterol was observed in hippocampal membranes of ApoE4 (apolipoprotein E4) AD cases. This effect was also evidenced by altered cholesterol-rich membrane domains (rafts) and raft-mediated functions, such as diminished generation of the Aβ-degrading enzyme plasmin. Thirdly, numerous genetic defects that cause neurodegeneration are due to defective cholesterol metabolism. Fourthly, in female mice, the most brain-permeant statin induces neurodegeneration and high amyloid production. Altogether, this evidence makes it difficult to accept that statins are beneficial through acting as brain cholesterol-synthesis inhibitors. It appears more likely that their advantageous role arises from improved brain oxygenation

[1]To whom correspondence should be addressed (email lola.ledesma@unito.it or carlos.dotti@unito.it)

and/or because of their anti-inflammatory properties. At this point, the true role of cholesterol and statins in AD is still to be determined and the final aim of statin-based treatments should be evaluated carefully.

Amyloid accumulation can occur through a loss of function event such as inefficient degradation

Abnormal protein aggregates, whether in the extracellular or intracellular space, can certainly affect cell function. The severity of the effect will depend on the composition of the cumuli, the size and the place of formation. From this basic and obvious standpoint, it is more than justified that researchers who are interested in understanding the cause of AD (Alzheimer's disease) have put so much effort in the analysis of the molecular and biochemical mechanisms responsible for the anomalous Aβ (β-amyloid peptide) and tau protein aggregates that invariably appear in AD patients' brains [1]. In the case of the amyloid cumuli, these can occur by a combination of factors: higher production of the peptide, generation of an abnormal, aggregation-prone form, or decreased rate of degradation. Thus, in some cases, aggregation can be seen as a gain-of-function event, whereas, in others, it reflects a loss of function. In the following sections, we will summarize data from our laboratory showing that loss of function of the brain plasminogen system in AD brains, resulting in a reduction of the levels and activity of the amyloid degrading enzyme plasmin, can account for amyloid accumulation. Yet this is not the only enzyme involved in Aβ clearance, and we recommend to the reader interested in knowing more about this type of mechanism to consult the excellent publications analysing the amyloid-degrading function of other enzymes such as neprilysin, IDE (insulin-degrading enzyme), MMP9 (matrix metalloproteinase 9), endothelin-converting enzyme and elastase (see [2] and references therein).

Among its various roles, the brain plasminogen system is involved in Aβ degradation

The active serine protease plasmin arises following the binding to the cell surface and subsequent cleavage of the inactive pro-enzyme, plasminogen [3]. The cleavage cascade is balanced through activators: tPA (tissue-type plasminogen activator) and uPA (urokinase-type plasminogen activator), and inhibitors: plasminogen-activator inhibitors (i.e. PAI1, PAI2) and plasmin inhibitors (i.e. α-2-antiplasmin, α-2-macroglobulin) [4]. Although the plasminogen–plasmin system is most known in relation to blood coagulation, because of the involvement in fibrin-clot clearance [4], it has been also recognized for years to exist in the brain. In this organ, tPA is the main plasminogen activator, appearing widely expressed in the hippocampus, hypothalamus, cerebellum and amygdala [5]. Interestingly, fibrin has not been found in the brain, indicating that, in this organ, the plasminogen system has different target(s). In this regard, various evidence has demonstrated that the brain tPA–plasminogen cascade is important for proper extracellular matrix degradation during neuronal development [6] and for events that require synaptic plasticity, such as long-term potentiation [7], memory [8] and motor learning [9]. Although plasminogen is potentially the most likely substrate for tPA, thus making plasmin the putative main modulator of such functions, it is now clear that tPA itself can exert certain roles such as microglial activation [10], mossy fibre outgrowth stimulation [11] and enhancement of NMDA (*N*-methyl-D-aspartate) receptor-mediated signalling [12]. More work is needed to elucidate which of the numerous roles of tPA are plasmin-dependent or -independent. In addition to the physiological functions, the brain plasminogen system has been implicated in neuronal death in conditions such as seizure and stroke [13,14]. In ischaemia and excitotoxicity models, neurotoxicity is mediated by increased tPA synthesis, followed by plasmin generation leading to laminin degradation [14,15]. In contrast, plasmin-independent neurotoxic effects triggered by excess tPA are thought to mediate the neuronal death that occurs after stroke [16].

Recent evidence points to the plasminogen system as a key player in AD pathogenesis. Thus it has been observed that plasmin can efficiently degrade Aβ *in vitro* [17,18] and that the degradation of Aβ in the brain, after injection of the peptide, is significantly less efficient in tPA or plasminogen-knockout mice [19]. That a defect in plasmin activity could lead to reduced clearance of Aβ, and therefore enhanced accumulation in the extracellular milieu in the human disease, came from the finding that brain plasmin levels and activity are significantly reduced in a relevant percentage of AD cases [20]. These three pieces of evidence add the plasminogen system to the complex picture of the molecular defects that can lead to the occurrence of abnormal amyloid aggregates contributing to neurodegeneration. This is supported further by the fact that the Aβ is a strong inducer of tPA expression [18], which, as described above, can lead to cell death in plasmin-dependent or -independent manners [14–16]. We envision that Aβ aggregates and low plasmin would trigger compensatory mechanisms to produce more plasmin relying on increased tPA or uPA production. The involvement of both molecules in cell signalling [12] raises the concern of excessive activation of signalling cascades.

Because of all the above, a reasonable scenario for AD is that an age-dependent down-regulation of the brain plasminogen system could contribute to the appearance of the typical signs of the disease, such as memory loss and learning impairment, because of amyloid accumulation on one hand and increased tPA signalling on the other. Together with our data, the demonstration that a short-term exposure to exogenous tPA and plasminogen can reduce the cytotoxic effects induced by amyloid aggregates [21] indicates further how delicate the balance of these molecules is to play beneficial or noxious roles.

Efficient plasminogen–plasmin conversion is impaired in situations of low neuronal membrane cholesterol

The observation of reduced plasmin levels and activity in certain AD brains raises the question of whether this can be a cause for abnormal amyloid aggregation or simply a consequence of the cell death or other signs that accompany this disease. Against the second premise is the fact that not all AD brains present low plasmin, despite all of them having similar neurodegeneration and disease signs [20]. Furthermore, the analysis of the level of the precursor plasminogen in the low-plasmin brains revealed that total plasminogen levels are comparable with those of non-AD cases and of the AD cases with normal plasmin [20], pinpointing that disease-associated features, such as inflammation or cell death, cannot be considered in the main responsible for plasmin paucity. Then what could the cause be for such low levels and activity?

The amount and activity of plasmin depends not only on the levels of the pro-enzyme plasminogen, but also on the capacity of the latter to bind to the cell surface [3], an event that requires the proper expression and distribution of different acceptor molecules. Thus defects in the organization of the neurons' membrane and/or on the expression of any of the acceptors may account for low plasmin. In fact, it was shown that exogenous plasminogen binds poorly to the membranes from low-plasmin AD brains [20]. Yet, the levels of expression of the plasminogen membrane-binding acceptors, the ganglioside GM1, and the C-terminal lysine-containing proteins annexin, amphoterin and α-enolase were found to be comparable in all three types of human brain samples: AD with normal and low plasmin levels and non-AD individuals [20]. From these two series of data, it appears that low plasmin in this peculiar AD group is due to a defect at the level of the neuronal plasma membrane.

Early studies from our laboratory demonstrated that the plasma membrane domains responsible for plasminogen binding and activation are the raft domains [22]. Rafts are small domains of cell membranes, characterized by the enrichment of cholesterol molecules in the exoplasmic leaflet at the hydrophobic–hydrophilic interface of phosphatidylcholine, as well as sphingomyelin that, depending on the cell type and stage of differentiation, have specific membrane and membrane-associated proteins [23]. The presence of cholesterol confers higher rigidity to these areas of the membrane [24]. Proteins that have an especial affinity for this type of lipidic environment perform different functions, either qualitatively or quantitatively, to those when in non-raft domains [25]. Biochemically, such types of proteins can be seen by their flotation to the buoyant density of cholesterol after extraction in cold non-ionic detergents [23]. Such an approach using membranes from cultured hippocampal neurons revealed that the plasminogen-binding molecules α-enolase, amphoterin, annexin and GM1 are enriched in rafts [22]. In agreement with these domains being the precise site of the membrane where plasminogen binding and cleavage occurs, plasmin was found exclusively in the raft fraction of both human and rodent hippocampal membranes [17,20,22]. More importantly, a statin-based reduction of less than 30% of membrane cholesterol in rodent hippocampal neurons that sufficed to cause raft disorganization resulted in a dramatic reduction of plasminogen binding and consequently in low plasmin levels [22]. To us, this last result acquires special relevance when it comes to the use of drugs that may lead to the reduction of cholesterol levels in neurons.

Hippocampal membranes of the AD brains with low plasmin have reduced membrane cholesterol

The reported observation that plasminogen binding is reduced in the hippocampal membranes of the AD patients with low plasmin, added to the demonstration that plasmin is essentially made in the raft membrane domains in a cholesterol-level-dependent manner, led us to the hypothesis that these patients' defect might be a paucity in membrane cholesterol. Indeed, when we measured the cholesterol content of the different samples, we found similar levels of total cholesterol, yet the membrane cholesterol was reduced significantly (30%) in the low-plasmin hippocampi compared to the AD cases with normal plasmin levels or with non-AD samples [20]. Consistently, specific and significant changes in the protein content were evident in the raft fractions of the AD low-plasmin membranes compared with those of the age-matched control individuals or AD patients with normal levels of the protease [20]. The changes in raft organization affected the membrane distribution of plasminogen binding and activating molecules, such as GM1 or uPA receptor. Notably, these alterations were not evident in brain areas that are normally spared from amyloid deposition in AD, such as the cerebellum. Membrane cholesterol measurement in the cerebellum revealed a reduction not higher than 19% in low-plasmin AD samples (M.D. Ledesma and C.G. Dotti, unpublished work), suggesting that cholesterol loss of more than 20% from the membrane would be required to produce defects in rafts and raft-mediated functions. These data also suggest that different brain areas can be more predisposed to cholesterol loss than others and/or that plasmin-independent proteolytic pathways might have a variable relevance in amyloid clearance depending on the brain area.

Low-membrane-cholesterol AD brains: possible relationship with the ApoE4 (apolipoprotein E4) allele

The existence of AD cases with low cholesterol/low plasmin and normal cholesterol/normal plasmin could not be correlated with the amount of amyloid plaques nor tau filament aggregation. They could not be associated either with differences in age of occurrence of the disease, severity of progression, sex, post-mortem brain-sampling delay, hippocampal area under scrutiny or cause of death. The only correlation we observed was the presence of the $\epsilon 4$ allele of the ApoE in AD brains with low plasmin and altered rafts and membrane cholesterol, while AD brains without these alterations all had the $\epsilon 2$ or $\epsilon 3$ alleles [20]. Although a study with a larger population is needed to confirm this correlation, it opens the possibility that the presence of the ApoE4 allele might be relevant in these cases. The fact that ApoE is a protein that participates in the homoeostasis of cholesterol and that the inheritance of the $\epsilon 4$ allele has been confirmed as a risk factor from sporadic AD [26] also supports this notion.

We have no scientific data yet to explain how the $\epsilon 4$ allele of ApoE could favour the cholesterol loss in these patients. It is possible that the astrocytes from these individuals are less efficient in providing cholesterol to neurons, as has been described in mice expressing this allele [27]. However, this might not be the case if it is considered that neuronal cholesterol levels are largely independent from outside cholesterol uptake [28]. In this situation, the presence of the ApoE4 allele might contribute to a more general defect, such as impaired brain oxygenation because of high blood cholesterol that indirectly affects the cholesterol synthesis machinery in the neurons. Still, the question of why not all ApoE4 bearers develop the disease remains. One explanation would be the existence of environmental or associated genetic factors that make some of the bearers more resistant to such indirect effects.

Considerations about membrane cholesterol and rafts, and how statins might help to fight against AD

Besides the possible relevance of the plasminogen system in the prevention of amyloid aggregation, the results detailed above serve to raise the conclusion that a deficit in brain cholesterol can be of extreme danger for proper neuronal function. As an example, cells with reduced cholesterol synthesis might no longer be able to respond efficiently to growth/survival factors, which are known to signal via receptors clustered in cholesterol-rich membrane domains [25].

Different from the above prediction, the opinion exists today that reduction of brain cholesterol can be beneficial to prevent an excess generation of Aβ and thus its accumulation. Indeed, cholesterol-synthesis inhibitors are being proposed for clinical trials for the prevention and cure of AD [29]. Two pillars support this opinion: clinical retrospective studies and basic cell biology/biochemical work in cells and animal models.

Retrospective clinical studies showed that patients with elevated serum cholesterol levels chronically treated with blood-cholesterol-lowering drugs, such as statins, present a decreased incidence of AD [30,31]. Since these drugs have as their basic mechanism of action the capacity to inhibit the early steps of cholesterol synthesis [32], the association between high brain cholesterol predisposing to AD was immediate. Yet, it is predictable that, in addition to a direct effect on the inhibition of cholesterol synthesis, statins might have had a beneficial role through other effects, secondary to the main mechanism of action (lowering blood cholesterol resulting in improved blood circulation and lower blood pressure) or unrelated to it, such as their antioxidant or anti-inflammatory roles [32]. The clinical retrospective studies themselves cannot distinguish between these different possibilities.

Albeit indirectly, cell biological/biochemical data in cells *in vitro* support the view that high brain cholesterol may be at the base of the anomalous amyloid production that occurs in AD. These studies showed that reduction of cholesterol via a combined treatment of cholesterol-synthesis inhibitors, statins, and cholesterol-extracting drugs in cultured cells overexpressing the precursor of Aβ, APP (amyloid precursor protein), or some of the enzymes involved in amyloid generation, inhibited Aβ release [33–36]. Hence, obvious conclusions were that excess amyloid is due to high cholesterol and that the Aβ is produced in cholesterol-rich raft domains. A first disadvantage of this type of approach concerns the fate of the overexpressed protein, whether or not it is targeted and accumulates in the same compartment as the endogenous counterpart. In fact, the presence of APP in rafts is a controversial issue, with several studies indicating that none or only a minor amount of the protein is in these membrane domains, with variation depending on the experimental approach (i.e. different detergents used in the extraction protocols and conditions of overexpression) [35,37]. Thus, if contrary to the endogenous protein, the overexpressed APP is delivered to rafts, the reduction of cholesterol could result in reduced amyloid generation, but this will reflect an extraordinary situation, not a physiological one. The second drawback is that the cholesterol-reduction treatments produced in those studies resulted in membrane cholesterol loss as high as 70%, raising the concern of disruption not only of raft domains, but also of the membrane as a whole that would affect cell viability. Thus, while useful to raise awareness about the possible role of neuronal cholesterol, we believe that these data do not prove that rafts and high neuronal cholesterol are involved in high amyloid production.

Conceivably, high brain cholesterol could be at the base of excessive amyloid production if increased amyloid production would occur in neurons induced to express high cholesterol. Although many of the enzymes participating in the biosynthesis of cholesterol are known and cloned, this point has not yet been addressed. Support would also come if the brains of AD patients presented higher cholesterol levels than those of non-AD individuals. Contrary to such a prediction, both the mRNA and protein levels of seladin-1, a protein involved in cholesterol synthesis, are down-regulated in the brain areas of AD patients where amyloid plaques form most frequently [38]. In agreement with AD being accompanied by low brain cholesterol, not high, we observed low membrane cholesterol in the hippocampi of a certain population of AD individuals [20]. On the opposite side, it has been claimed that AD brains have higher cholesterol than controls based on the finding that the levels of 24*S*-hydroxycholesterol are high in blood and CSF (cerebrospinal fluid) of AD patients [39]. The major drawback of using this catabolite as a marker for high brain cholesterol content is that it reflects the degree of cell damage and not the level of neuronal synthesis. In fact, its measurement constitutes a standard parameter of analysis to evaluate the degree of neuronal loss in situations such as stroke [40].

Although the lack of clearly indicative data showing that high brain cholesterol occurs in AD brains or that high neuronal cholesterol increases amyloid production weakens the need for using a cholesterol-synthesis inhibitor to treat this disease, it can be argued that what statins do in hypercholesterolaemic people is reduce the circulating cholesterol that is later exposed to the neurons. While this cannot be ruled out completely, it appears unlikely because neuronal cholesterol content is mostly, if not exclusively, dependent on internal synthesis, not from extracellular uptake [28]. This implies that neurons have a very strict control of their own cholesterol levels that, in principle, would make these cells able to cope with an extrinsic increase of the lipid with transcriptional shut-down. Although it can certainly be that hypercholesterolaemia perturbs such a feedback loop, this would be reflected in the presence of high circulating cholesterol in AD brains, an event that has not been documented.

All the above data lead necessarily to question whether it is worth trying to reduce cholesterol synthesis with statins in neurons to prevent AD and how dangerous it could be if neuronal cholesterol synthesis is impaired.

Animals treated with different types of statins should be useful to address these questions. Unfortunately, the data so far are contradictory. Thus guinea pigs treated with high doses of simvastatin have less Aβ in the CSF and in brain homogenates [34]. Since these are not hypercholesterolaemic animals, there is no reason to think that the anti-amyloidogenic effect involved inhibition of neuronal cholesterol synthesis. In fact, brain cholesterol levels are unchanged compared with control animals, suggesting that the statins utilized do not cross the blood–brain barrier efficiently. Quite the contrary: when Park et al. [41] used a blood–brain-barrier-permeant statin, an enhancement of amyloid production and senile plaque deposition was observed in female, but not male, mouse brains. From these results, it comes that statins can have noxious effects. This is in agreement with our data on raft disturbance due to moderate cholesterol loss in AD patient brains [20].

A number of genetic disorders also support the view that low, and not high, brain cholesterol is harmful to brain function. Thus mice bearing a defect in cholesterol homoeostasis due to mutations in the NPC (Niemann–Pick type C) 1 gene, that in humans causes NPC disease, present neurodegeneration. In NPC cells, cholesterol accumulates in late endosomes and lysosomes, while its synthesis is reduced as well as its anterograde transport to the axonal membrane [42]. Interestingly, an enhanced generation of Aβ, like in AD, has been reported in this disease [43]. Another example is the genetic disorder RSH/Smith–Lemli–Opitz syndrome, where the accumulation of 7-DHC (7-dehydrocholesterol) and paucity of cholesterol are thought to be responsible for neurodegeneration and early death [44]. Importantly, the gene responsible for this disease is homologous with seladin-1, which, as mentioned above, is abnormally down-regulated in the neurons of patients with AD in the areas of the brain that show the highest levels of amyloid deposition [38]. The obvious conclusion from these last series of observations is that the maintenance of appropriate cholesterol levels is essential for proper neuronal function.

Conclusion

The lack of a clear demonstration that amyloid plaques in AD form because of anomalously high brain cholesterol, added to the numerous data, biochemical and genetic, showing that a reduction of neuronal membrane cholesterol can be deleterious, make us conclude that only statins that do not cross the blood–brain barrier should be used. These drugs can still be extremely useful to fight the disease, either because of the improvement of brain oxygenation that a reduction in circulating cholesterol would produce, or because of their anti-inflammatory roles. Because of the last possibility, it would be interesting to establish precisely if chronic treatments with non-steroidal anti-inflammatory compounds, which were shown to also be beneficial against the occurrence of the disease in retrospective clinical studies [45], could replace statins, as these would not have the risk of producing undesired cholesterol synthesis inhibition, as has been the case for similar drugs in the recent past.

We thank the financial support of the EU grants DIADEM (Early Diagnosin of Alzheimer's Disease and Related Dementia) (QLK-3-CT-2001-02362) and APOPIS (Abnormal Proteins in the Pathogenesis of Neurodegenerative Disorders) to C.G.D. and M.D.L.

References

1. Selkoe, D.J. (2001) Physiol. Rev. **81**, 741–766
2. Selkoe, D.J. (2001) Neuron **32**, 177–180
3. Plow, E.F., Herren, T., Redlitz, A., Miles, L.A. and Hoover-Plow, J.L. (1995) FASEB J. **9**, 939–945
4. Collen, D. (1999) Thromb. Haemostasis **82**, 259–270
5. Sappino, A.P., Madani, R., Huarte, J., Belin, D., Kiss, J.Z., Wohlwend, A. and Vassalli, J.D. (1993) J. Clin. Invest. **92**, 679–685
6. Friedman, G.C. and Seeds, N.W. (1994) Dev. Brain Res. **81**, 41–49
7. Madani, R., Hulo, S., Toni, N., Madani, H., Steimer, T., Muller, D. and Vassalli, J.D. (1999) EMBO J. **18**, 3007–3012
8. Calabresi, P., Napolitano, M., Centonze, D., Marfia, G.A., Gubellini, P., Teule, M.A., Berretta, N., Bernardi, G., Frati, L., Tolu, M. and Gulino, A. (2000) Eur. J. Neurosci. **12**, 1002–1012
9. Pawlak, R., Nagai, N., Urano, T., Napiorkowska-Pawlak, D., Ihara, H., Takada, Y., Collen, D. and Takada, A. (2002) Neuroscience **113**, 995–1001
10. Rogove, A.D., Siao, C., Keyt, B., Strickland, S. and Tsirka, S.E. (1999) J. Cell Sci. **112**, 4007–4016
11. Wu, Y.P., Siao, C.J., Lu, W., Sung, T.C., Frohman, M.A., Milev, P., Bugge, T.H., Degen, J.L., Levine, J.M., Margolis, R.U. and Tsirka, S.E. (2000) J. Cell Biol. **148**, 1295–1304
12. Nicole, O., Docagne, F., Ali, C., Margaill. I., Carmeliet, P., MacKenzie, E.T., Vivien, D. and Buisson, A. (2001) Nat. Med. **7**, 59–64
13. Qian, Z., Gilbert, M.E., Colicos, M.A., Kandel, E.R. and Kuhl, D. (1993) Nature (London) **361**, 453–457
14. Tsirka, S.E., Rogove, A.D. and Strickland, S. (1996) Nature (London) **384**, 123–124
15. Chen, Z.L. and Strickland, S. (1997) Cell **91**, 917–925
16. Wang, Y.F., Tsirka, S.E., Strickland, S., Stieg, P.E., Soriano, S.G. and Lipton, S.A. (1998) Nat. Med. **4**, 228–231
17. Ledesma, M.D., Da Silva, J.S., Crassaerts, K., Delacourte, A., De Strooper, B. and Dotti, C.G. (2000) EMBO Rep. **1**, 530–535
18. Tucker, H.M., Kihiko, M., Caldwell, J.N., Wright, S., Kawarabayashi, T., Price, D., Walker, D., Scheff, S., McGillis, J.P., Rydel, R.E. and Estus, S. (2000) J. Neurosci. **20**, 3937–3946
19. Melchor, J.P., Pawlak, R. and Strickland, S. (2003) J. Neurosci. **23**, 8867–8871
20. Ledesma, M.D., Abad-Rodriguez, J., Galvan, C., Biondi, E., Navarro, P., Delacourte A., Dingwall, C. and Dotti, C.G. (2003) EMBO Rep. **4**, 1190–1196
21. Tucker, H.M., Kihiko-Ehmann, M., Wright, S., Rydel, R.E. and Estus, S. (2000) J. Neurochem. **75**, 2172–2177
22. Ledesma, M.D., Da Silva, J.S., Schevchenko, A., Wilm, M. and Dotti, C.G. (2003) Brain Res. **987**, 107–116
23. Simons, K. and Ikonen, E. (1997) Nature (London) **387**, 569–572
24. Zurzolo, C., van Meer, G. and Mayor, S. (2003) EMBO Rep. **4**, 1117–1121
25. Simons, K. and Toomre, D. (2000) Nat. Rev. Mol. Cell Biol. **1**, 31–39

26. Corder, E.H., Saunders, A.M., Strittmatter, W.J., Schmechel, D.E., Gaskell, P.C., Small, G.W., Roses, A.D., Haines, J.L. and Pericak-Vance, M.A. (1993) Science **261**, 921–923
27. Gong, J.S., Kobayashi, M., Hayashi, H., Zou, K., Sawamura, N., Fujita, S.C., Yanagisawa, K. and Michikawa, M. (2002) J. Biol. Chem. **277**, 29919–29926
28. Dietschy, J.M. and Turley, S.D. (2001) Curr. Opin. Lipidol. **12**, 105–112
29. Scarpini, E., Scheltens, P. and Feldman, H. (2003) Lancet Neurol. **2**, 539–547
30. Wolozin, B., Kellman, W., Ruosseau, P., Celesia, G.G. and Siegel, G. (2000) Arch. Neurol. **57**, 1439–1443
31. Jick, H., Zornberg, G.L., Jick, S.S., Seshadri, S. and Drachman, D.A. (2000) Lancet **356**, 1627–1631
32. Cucchiara, B. and Kasner, S.E. (2001) J. Neurol. Sci. **187**, 81–89
33. Simons, M., Keller, P., De Strooper, B., Beyreuther, K., Dotti, C.G. and Simons, K. (1998) Proc. Natl. Acad. Sci. U.S.A. **95**, 6460–6464
34. Fassbender, K., Simons, M., Bergmann, C., Stroick, M., Lutjohann, D., Keller, P., Runz, H., Kuhl, S., Bertsch, T., von Bergmann, K. et al. (2001) Proc. Natl. Acad. Sci. U.S.A. **98**, 5856–5861
35. Ehehalt, R., Keller, P., Haass, C., Thiele, C. and Simons, K. (2003) J. Cell Biol. **160**, 113–123
36. Cordy, J.M., Hussain, I., Dingwall, C., Hooper, N.M, and Turner, A.J. (2003) Proc. Natl. Acad. Sci. U.S.A. **100**, 11735–11740
37. Parkin, E.T., Turner, A.J. and Hooper, N.M. (1999) Biochem. J. **344**, 23–30
38. Greeve, I., Hermans-Borgmeyer, I., Brellinger, C., Kasper, D., Gomez-Isla, T., Behl, C., Levkau, B. and Nitsch, R.M. (2000) J. Neurosci. **20**, 7345–7352
39. Schonknecht, P., Lutjohann, D., Pantel, J., Bardenheuer, H., Hartmann, T., von Bergmann, K., Beyreuther, K. and Schroder, J. (2002) Neurosci. Lett. **324**, 83–85
40. Bretillon, L., Siden, A., Wahlund, L.O., Lutjohann, D., Minthon, L., Crisby, M., Hillert, J., Groth, C.G., Diczfalusy, U. and Bjorkhem, I. (2000) Neurosci. Lett. **293**, 87–90
41. Park, I.H., Hwang, E.M., Hong, H.S., Boo, J.H., Oh, S.S., Lee, J., Jung, M.W., Bang, O.Y., Kim, S.U. and Mook-Jung, I. (2003) Neurobiol. Aging **24**, 637–643
42. Karten, B., Vance, D.E., Campenot, R.B. and Vance, J.E. (2002) J. Neurochem. **83**, 1154–1163
43. Burns, M. and Duff, K. (2002) Ann. N.Y. Acad. Sci. **977**, 367–375
44. Porter, F.D. (2000) Mol. Genet. Metab. **71**, 163–174
45. Wyss-Coray, T. and Mucke, L. (2000) Nat. Med. **6**, 973–974

Biochem. Soc. Symp. **72**, 139–150
(Printed in Great Britain)

14

Endocytosis and retrograde axonal traffic in motor neurons

Katrin Deinhardt and Giampietro Schiavo[1]

Molecular NeuroPathobiology Laboratory, Cancer Research UK London Research Institute, 44 Lincoln's Inn Fields, London WC2A 3PX, U.K.

Abstract

Spinal cord motor neurons control voluntary movement by relaying messages that arrive from upper brain centres to the innervated muscles. Despite the importance of motor neurons in human health and disease, the precise control of their membrane dynamics and its effect on motor neuron homoeostasis and survival are poorly understood. In particular, the molecular basis of the co-ordination of specific endocytic events with the axonal retrograde transport pathway is largely unknown. To study these important vesicular trafficking events, we pioneered the use of atoxic fragments of tetanus and botulinum neurotoxins to follow endocytosis and retrograde axonal transport in motor neurons. These neurotoxins bind specifically to pre-synaptic nerve terminals, where they are internalized. Whereas botulinum neurotoxins remain at the neuromuscular junction, tetanus toxin is retrogradely transported along the axon to the cell body, where it is released into the intersynaptic space and is internalized by adjacent inhibitory interneurons. The high neurospecificity and the differential intracellular sorting make tetanus and botulinum neurotoxins ideal tools to study neuronal physiology. In the present review, we discuss recent developments in our understanding of the internalization and trafficking of these molecules in spinal cord motor neurons. Furthermore, we describe the development of a reliable transfection method for motor neurons based on microinjection, which will be extremely useful for dissecting further the molecular basis of membrane dynamics and axonal transport in these cells.

Introduction

Membrane dynamics is a process of paramount importance in all eukaryotic cells. It plays a role in multiple pathways, controlling, among other events, cellular homoeostasis, motility and survival. Due to their specialized functions

[1]To whom correspondence should be addressed (email giampietro.schiavo@cancer.org.uk).

and topology, neurons have developed unique mechanisms for the regulation of their membrane dynamics, which are responsible for neurite outgrowth and remodelling, synaptogenesis and SV (synaptic vesicle) exo-/endocytosis. The precise co-ordination of these processes is ensured by fine-tuned vesicular traffic, leading to the transport of many different ligands and organelles from and to the cell body. Newly synthesized proteins, RNAs and vesicular cargos are delivered to distal neuronal sites via the anterograde transport pathway, whereas retrograde transport is required for the retrieval of organelles and ligands from the cell periphery to the soma [1]. Pathogens and virulence factors, including TeNT (tetanus neurotoxin) and several neurotrophic viruses, induce these transport pathways to enter the central nervous system and to spread to neighbouring cells [2].

Despite the importance of axonal retrograde transport in health and disease, the molecular machinery responsible for this process and its coupling to specific endocytic events at the nerve terminal is poorly characterized. Whereas our understanding of SV dynamics within mammalian synapses has witnessed remarkable progress in the last decade [3], little is known about other endocytic events occurring at the same cellular site. Moreover, the relationship between the rate and co-ordination of SV exo-/endocytosis and other synaptic endocytic events is presently unknown. We sought to fill this gap by using a non-toxic fragment of TeNT (H_C) as a probe to monitor high-affinity binding, internalization and retrograde axonal transport in spinal cord MNs (motor neurons) [4,5].

TeNT H_C as a probe for neuronal membrane dynamics

TeNT is expressed by toxigenic strains of *Clostridium tetani*, an anaerobic bacterium, and belongs to the CNT (clostridial neurotoxin) family, which also comprises seven BoNT (botulinum neurotoxin) serotypes. TeNT is the causative agent of tetanus, a life-threatening disease characterized by an involuntary general contraction of skeletal muscles due to the unbalanced activity of the NMJs (neuromuscular junctions). CNTs are structurally organized into heavy and light chains linked by a disulphide bond (Figure 1a) [6]. The light chains are zinc-endopeptidases specific for synaptic members of the SNARE (soluble *N*-ethylmaleimide-sensitive fusion protein attachment protein receptor) family, which are core components of the neurotransmitter release apparatus [7]. These proteins [VAMP (vesicle-associated membrane protein)/synaptobrevin, SNAP-25 and syntaxin 1] are necessary for the fusion of SVs with the synaptic plasma membrane; their selective proteolysis accounts fully for the inhibition of neurotransmitter release caused by CNTs *in vivo* [7,8]. As a consequence, CNTs are the tools of choice to study the functions of synaptic SNARE proteins in neurons and secretory cells both *in vitro* and *in vivo* [7,8]. TeNT and BoNT have also been exploited to ascertain additional functions of synaptic SNAREs, such as their role in membrane repair and axonal maintenance [5]. In addition, TeNT is used in experimental models of chronic epilepsy to highlight the effects of repetitive seizures on brain functions [9].

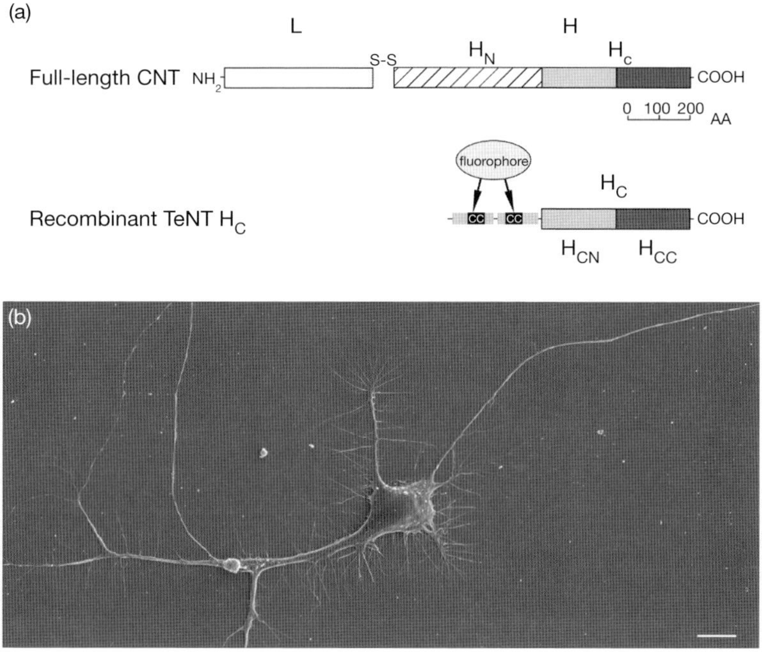

Figure 1 Scheme of clostridial neurotoxins and their domains. (a) TeNT and BoNT are composed of a heavy (H) and a light (L) chain, held together by a disulphide bridge and other non-covalent interactions. The L chain (in white) is a zinc-endopeptidase specific for SNARE proteins. The H chain is composed of a 50 kDa N-terminal domain (H_N, hatched), which is involved in membrane translocation and a C-terminal portion (H_C) responsible for membrane binding and retrograde transport. This domain is formed by the H_{CN} (light grey) and H_{CC} (dark grey) subdomains. H_C fragments are expressed as recombinant proteins tagged at the N-terminus [11]. These tags include a cysteine-rich motif suitable for Alexa-maleimide modification [42]. AA, amino acids. Fluorescent TeNT H_C is used as a probe to monitor axonal retrograde transport in spinal cord MNs (**b**), here imaged by field emission scanning electron microscopy. Notice the typical large soma, multiple dendritic arborizations and a single long axon with constant diameter. Scale bar, 10 μm.

Alongside these applications, TeNT and BoNT are also ideal tools to study membrane dynamics at the nerve terminal due to their neurospecificity and unique intracellular trafficking features. All CNTs are recruited to the NMJ plasma membrane, where they undergo distinct sorting events. BoNT enters the NMJ and blocks acetylcholine release, causing the flaccid paralysis, which is the clinical trademark of botulism. TeNT is instead recruited to the axonal retrograde transport pathway and reaches the soma of MNs, which is located in the spinal cord [7]. TeNT is then targeted to inhibitory interneurons, where it blocks the release of glycine, thus leading to spastic paralysis. This distinct sorting of TeNT and BoNT is saturable, since at high concentrations BoNT is also transported to the spinal cord, causing central nervous system effects [10].

However, at concentrations relevant for the pathophysiology of tetanus and botulism (picomolar and below), TeNT and BoNT are characterized by a specific targeting to distinct endocytic routes and constitute useful tools for the analysis and identification of these pathways in neurons [5].

The main determinants of the neurospecificity and intracellular targeting of CNTs are encoded by their C-terminal portions, termed H_C (Figure 1a) [5]. Isolated H_C fragments bind to the functional receptors of TeNT and BoNT, since they counteract the paralysis induced by the parental neurotoxins [11]. H_C is composed of two distinct subdomains of almost identical size with different binding specificity. The foremost C-terminal domain of H_C (H_{CC}) contains four distinct carbohydrate-binding regions, which mediate the interaction with surface lipids containing sialic acid, such as the polysialoganglioside GT1b [12,13], and, possibly, with neuronal glycoproteins [14,15]. The structural differences detected in the binding of ganglioside GT1b to the H_{CC} domain of TeNT and BoNT suggest that this domain might contribute to the specificity of TeNT and BoNT recruitment to the NMJ and their selective trafficking [16,17]. Accordingly, deletions in H_{CC} cause loss of binding to MNs, the impairment of TeNT to undergo retrograde transport *in vivo* [18], and a decrease in the neurotoxicity of CNT [15,16]. The N-terminal portion (H_{CN}) of H_C structurally resembles the carbohydrate-binding moiety found in lectins [19]. Despite the fact that no ligand for this portion of the molecule has been reported to date, it has been suggested that H_{CN} might play a role in a later step of the intoxication process, such as the intracellular sorting of TeNT [5].

Specificity of the binding of clostridial neurotoxins to the neuronal membrane

Multiple receptors participate in the high-affinity interaction of CNTs and their H_C fragments to the neuronal membrane. TeNT and BoNT bind to polysialogangliosides [7]. This interaction is biologically important, since CNTs display a reduced activity in neurons in which ganglioside biosynthesis has been impaired [20,21]. However, several considerations, including the limited binding competition among CNTs, suggest that other determinants play a role in the high-affinity interaction of these molecules with the neuronal surface [22]. In addition to acidic phospholipids, specific protein co-receptors have been characterized. BoNT/B and G interact with members of the synaptotagmin family in a GT1b-dependent manner [23–26]. These SV proteins, which are involved in calcium sensing at the synapse, expose their intraluminal domain to the extracellular environment following SV fusion with the plasma membrane [27], potentially allowing CNT binding. These findings suggest that the internalization of at least these two BoNT serotypes may occur via SV endocytosis.

Despite the efforts from different laboratories, less is known regarding the specific receptors for TeNT and the remaining BoNT serotypes. One or more GPI (glycosylphosphatidylinositol)-anchored proteins of 15 kDa have been shown to bind selectively to TeNT and its H_{CC} fragment, but not BoNT [28,29]. In NGF (nerve growth factor) differentiated PC12 cells, one of these TeNT-interacting factors has been identified as Thy-1 [30], a GPI-anchored glycoprotein involved in multiple cellular functions, including neurite outgrowth [31]. Although Thy-1 is unlikely to be the main protein receptor for TeNT [5], GPI-anchored proteins play a major role in TeNT binding and uptake, since pre-treating MNs with a phospholipase which cleaves their lipid anchor protects neurons from TeNT intoxication [32]. Altogether, these findings support the hypothesis that one or more still unidentified GPI-anchored glycoproteins act as a physiological TeNT receptor.

GPI-anchored proteins, together with cholesterol, gangliosides and other sphingolipids (but notably not synaptotagmins), are enriched in microdomains of the plasma membrane termed lipid rafts [33,34]. Lipid rafts act as functional platforms for signalling, endocytosis and sorting [34], and might be exploited by pathogens to promote infection or intoxication [35]. The binding of TeNT H_C to multiple raft-associated components suggests that the recruitment of TeNT on the neuronal surface is mediated by lipid microdomains. Accordingly, CNTs associate with detergent-resistant glycolipid-enriched membranes in a cholesterol-dependent manner [30,36]. Disruption of lipid microdomains with cholesterol-sequestering drugs abolishes TeNT H_C internalization [30] and VAMP/synaptobrevin cleavage [30,32], suggesting that lipid rafts play a central role in TeNT trafficking and represent a common requisite for the recruitment of CNTs to the neuronal plasma membrane. Furthermore, the distinct protein receptors and trafficking of TeNT and BoNT imply that these neurotoxins bind to different lipid rafts with independent intrasynaptic sorting. Strong evidence supporting the presence of pools of membrane rafts undergoing distinct regulation is accumulating rapidly, together with the notion that both the composition and distribution of lipid microdomains are likely to determine their distinct signal outputs and, ultimately, their cellular fate [34].

The high affinity of CNTs, their multiple receptors and the apparent redundancy of some of these molecules in CNT binding to the neuronal membrane have led to the recent proposal that APRs (arrays of pre-synaptic receptors) are involved in this process [22]. APRs are suggested to be dynamic entities sharing several functional features with lipid rafts. Although the initial CNT binding may involve any of their receptor molecules, the polysialoganglioside headgroups, rather than the carbohydrate moieties of glycoproteins, are likely to act as preferred initial binding factors due to their high lateral mobility and high concentration at the NMJ. Polysialogangliosides would act as ‘antennas’ in capturing CNT molecules present in the intersynaptic fluid and allow their clustering in APRs on the pre-synaptic membrane [22]. The presence of different receptor molecules within an APR, each one in several copies, would transform the initial low-affinity reversible membrane interaction [37] to a multivalent high-affinity one, which is virtually irreversible [8,38]. The recruitment of CNTs to APRs is expected to favour the delivery of CNTs to specialized sites for endocytosis and sorting, where these neurotoxins may exploit pre-assembled signalling complexes for their internalization, such as the Trk-dependent neurotrophin signalling cascade, which has been shown to be activated by TeNT H_C in cortical neurons [39]. Therefore specific APR signalling might functionally engage a short-range pathway that is responsible for

BoNT localization at the NMJ or, alternatively, couple TeNT endocytosis with the sorting machinery responsible for the long-haul retrograde transport routes to the spinal cord followed by neurotrophins.

Synaptic sorting and axonal transport in MNs

Axonal retrograde transport is essential for neuronal homoeostasis, and its impairment is incompatible with neuronal survival [40]. Despite its importance in neuronal physiology and disease, a limited number of studies have focused on the identification of the machinery responsible for the sorting of endocytic vesicles to the axonal retrograde transport pathway in mammalian neurons. Furthermore, the direct visualization of retrograde transport and its quantitative analysis have been hampered by the paucity of reliable assays in living cells. We have recently established an axonal retrograde transport assay in differentiated spinal cord MNs, which exploits a fluorescent derivative of TeNT H_C as a probe [4]. MNs are isolated from embryonal rat spinal cords (E14) by a purification method that combines metrizamide density centrifugation and magnetic immuno-isolation using antibodies against p75 low-affinity NGF receptors ($p75^{NTR}$). This procedure exploits both the relative large size of MN somas and the selective early expression of $p75^{NTR}$ in these cells [41]. The same procedure can be adapted to murine MNs by using embryos at E13 (embryonic day 13) and omitting the magnetic purification step [42]. Differentiated MNs [DIV5 (5 days *in vitro*) onwards] exhibit an extensive dendritic network and a single axon (Figure 1b), which may contain lateral branching.

Differentiated MNs internalize both fluorescently labelled TeNT and TeNT H_C in morphologically identical endocytic structures [4], which display an overlapping speed distribution [5]. These carriers have an average speed that agrees well with the rates observed for TeNT retrograde transport *in vivo* [8]. Two main groups of organelles are responsible for the retrograde transport of TeNT H_C: round vesicles and tubules. Both types of carriers are characterized by distinct kinetic behaviour; vesicles move discontinuously, alternating frequent pauses with phases of movement, whereas tubules are faster and display a more continuous transport [4]. Both types of carriers are likely to originate from coated invaginations, which appear on the axonal plasma membrane upon incubation of TeNT H_C at 4°C. When neurons were subsequently shifted to 37°C, TeNT H_C was initially found in deep pits and coated vesicles, and was detected later on in uncoated tubulovesicular and round endocytic structures, as seen by electron microscopy [42]. The diameter of vesicular compartments varies between 50 and 100 nm, whereas the average length of tubular structures is 2–3 μm. Analogous coated structures have been observed independently for TeNT internalization in mixed spinal cord cells [43]. However, the nature of the TeNT H_C compartment and its internalization mechanism have been elusive due to, among other factors, the paucity of systems for efficient transfection of fluorescently tagged and dominant-negative mutant proteins in MNs.

Optimization of transfection protocols in spinal cord MNs

We sought to fill this gap by testing different chemical and physical protocols for MN transfection. Lipofectamine 2000 is a cationic lipid-based transfection reagent designed to give high transfection efficiencies coupled to high protein expression levels in different mammalian cell types. We tested a variety of conditions in terms of Lipofectamine 2000 concentration (0.5–1.5%, v/v), amount of DNA (0.5–3 μg), incubation time (2–24 h), expression times (20–120 h) and differentiation state of the transfected MNs (DIV0–DIV4). As a transfection probe, we chose the vector pEGFP-C2, which drives the expression of EGFP (enhanced green fluorescent protein) under the strong human CMV (cytomegalovirus) immediate early promoter. Despite our attempts, we could observe medium–strong cytoplasmic EGFP expression only in a low percentage of MNs: up to 5–10%, if transfected at DIV1, but falling to well below 1% at later differentiation stages. Furthermore, when tested for TeNT H_C internalization and axonal retrograde transport [4], cells expressing EGFP did not show many moving TeNT H_C carriers. In addition, untransfected cells in the same culture dish, or in control dishes where the exogenous DNA was omitted, displayed only a few slow-moving endosomes, similar to pEGFP-C2 transfected cells, suggesting that Lipofectamine 2000 treatment itself has a deleterious effect on the ability of MNs to sustain an efficient axonal retrograde transport (K. Deinhardt and G. Schiavo, unpublished work). Although the precise explanation of this inhibitory effect is presently unclear, it is possible that the permeability of the MN plasma membrane is altered by the cationic lipid formulation of the transfection reagent, leading to a large and long-term imbalance of the ionic equilibrium and membrane potential of the neurite network, which in turn negatively affects axonal transport.

On the basis of these early findings, we then chose microinjection as the elected protocol to introduce foreign DNA into differentiated MNs. Despite the relative low throughput of this technique, it is generally compatible with viability and functions of fully differentiated neuronal cultures [44], and its efficiency is independent of the maturation stage of the neurons. Upon microinjection of the same EGFP-based vector, MNs were left to recover for 4–72 h and were then tested for TeNT H_C internalization and axonal retrograde transport as before. Differentiated MNs expressing high levels of EGFP are able to bind and internalize TeNT H_C to the same levels as untransfected cells (Figures 2a and 2b). Moreover, the percentage of still and moving TeNT H_C-containing structures in EGFP-expressing MNs is identical with that of control cells (Figure 2e), as well as the speed distribution of retrograde transport carriers (Figure 2f). This last parameter has been calculated by kinetic analysis of time-series images (Figure 2c), as described previously [42]. Therefore microinjection is the method of choice for the expression of high levels of exogenous proteins in differentiated MNs. In addition, microinjection allows the introduction of other cell-impermeant molecules into MNs, such as purified proteins or labelled antibodies.

Molecular motors and fast axonal retrograde transport

Fast retrograde transport of TeNT H_C in MNs requires both F-actin (filamentous actin) and MTs (microtubules) [42], suggesting that both MT-dependent and microfilament-dependent motors are involved in this process. To impair MT- and F-actin-based motors selectively, MNs were incubated with the dynein inhibitor EHNA [erythro-9-(2-hydroxy-3-nonyl) adenine] [45], the kinesin inhibitor aurintricarboxylic acid [46] and the broad-spectrum myosin inhibitor 2,3-butanedione monoxime [47] following TeNT H_C internalization. Only the simultaneous treatment of MNs with the three inhibitors stopped TeNT H_C carriers completely [42], suggesting that multiple classes of molecular motors are involved in fast axonal retrograde transport.

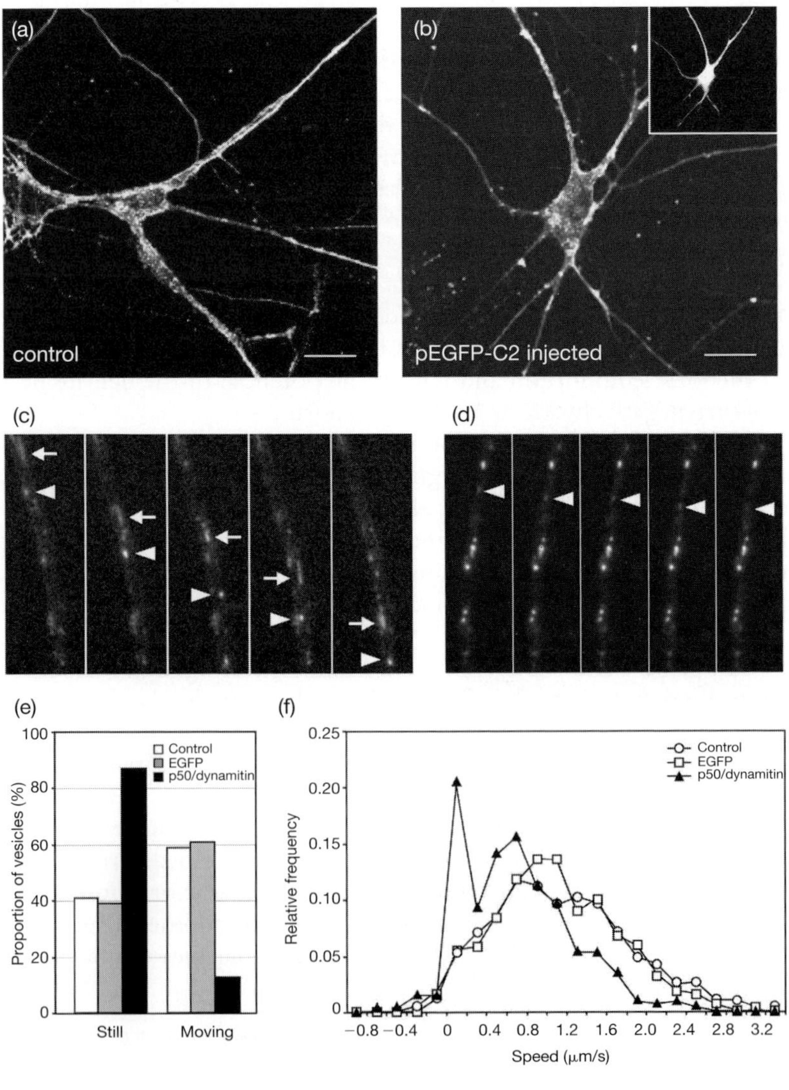

To overcome the specificity concerns of this pharmacological analysis, the function of dynein and axonal myosins in retrograde transport have been investigated further by using a combination of transfection and genetic approaches. To this end, MNs have been microinjected with p50/dynamitin, a component of the dynein/dynactin complex [48]. Overexpression of p50/dynamitin in mammalian cells disrupts the complex, resulting in dissociation of cytoplasmic dynein from several membrane organelles, including early and late endosomes [49]. MNs expressing p50/dynamitin display an impaired motility of the TeNT H_C carriers (Figure 2d), with a pronounced increase of still TeNT H_C-positive endosomes (Figure 2e) and a significant shift of the speed-distribution profile towards lower speed values (Figure 2f). Altogether, these findings indicate that cytoplasmic dynein acts as the main motor unit for retrograde transport in differentiated MNs, which is in agreement with the results obtained by overexpressing p50/dynamitin in postnatal MNs [50]. However, recent findings describing a direct link between dynactin and both kinesin II and cytoplasmic dynein [51] complicate the interpretation of the effects of p50/dynamitin overexpression in these cells. This notion is strengthened by the discovery that a point mutation in the MT-binding domain of dynactin causes MN disease in humans [52]. Evidence supporting a direct role for dynein in MN survival has been recently provided by the finding that mutations in dynein heavy chain generate pathological conditions closely resembling amyotrophic lateral sclerosis [53]. Interestingly, fast retrograde transport of TeNT H_C in MNs isolated from these mice is impaired, confirming a central role of cytoplasmic dynein in this process.

As predicted by the need for F-actin, TeNT transport requires members of the myosin superfamily and, in particular, myosin Va. MNs from myosin-Va-null embryos displayed slower retrograde transport compared with wild-type cells [42]. Therefore the co-ordination of myosin Va and MT-dependent motors is necessary for fast axonal retrograde transport in spinal cord MNs.

Figure 2 p50/dynamitin blocks axonal retrograde transport of TeNT H_C in differentiated MNs. (**a**) MNs bind and internalize TeNT H_C–Alexa Fluor 555. (**b**) Overexpression of EGFP by microinjection of the pEGFP-C2 vector (inset, 26 h post-injection) does not affect TeNT H_C–Alexa Fluor 555 binding and internalization (scale bars, 20 μm) nor its retrograde transport (**c**). This panel shows time series images of an MN axon (from left to right) containing TeNT H_C-positive carriers (arrowhead, arrow; frames=9×29 μm). In contrast, MNs overexpressing p50/dynamitin are characterized by slower TeNT H_C-positive carriers and a higher number of stationary TeNT H_C-positive vesicles (**d**) (arrowhead; frames =7.6×29 μm). In both (**c**) and (**d**), MN somas were located out of view at the bottom of the image. Intervals between frames are 5 s. (**e**) Percentage of still and moving TeNT H_C-positive structures in non-injected MNs (empty bars, n=265) or MNs injected with pEGFP-C2 (grey bars, n=289) or pCMV-myc encoding p50/dynamitin [49] (solid bars, n=333). (**f**) Kinetic analysis of TeNT H_C carriers in control (○; 71 carriers, 729 single movements), EGFP-expressing (□; 68 carriers, 674 single movements) and p50/dynamitin microinjected MNs (▲; 56 carriers, 679 single movements). Relative frequencies of speed values for TeNT H_C–Alexa Fluor 555 carriers observed between two consecutive frames (interval=5 s) are reported. Retrograde movement is conventionally shown as positive.

Physiological role of TeNT H_C carriers

TeNT H_C carriers are not acidified during transport, a condition necessary to retain TeNT within their lumen [8]. SV40 (simian virus 40) and *Escherichia coli* are also internalized into neutral compartments that do not belong to the classical endosomal system, which contain glycosphingolipid-rich caveolae [54]. Despite the absence of caveolin in MNs, TeNT could use a distinct endocytic pathway to bypass the classical endosomal–lysosomal system and escape degradation, as suggested by the long half-life of TeNT in spinal neurons [8]. Similarly to TeNT, endogenous ligands might use the same pathway to avoid proteolysis, reaching the neuronal cell body in a biologically active form. TeNT H_C and NGF share the same retrograde organelles in MNs, a subset of which contains $p75^{NTR}$ [4]. NGF is retrogradely transported in MNs and accumulates in their somas without being degraded [55]. Although the endocytic compartment responsible for $p75^{NTR}$ trafficking in MNs is not fully characterized [56], recent studies highlighted $p75^{NTR}$ association with GT1b [57] and lipid raft components [58]. On this basis, lipid microdomains involved in the retrograde transport of $p75^{NTR}$ might play an important role in the trafficking of TeNT and other endogenous and exogenous factors through a specialized endocytic route.

We apologize to the many colleagues whose work could not be cited here due to space limitations. We thank S. Gschmeissner for the field emission scanning electron microscopy image of spinal cord MNs in culture, S. Salinas and S. Tooze for reading the manuscript and for helpful comments. Laboratory work by the authors is supported by Cancer Research UK.

References

1. Brown, A. (2003) J. Cell Biol. **160**, 817–821
2. von Bartheld, C.S. (2004) J. Neurobiol. **58**, 295–314
3. Murthy, V.N. and De Camilli, P. (2003) Annu. Rev. Neurosci. **26**, 701–728
4. Lalli, G. and Schiavo, G. (2002) J. Cell Biol. **156**, 233–240
5. Lalli, G., Bohnert, S., Deinhardt, K., Verastegui, C. and Schiavo, G. (2003) Trends Microbiol. **11**, 431–437
6. Turton, K., Chaddock, J.A. and Acharya, K.R. (2002) Trends Biochem. Sci. **27**, 552–558
7. Humeau, Y., Doussau, F., Grant, N.J. and Poulain, B. (2000) Biochimie **82**, 427–446
8. Schiavo, G., Matteoli, M. and Montecucco, C. (2000) Physiol. Rev. **80**, 717–766
9. Benke, T.A. and Swann, J. (2004) Adv. Exp. Med. Biol. **548**, 226–238
10. Moreno-Lopez, B., de la Cruz, R.R., Pastor, A.M. and Delgado-Garcia, J.M. (1997) Neuroscience **81**, 437–455
11. Lalli, G., Herreros, J., Osborne, S.L., Montecucco, C., Rossetto, R. and Schiavo, G. (1999) J. Cell Sci. **112**, 2715–2724
12. Emsley, P., Fotinou, C., Black, I., Fairweather, N.F., Charles, I.G., Watts, C., Hewitt, E. and Isaacs, N.W. (2000) J. Biol. Chem. **275**, 8889–8894
13. Fotinou, C., Emsley, P., Black, I., Ando, H., Ishida, H., Kiso, M., Sinha, K.A., Fairweather, N.F. and Isaacs, N.W. (2001) J. Biol. Chem. **276**, 32274–32281
14. Louch, H.A., Buczko, E.S., Woody, M.A., Venable, R.M. and Vann, W.F. (2002) Biochemistry **41**, 13644–13652

15. Rummel, A., Bade, S., Alves, J., Bigalke, H. and Binz, T. (2003) J. Mol. Biol. **326**, 835–847
16. Rummel, A., Mahrhold, S., Bigalke, H. and Binz, T. (2004) Mol. Microbiol. **51**, 631–643
17. Yowler, B.C. and Schengrund, C.L. (2004) Biochemistry **43**, 9725–9731
18. Sinha, K., Box, M., Lalli, G., Schiavo, G., Schneider, H., Groves, M., Siligardi, G. and Fairweather, N. (2000) Mol. Microbiol. **37**, 1041–1051
19. Lacy, D.B. and Stevens, R.C. (1999) J. Mol. Biol. **291**, 1091–1104
20. Williamson, L.C., Bateman, K.E., Clifford, J.C.M. and Neale, E.A. (1999) J. Biol. Chem. **274**, 25173–25180
21. Kitamura, M., Takamiya, K., Aizawa, S. and Furukawa, K. (1999) Biochim. Biophys. Acta **1441**, 1–3
22. Montecucco, C., Rossetto, R. and Schiavo, G. (2004) Trends Microbiol. **12**, 442–446
23. Nishiki, T., Tokuyama, Y., Kamata, Y., Nemoto, Y., Yoshida, A., Sato, K., Sekiguchi, M., Takahashi, M. and Kozaki, S. (1996) FEBS Lett. **378**, 253–257
24. Yowler, B.C., Kensinger, R.D. and Schengrund, C.L. (2002) J. Biol. Chem. **277**, 32815–32819
25. Ihara, H., Kohda, T., Morimoto, F., Tsukamoto, K., Karasawa, T., Nakamura, S., Mukamoto, M. and Kozaki, S. (2003) Biochim. Biophys. Acta **1625**, 19–26
26. Dong, M., Richards, D.A., Goodnough, M.C., Tepp, W.H., Johnson, E.A. and Chapman, E.R. (2003) J. Cell Biol. **162**, 1293–1303
27. Matteoli, M., Takei, K., Perin, M.S., Sudhof, T.C. and De Camilli, P. (1992) J. Cell Biol. **117**, 849–861
28. Herreros, J., Lalli, G., Montecucco, C. and Schiavo, G. (2000) J. Neurochem. **74**, 1941–1950
29. Herreros, J., Lalli, G. and Schiavo, G. (2000) Biochem. J. **347**, 199–204
30. Herreros, J., Ng, T. and Schiavo, G. (2001) Mol. Biol. Cell **12**, 2947–2960
31. Simon, P.D., McConnell, J., Zurakowski, D., Vorwerk, C.K., Naskar, R., Grosskreutz, C.L. and Dreyer, E.B. (1999) Dev. Brain Res. **117**, 219–223
32. Munro, P., Kojima, H., Dupont, J.L., Bossu, J.L., Poulain, B. and Boquet, P. (2001) Biochem. Biophys. Res. Commun. **289**, 623–629
33. Chamberlain, L.H., Burgoyne, R.D. and Gould, G.W. (2001) Proc. Natl. Acad. Sci. U.S.A. **98**, 5619–5624
34. Pike, L.J. (2004) Biochem. J. **378**, 281–292
35. Lafont, F., Abrami, L. and van der Goot, F.G. (2004) Curr. Opin. Microbiol. **7**, 4–10
36. Herreros, J. and Schiavo, G. (2002) Int. J. Med. Microbiol. **291**, 447–453
37. Keller, J.E., Cai, F. and Neale, E.A. (2004) Biochemistry **43**, 526–532
38. Habermann, E. and Dreyer, F. (1986) Curr. Top. Microbiol. Immunol. **129**, 93–179
39. Gil, C., Chaib-Oukadour, I. and Aguilera, J. (2003) Biochem. J. **373**, 613–620
40. Goldstein, L.S. and Yang, Z. (2000) Annu. Rev. Neurosci. **23**, 39–71
41. Yan, Q. and Johnson, E.M., Jr. (1988) J. Neurosci. **8**, 3481–3498
42. Lalli, G., Gschmeissner, S. and Schiavo, G. (2003) J. Cell Sci. **116**, 4639–4650
43. Parton, R.G., Ockleford, C.D. and Critchley, D.R. (1987) J. Neurochem. **49**, 1057–1068
44. Ikeda, S.R. (1997) Methods Mol. Biol. **83**, 191–202
45. Penningroth, S.M., Cheung, A., Bouchard, P., Gagnon, C. and Bardin, C.W. (1982) Biochem. Biophys. Res. Commun. **104**, 234–240
46. Hopkins, S.C., Vale, R.D. and Kuntz, I.D. (2000) Biochemistry **39**, 2805–2814
47. Cramer, L.P. and Mitchison, T.J. (1995) J. Cell Biol. **131**, 179–189
48. Wittmann, T. and Hyman, T. (1999) Methods Cell Biol. **61**, 137–143
49. Burkhardt, J.K., Echeverri, C.J., Nilsson, T. and Vallee, R.B. (1997) J. Cell Biol. **139**, 469–484
50. LaMonte, B.H., Wallace, K.E., Holloway, B.A., Shelly, S.S., Ascano, J., Tokito, M., Van Winkle, T., Howland, D.S. and Holzbaur, E.L. (2002) Neuron **34**, 715–727
51. Deacon, S.W., Serpinskaya, A.S., Vaughan, P.S., Lopez Fanarraga, M., Vernos, I., Vaughan, K.T. and Gelfand, V.I. (2003) J. Cell Biol. **160**, 297–301
52. Puls, I., Jonnakuty, C., LaMonte, B.H., Holzbaur, E.L., Tokito, M., Mann, E., Floeter, M.K., Bidus, K., Drayna, D., Oh, S.J. et al. (2003) Nat. Genet. **33**, 455–456

53. Hafezparast, M., Klocke, R., Ruhrberg, C., Marquardt, A., Ahmad-Annuar, A., Bowen, S., Lalli, G., Witherden, A.S., Hummerich, H., Nicholson, S. et al. (2003) Science **300**, 808–812
54. Duncan, M.J., Shin, J.S. and Abraham, S.N. (2002) Cell. Microbiol. **4**, 783–791
55. Butowt, R. and Von Bartheld, C.S. (2003) Eur. J. Neurosci. **17**, 673–680
56. Kuruvilla, R., Zweifel, L.S., Glebova, N.O., Lonze, B.E., Valdez, G., Ye, H. and Ginty, D.D. (2004) Cell **118**, 243–255
57. Yamashita, T., Higuchi, H. and Tohyama, M. (2002) J. Cell Biol. **157**, 565–570
58. Higuchi, H., Yamashita, T., Yoshikawa, H. and Tohyama, M. (2003) EMBO J. **22**, 1790–1800

Biochem. Soc. Symp. **72**, 151–164
(Printed in Great Britain)

15

Consequences of lipid raft association on G-protein-coupled receptor function

Anja Becher[1] and R.A. Jeffrey McIlhinney

MRC Anatomical Neuropharmacology Unit, University of Oxford, Mansfield Road, Oxford OX1 3TH, U.K.

Abstract

GPCRs (G-protein-coupled receptors) play key roles in many cellular processes, and malfunction may lead to a range of pathologies, including psychiatric and neurological disorders. It is therefore not surprising that this group of receptors supplies a majority of the targets for pharmaceutical drug development. Despite their importance, the mechanisms that regulate their function and signalling still remain only partially understood. Recently, it has become evident that a subset of GPCRs is not homogeneously distributed in the plasma membrane, but localizes instead to specific membrane microdomains known as lipid rafts. Lipid rafts are characterized by their enrichment in cholesterol and sphingolipids, and have been suggested to serve as platforms for a range of cellular signalling complexes. In the present review, we will be discussing the effects of the lipid raft environment on trafficking, signalling and internalization of raft-associated GPCRs.

Introduction

The large majority of receptors belongs to the GPCR (G-protein-coupled receptor) superfamily, which means that they couple to heterotrimeric G-proteins. Stimuli as diverse as light photons, taste ligands, neurotransmitters and hormones signal through GPCRs [1]. It is therefore not surprising that these receptors represent the third largest family of genes in the human genome, with an estimated 1% of the genome coding for them.

Even though they are functionally diverse, GPCRs share a range of structural features [1]. Apart from their G-protein-binding properties, these include an extracellular receptor domain involved in ligand binding and a cluster of seven

[1]To whom correspondence should be addressed (email anja.becher@pharm.ox.ac.uk).

transmembrane helical domains linking the extracellular region with the cytoplasmic tail (Figure 1). Upon receptor activation, a conformational change is triggered that leads to separation of the associated G-protein into its constituent α and βγ subunits, which then interact with and regulate distinct intracellular second messenger cascades depending upon the type of G-protein involved.

Recent studies have shown that many G-proteins do not diffuse freely at the plasma membrane, but are instead localized to specialized sphingolipid- and cholesterol-enriched platforms known as lipid raft domains [2–4]. Accordingly, a number of GPCRs have also been found to localize to lipid rafts (Table 1), and we will be discussing examples of GPCRs whose trafficking and/or signalling may be altered by raft localization. First, we will give a brief summary of lipid raft structure and the main methods used to isolate them. For a broader examination of lipid rafts and their role in cell function, the interested reader is referred to some recent more general reviews [5–9].

Lipid raft structure

The saturated hydrocarbon chains of sphingolipids allow for tight packaging of sphingolipid and cholesterol molecules in lipid raft domains. This makes rafts more ordered and less fluid than non-raft membranes, which mostly contain unsaturated, and therefore less densely packed, phospholipids. It also gives raft domains a slightly increased membrane thickness compared with non-raft domains. Because of these physical properties, lipid raft domains concentrate a specific subset of membrane proteins, while excluding most others. In addition, the unique lipid environment may also influence the functional properties of its resident proteins directly.

Structurally, two main types of lipid raft can be identified, namely caveolin-free rafts and caveolae, and a number of GPCRs have been identified in

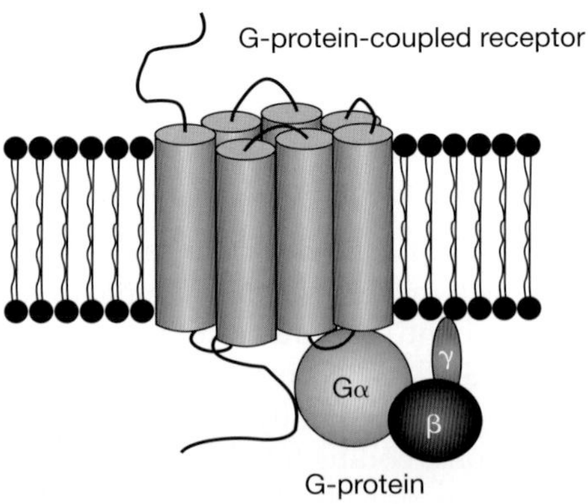

Figure 1 Schematic diagram of GPCR. GPCRs share structural similarities including an extracellular N-terminal ligand-binding domain, seven helical transmembrane domains linked by three extracellular and three cytoplasmic loops, and a cytoplasmic C-terminal tail.

these raft types [10–40]. Caveolae are named after their constituent structural proteins, the caveolins, which give caveolae their distinct morphology. When visualized by electron microscopy, caveolae are identifiable as flask-shaped plasma membrane invaginations that open up towards the extracellular surface [41]. Caveolin-free rafts lack the distinct structural architecture of caveolae. In fact, caveolin-free rafts have a similar curvature to the surrounding non-raft membrane, and are therefore difficult to study by visual means. Most cell types probably contain both caveolae and caveolin-free lipid rafts, while some, such as neurons, contain only caveolin-free rafts [42].

The reasons why proteins segregate to lipid rafts are not fully understood. Several mechanisms may contribute to raft association, including post-translational modifications, such as GPI (glycosylphosphatidylinositol) anchors and acylation with myristate and/or palmitate. The G-protein $G_{\alpha i}$ subunit, for example, needs to be both myristoylated and palmitoylated in order to localize to lipid rafts [43]. Some proteins may also associate with rafts via a direct protein–protein interaction with a raft-resident protein. Examples of GPCRs that are able to interact directly with the lipid raft protein caveolin include the adenosine A_1 receptor [10,12], AT_1 (angiotensin II type 1 receptor) [17], and the β_2-adrenoceptor [16]. For a few transmembrane proteins, the lengths and amino acid sequences of their membrane-spanning domains, as well as the sequences of their membrane-proximal extracellular domains, have been shown to influence lipid raft association [5]. Although transmembrane domains have been implicated in the lipid raft association of the latent membrane protein 1 of Epstein–Barr virus [44] and the influenza virus neuraminidase [45], we are unaware at time of writing (July 2004) of any GPCRs where this has been conclusively shown to be the case.

Studying lipid rafts

Lipid rafts, because of their specific lipid composition, are insoluble in cold non-ionic detergents, and this characteristic has been used historically both to define lipid raft domains and to isolate them [5–9]. Briefly, detergent-dependent raft isolation methods require that detergent-treated cell membranes are loaded to the bottom of a density gradient and centrifuged at high speed. Because of their low buoyant density and detergent-insolubility, lipid rafts will float up the density gradient, away from the bulk of the cellular material. The detergent-soluble proteins will remain in the lower fractions of the gradient, and the detergent-insoluble cytoskeletal components will pellet at the bottom.

Recently, detergent-free methods of raft isolation have gained some popularity, especially for the enrichment of caveolin-containing rafts [5–9]. These methods mostly replace the non-ionic detergents used during conventional isolation procedures with a high pH sodium carbonate buffer and rigorous homogenization steps to rupture cell membranes. The resulting membrane fragments are separated by virtue of their different buoyant densities by density-gradient centrifugation. Fractions isolated by detergent-free methods elude the definition of rafts as detergent-insoluble, and may differ in their lipid and protein content from fractions isolated using detergent-insolubility criteria [5,46].

Table 1 GPCRs that associate with lipid rafts. 'Detergent-dependent' and 'detergent-free' refer to the biochemical isolation methods of lipid rafts either using non-ionic detergents or not using detergent respectively; CM and EM refer to co-localization studies of receptors with lipid rafts by confocal microscopy or electron microscopy respectively; n.a., not assessed.

Receptor	Methods used to detect raft-association	Mechanism of raft association	Functional implications
Adenosine A_1	Detergent-free [10,11], CM [12], EM [12]	Bound caveolin-3 [10] and caveolin-1 [12]	Variable results: agonist stimulation led to movement out of [10] or into [11,12] lipid rafts
α_1-Adrenoceptor	Detergent-free [13]	Bound caveolin-3 [13]	n.a.
β_1-Adrenoceptor	Detergent-free [14–16]	n.a.	No functional effects were observed upon agonist treatment or raft disruption [15,16]
β_2-Adrenoceptor	Detergent-free [14–16]	Bound caveolin-3 [16]	Lipid raft disruption by cyclodextrin and filipin increased signalling [15,16]
AT_1	Detergent-free [17–19], detergent-dependent [18], EM [19]	Bound caveolin-1 [17]	Variable results: agonist stimulation either led to movement into rafts [17,19] or had no effect on localization [18]
Bradykinin B1	CM [20]	Putative caveolin binding domain [21]	n.a.
Bradykinin B2	CM [20], detergent-free [18], detergent-dependent [22]	Putative caveolin binding domain [21]	Agonist stimulation led to movement into rafts [18]; internalization via a raft-dependent mechanism [20]
Calcium-sensing	Detergent-free [23]	n.a.	n.a.
CCR5	Detergent-free [25], detergent-dependent [24]	Three palmitoylation sites in cytoplasmic tail [26]	Cyclodextrin treatment led to loss of receptor signalling [25]
Endothelin A	CM [27], detergent-dependent [29]	Bound caveolin [27,29]	Filipin inhibited agonist-induced endocytosis [28]
Endothelin B	Detergent-free [30], detergent-dependent [29]	Bound caveolin-1 only in absence of agonist [29]	Filipin inhibited signalling [29,30], agonist stimulation resulted in movement out of rafts [29]
$GABA_B$ (γ-aminobutyric acid B)	Detergent-dependent [31,32]	n.a.	Decreased signalling in raft-enriched fractions [32]

Table 1 (contd.)

Receptor	Methods used to detect raft-association	Mechanism of raft association	Functional implications
GnRH	Detergent-free, detergent-dependent [33]	n.a.	Cyclodextrin led to loss of agonist-induced responses [33]
Muscarinic M_2	Detergent-free [34]	n.a.	Agonist stimulation led to movement into rafts [34]; possibly internalization by raft-dependent mechanism [35]
Oxytocin	Detergent-free [36,37]	Cholesterol-binding	Agonist stimulation of raft-associated compared with non-raft receptors may increase or inhibit cell proliferation respectively [36]
Serotonin	Detergent-free [38]	n.a.	Reduced agonist response upon cyclodextrin treatment [38]
Sphingosine 1-phosphate (EDG-1)	Detergent-free [39]	Bound caveolin 1 [39]	Agonist-induced movement into rafts; caveolin overexpression inhibited receptor function [39]
Thyrotropin	CM [40]	n.a.	Receptor moved out of rafts upon stimulation [40]

The different methods that exist to isolate lipid raft domains can make it difficult to compare and assess the data obtained, especially when results are contradictory [46,47]. Even when using only detergent-dependent isolation methods, results can differ depending on the specific isolation procedure, the protein to detergent ratio, and the type of detergent used [48]. Thus, while biochemical isolation procedures are useful tools for the study of lipid rafts, they are best used in combination with other techniques of raft study.

Additional techniques used to study lipid rafts include the microscopic visualization of lipid raft domains. Lipid rafts are thought to be too small and/or dynamic to be resolved by conventional light microscopy. Because caveolae have a distinct morphology, association of proteins with this raft type can be analysed by electron microscopy as long as suitable antibodies are available. Methods do not (yet) exist for detection of caveolin-free rafts by electron microscopy. Indirect methods of visualization by light microscopy are being developed, an example being FRET (fluorescence resonance energy transfer), which yields a fluorescent signal only when a tagged protein of interest is in very close proximity (<50 nm) to a tagged raft marker, and can thus be inferred to reside in lipid raft domains [47]. An advantage of FRET is that it can be combined with live-imaging techniques, which give an indication of the dynamic behaviour over time of lipid raft domains.

A widely used approach to study the functional implications of raft association is to manipulate the cholesterol content of membranes using tools such as the cholesterol depleting drug methyl-β-cyclodextrin, or the cholesterol-sequestering antibiotic filipin [49]. These treatments lead to the dissociation of proteins from lipid rafts into non-raft membranes and thereby allow for comparison of receptor signalling properties in raft and non-raft domains. While cholesterol depletion is a very useful tool for studying rafts, it does have pleiotropic effects on all membrane areas, and may potentially affect the functional properties of both raft-associated and non-raft proteins [50]. Functional changes observed upon cholesterol depletion are thus not necessarily due to prior raft-association, and additional methods should be employed to ensure that the protein under study does indeed reside in lipid raft domains.

The effects of lipid raft association on receptor function are increasingly being studied without altering membrane properties by removing or mutating known raft-targeting domains in the receptor. As discussed in more detail below, such an approach was used by Wyse et al. [19] to show, without altering cholesterol levels, that caveolae are necessary for trafficking of the AT_1 receptor to the plasma membrane. Of course, this approach requires that the raft-targeting domains are known and that their removal or mutation will not alter additional properties of the protein other than its raft association.

Table 1 lists GPCRs that have been shown to associate with lipid raft domains, and summarizes the mechanisms of raft association and the functional implications if known. From this list, we have chosen seven receptors as representative examples to illustrate the effects of lipid raft domains on GPCR signalling and trafficking.

Lipid rafts and GPCR signalling

Examples of GPCRs whose signalling seems to be affected by raft localization include the CCR5 (chemokine receptor 5), the GnRH (gonadotropin-releasing hormone) receptor and the β_2-adrenoceptor (CCR5 may also be an example of a receptor that is internalized via lipid rafts [51]). Cholesterol depletion decreases signalling of CCR5 and the mammalian GnRH receptor, while enhancing signalling of the β_2-adrenoceptor.

CCR5

The chemokine receptors are a family of GPCRs that mediate immune cell responses to soluble chemoattractants called chemokines. CCR5 has received much attention recently as a co-receptor of CD4 required for HIV entry into host cells.

CCR5 has been shown by detergent-dependent methods [24] and by confocal microscopy [25] to reside in lipid-raft-enriched domains, and this localization may affect receptor signalling. In studies performed by Nguyen and Taub [25], CCR5s lost their ligand-binding ability in response to cholesterol extraction using hydroxypropyl-β-cyclodextrin. Cholesterol extraction also led to a loss of receptor signalling, as measured by intracellular calcium mobilization. The majority of bound agonist appeared to be localized to lipid rafts, suggesting that rafts contained the high-affinity CCR5. These results suggest that localization to lipid rafts is a requirement for effective CCR5 signalling.

Raft localization and signalling seem to be enhanced by CCR5 palmitoylation. Three cysteine residues at the boundary between the seventh transmembrane domain and the cytoplasmic tail of the receptor have been shown to be palmitoylated [26,52], and palmitoylation-deficient mutants appear to have both a decreased raft-association, as assessed by confocal microscopy [26], and a decreased coupling to downstream signalling cascades [52].

Although lipid raft localization appears to be important for endogenous CCR5 signalling, it does not seem necessary for HIV entry into cells. Results obtained by Percherancier et al. [53] suggest that, when CCR5 resided exclusively in non-raft domains, HIV-1 could still efficiently enter into cells. However, even though lipid raft localization of the chemokine receptor was not essential for HIV cell entry, entry did nevertheless depend on a high cholesterol content of the host cell membrane. Depletion of plasma membrane cholesterol by the drug cyclodextrin inhibited HIV-1 cell entry even when the CCR5 and CD4 receptors were not localized to lipid rafts [53].

It therefore seems that lipid rafts are needed for endogenous agonist signalling of the CCR5, but probably not for HIV-1 entry into T-cells, which depends more on overall membrane cholesterol content.

Mammalian GnRH receptor

GnRH plays a pivotal role in reproductive processes. It is released by the hypothalamus and binds to receptors on the anterior pituitary, thereby stimulating the synthesis and release of LH (luteinizing hormone), and the release of FSH (follicle-stimulating hormone). Agonist-occupied GnRH receptors link via early second messenger pathways, such as phospholipase C and InsP_3 formation, to MAPK (mitogen-activated protein kinase) signalling. An important MAPK in this context is ERK (extracellular-signal-regulated kinase). Mammalian GnRH receptors are unusual in that the C-terminal cytoplasmic domain is only 1–2 amino acids in length.

Using both detergent and detergent-free isolation methods, Navratil et al. [33] showed that the GnRH receptor constitutively resided in lipid-raft-enriched domains isolated from gonadotropin-derived αT3-1 cells. This localization was observed both in the presence and absence of agonist. Similarly, the GnRH receptor resided in lipid-raft-enriched fractions from transfected CHO (Chinese-hamster ovary) cells independently of hormone treatment. Interestingly, while CHO cells express caveolin, this caveolae structural protein is absent in αT3-1 cells. Thus localization of the GnRH receptor to lipid rafts seemed independent of agonist stimulation, cell type or expression of caveolin.

While raft localization of the GnRH receptor did not change upon agonist treatment, the lipid raft environment was nevertheless required for effective receptor signalling. Results by Navratil et al. [33] showed that stimulation of the raft-localized receptor resulted in an InsP_3 response, leading to agonist-dependent ERK phosphorylation and c-fos gene expression. When rafts were disrupted by methyl-β-cyclodextrin, all of these agonist responses were lost.

Thus, while the GnRH receptors do not seem to change their raft localization in response to agonist stimulation, they do appear to require lipid rafts for effective signalling.

β-Adrenoceptor

Cardiac β_1- and β_2-adrenoceptors mediate the actions of catecholamines, which increase the rate, and force of contractions, of the heart. The β_1-adrenoceptor is the main receptor that mediates catecholamine actions under normal physiological conditions, while the β_2-adrenoceptor operates mainly in neonatal ventricles, and in aged and/or failing hearts, where the β_1 receptor is down-regulated. Both β_1- and β_2-adrenoceptors have very similar signalling properties, activating a G-protein-dependent pathway of adenylate cyclase stimulation leading to accumulation of cAMP and to PKA (protein kinase A)-dependent protein phosphorylation.

Both the β_1- and the β_2-adrenoceptors have been isolated in lipid-raft-enriched fractions, although the extent of raft-localization seems to differ between different cell types and developmental stages. β_1- and β_2-adrenoceptors were first studied by Schwencke et al. [14] in raft-enriched fractions obtained using detergent-free isolation methods. When overexpressing the receptors in a heterologous cell system, they found that both receptor types were localized almost exclusively to caveolin-enriched domains.

Subsequently, Rybin et al. [15] showed, using two separate detergent-free isolation methods, that only a subfraction of β_1-adrenoceptor localized to caveolae in rat neonatal myocyte cultures, and neither agonist treatment nor raft disruption greatly influenced the functional behaviour or distribution pattern of the receptor. In contrast, the β_2-adrenoceptor was located entirely in the caveolae-rich fractions isolated from both neonatal and adult mouse myocytes [15,16,54], and from quiescent fibroblasts [15]. Furthermore, co-immunoprecipitation of the β_2-adrenoceptor with caveolin-3 from neonatal mouse myocytes implied a direct interaction between the receptor and this structural protein of caveolae. Manipulation of lipid raft domains using the raft-disrupting agents filipin and methyl-β-cyclodextrin led to increased receptor signalling, as measured by an increased and prolonged myocyte contraction rate [54], and increased agonist-induced accumulation of the second messenger cAMP [15] respectively. Furthermore, filipin treatment also resulted in a decreased co-immunoprecipitation of the receptor with caveolin-3.

Thus signalling of the β_2-adrenoceptor appears to be inhibited by lipid raft localization, and is increased when lipid raft domains are disrupted. However, there may be cell- and expression-system-specific differences in β_2-adrenoceptor partitioning. The receptor seems largely excluded from caveolar membranes in vascular smooth muscle cells [55,56]. Furthermore, Odley et al. [57] failed to detect human β_2-adrenoceptor in lipid raft domains isolated from resting-state cardiac myocytes when overexpressed in transgenic mice.

Lipid rafts and GPCR trafficking and internalization

Internalization follows agonist-induced G-protein phosphorylation and contributes to prolonged receptor desensitization. The AT_1 receptor appears to utilize lipid rafts for trafficking, and may internalize via a raft-dependent mechanism. Other GPCRs that seem to be internalized by raft-dependent mechanisms include the adenosine A_1 receptor, the B_2 bradykinin receptor and the chicken GnRH receptor.

AT_1 receptor

Angiotensin II plays a key role in the regulation of cardiovascular homoeostasis by regulating blood volume and vascular resistance. The actions of angiotensin II are mediated by AT_1 and AT_2 (angiotensin II type 2) receptors. The AT_1 receptor is the main angiotensin receptor in adult tissue, and mediates most of the actions of angiotensin II. It acts mainly on phospholipases, phosphoinositides, calcium channels, and serine/threonine and tyrosine kinases.

Ishizaka et al. [17] and Ushio-Fukai et al. [58] showed that the AT_1 receptor was absent from lipid-raft-enriched fractions isolated by detergent-free methods from unstimulated vascular smooth muscle cells, but moved into raft-enriched fractions upon agonist activation. In addition, they also identified a consensus binding site for interaction with caveolin in the AT_1 receptor, and were furthermore able to co-immunoprecipitate the AT_1 receptor with caveolin-1 following agonist stimulation. Thus, according to these studies, the AT_1 receptor localizes to caveolae upon agonist activation, where it interacts with caveolin-1.

However, a later publication by Leclerc et al. [18], who used both detergent-dependent and detergent-free isolation methods, seemed to show that the proportion of AT_1 receptor in lipid-raft-enriched membrane fractions did not increase upon agonist stimulation, at least not in the heterologously transfected COS cell line used. Interestingly, however, the authors identified a caveolin scaffolding domain binding sequence at the C-terminus of the AT_1 receptor, a mutation of which led to decreased plasma membrane expression and decreased agonist-induced internalization of the receptor [18].

Wyse et al. [19] approached the questions of whether and when the AT_1 receptor resides in lipid rafts using a combination of methods. First, using the detergent-free method to isolate lipid-raft-enriched membrane domains, they showed that stimulation with angiotensin II caused a translocation of the AT_1 receptor to raft-enriched fractions. Secondly, using electron microscopy, they identified agonist-induced receptor clusters in membrane domains that were sensitive to the raft-disrupting drug cyclodextrin. However, these domains lacked the characteristic morphology of caveolae. Furthermore, identified caveolae did not contain AT_1. Thus the receptor probably resided in caveolin-free lipid raft domains. Surprisingly, in cells lacking caveolin-1 and caveolin-3, the AT_1 receptor failed to travel to the cell surface and remained instead in the endoplasmic reticulum. However, when caveolin-1 or caveolin-3 were ectopically expressed, this allowed for trafficking of the receptor to the plasma membrane. Furthermore, mutation of important residues within the caveolin scaffolding domain binding sequence of the AT_1 receptor similarly confined the receptor to the endoplasmic reticulum even in the presence of the caveolins.

These results suggest that caveolins are necessary for trafficking of the AT_1 receptor to the cell surface. Once at the cell surface, however, the receptor seems to localize to the caveolin-free type of lipid raft. Caveolins thus act as molecular chaperones rather than as plasma membrane scaffolds for the AT_1 receptor.

Adenosine A_1 receptor

The nucleotide adenosine plays an import role in coupling the metabolic rate of cells to their energy supply. To date, four adenosine receptors have been identified, namely A_1, A_{2a}, A_{2b} and A_3, and all are coupled to G-proteins.

The A_1 adenosine receptor is thought to mediate the effects of adenosine during pathophysiological conditions such as hypoxia. Caveolae have been shown to be enriched in the A_1 adenosine receptor. Reports differ, however, on the extent to which the receptor associates with caveolae in quiescent unstimulated cells, and on whether it moves into or out of rafts upon agonist treatment.

Lasley et al. [10] were the first to report an association of the adenosine A_1 receptor with caveolae. Using non-detergent methods, they isolated approx. two-thirds of the adenosine A_1 receptors in caveolin-enriched membrane fractions from unstimulated cardiac ventricular myocytes. The receptors were found to bind directly to caveolin, with an antibody against caveolin-3 co-immunoprecipitating A_1 adenosine receptors together with caveolin-3. Agonist stimulation resulted in a decrease of receptor in caveolin-enriched fractions. In particular, stimulation with the adenosine analogue CCPA (2-chloro-N^6-cyclopentyladenosine) in the presence of adenosine deaminase to sequester free adenosine resulted in A_1 receptor translocation from caveolin-enriched into non-caveolin-enriched membrane fractions, an effect that was blocked by a receptor antagonist. These observations suggest that, *in vivo* in cardiac myocytes, the inactive form of the receptor is located in lipid rafts, and moves into non-raft domains upon activation.

These results contrast with those obtained by Rafael Franco's research group [11,12]. Using a detergent-free method to enrich for lipid raft-containing membrane fractions from a cultured epithelial cell line, Ginés et al. [11] found that only a very small proportion of A_1 adenosine receptors was present in these fractions in unstimulated cells. Upon agonist treatment, the A_1 adenosine receptor was found to have translocated into caveolin-containing fractions, from where it was seen to be internalized [11,12]. Caveolin-1 was able to bind a C-terminal domain construct of the A_1 adenosine receptor, thereby indicating a possible *in vivo* interaction between the full-length receptor and caveolin.

The differences in lipid raft localization and agonist effects found in the above studies may be due to differences in the cell types under study. Different, cell-type-specific distribution patterns between raft and non-raft domains have been observed for other GPCRs, such as the β-adrenoceptor (discussed above). However, future studies need to test different agonist doses and exposure times with respect to lipid raft localization in the different cell types in order to determine the kinetics of receptor movement out of and (back) into rafts, and the effects on receptor function, internalization and desensitization. Furthermore, the effects on receptor raft localization of adenosine deaminase, which was included in the myocytes study, need to be determined.

B_1 and B_2 bradykinin receptors

Bradykinin is a vasoactive peptide that acts through the two homologous GPCRs, the bradykinin 1 (B_1) receptor and the bradykinin 2 (B_2) receptor.

The B_1 bradykinin receptor displays high basal agonist-independent signalling, no ligand-induced phosphorylation, and no internalization upon agonist stimulation. The B_2 receptor, on the other hand, shows ligand-induced phosphorylation, and is internalized upon agonist stimulation. Receptor signalling through G-proteins to phospholipase C leads to phosphoinositide hydrolysis and subsequent intracellular calcium mobilization. Both the B_1 and the B_2 bradykinin receptors have a putative caveolin-binding domain [21].

Using confocal microscopy of fluorescently tagged proteins, Sabourin et al. [20] showed that the B_1 bradykinin receptor co-localized with caveolin-1 upon agonist treatment, but was not internalized, while the B_2 receptor seemed to internalize via a raft-dependent process upon agonist treatment. Employing detergent and detergent-free biochemical enrichment methods of lipid raft membranes together with electron microscopy analysis, de Weerd and Leeb-Lundberg [22] studied raft-association of the bradykinin B_2 receptor in a smooth muscle cell line. In these cells, the basal amount of receptor in rafts was very low. However, when bradykinin was bound to the B_2 receptor, the receptor and its associated G-proteins were recruited to caveolae. Although bradykinin-induced hydrolysis of phosphoinositides occurred within just a few seconds of receptor stimulation ($t_{1/2}$ approx. 15 s), the enrichment of G-proteins in lipid rafts was only observed after several minutes ($t_{1/2}$ 2–5 min).

These results suggest that localization to caveolae is not required for early signalling through the B_2 bradykinin receptor. Instead, the time frame of G-protein enrichment in rafts parallels that of receptor desensitization [59] and internalization, suggesting that lipid rafts are involved in B_2 receptor internalization and possibly sequestration.

Non-mammalian GnRH receptor

As mentioned above, GnRH plays a key role in reproduction and its receptor localizes to lipid raft domains [33]. Interestingly, while the mammalian GnRH receptor displays slow internalization kinetics in response to agonist binding, the chicken GnRH receptor internalizes rapidly in a process that depends on its long C-terminal tail. Long intracellular tails are a common feature of non-mammalian GnRH receptors, but are absent in the mammalian receptors [60].

Pawson et al. [60] have shown that the fast internalization kinetics of the chicken GnRH receptor depend on caveolae. The fast kinetics were lost upon disruption of caveolae by the drug filipin or by overexpressing a dominant-negative caveolin-1 mutant. Furthermore, internalization was also inhibited when a putative C-terminal membrane-proximal palmitoylation site was mutated out (Cys^{328}→Ala). Although the authors did not show direct evidence for palmitoylation at this site, it corresponds to a similar site in the β_2-adrenoceptor that is palmitoylated [60]. A possible effect of this mutation on raft association was not reported.

These results suggest that the fast internalization kinetics of the chicken GnRH receptor depend on caveolae and possibly on receptor palmitoylation.

Concluding remarks

GPCRs play key roles both in physiological and pathological processes, and lipid rafts provide an important mechanism for modulating some of these functions. Not all GPCRs localize to lipid rafts. Of those that do, most associate in a partial and reversible fashion, thus providing a mechanism for influencing receptor behaviour depending on the membrane localization. The superfamily of GPCRs belongs to one of the largest family of genes in the human genome and is made up of eight evolutionarily unrelated, but convergent families of GPCRs. It is not surprising, therefore, that different lipid raft effects are observed for different GPCRs. Further research may allow us to identify a unifying pattern, such as specific receptor subgroups that associate with and are modulated by lipid rafts in similar ways.

References

1. Bockaert, J., Claeysen, S., Becamel, C., Pinloche, S. and Dumuis, A. (2002) Int. Rev. Cytol. **212**, 63–132
2. Oh, P. and Schnitzer, J.E. (2001) Mol. Biol. Cell **12**, 685–698
3. Waheed, A.A. and Jones, T.L. (2002) J. Biol. Chem. **277**, 32409–32412
4. Foster, L.J., De Hoog, C.L. and Mann, M. (2003) Proc. Natl. Acad. Sci. U.S.A. **100**, 5813–5818
5. Pike, L.J. (2004) Biochem. J. **378**, 281–292

6. Brown, D. (2002) Int. J. Med. Microbiol. **291**, 433–437
7. Ikonen, E. (2001) Curr. Opin. Cell Biol. **13**, 470–477
8. Kenworthy, A. (2002) Trends Biochem. Sci. **27**, 435–437
9. Simons, K. and Ehehalt, R. (2002) J. Clin. Invest. **110**, 597–603
10. Lasley, R.D., Narayan, P., Uittenbogaard, A. and Smart, E.J. (2000) J. Biol. Chem. **275**, 4417–4421
11. Ginés, S., Ciruela, F., Burgueno, J., Casado, V., Canela, E.I., Mallol, J., Lluis, C. and Franco, R. (2001) Mol. Pharmacol. **59**, 1314–1323
12. Escriche, M., Burgueno, J., Ciruela, F., Canela, E.I., Mallol, J., Enrich, C., Lluis, C. and Franco, R. (2003) Exp. Cell Res. **285**, 72–90
13. Fujita, T., Toya, Y., Iwatsubo, K., Onda, T., Kimura, K., Umemura, S. and Ishikawa, Y. (2001) Cardiovasc. Res. **51**, 709–716
14. Schwencke, C., Okumura, S., Yamamoto, M., Geng, Y.J. and Ishikawa, Y. (1999) J. Cell. Biochem. **75**, 64–72
15. Rybin, V.O., Xu, X., Lisanti, M.P. and Steinberg, S.F. (2000) J. Biol. Chem. **275**, 41447–41457
16. Rybin, V.O., Pak, E., Alcott, S. and Steinberg, S.F. (2003) Mol. Pharmacol. **63**, 1338–1348
17. Ishizaka, N., Griendling, K.K., Lassegue, B. and Alexander, R.W. (1998) Hypertension **32**, 459–466
18. Leclerc, P.C., Auger-Messier, M., Lanctot, P.M., Escher, E., Leduc, R. and Guillemette, G. (2002) Endocrinology **143**, 4702–4710
19. Wyse, B.D., Prior, I.A., Qian, H., Morrow, I.C., Nixon, S., Muncke, C., Kurzchalia, T.V., Thomas, W.G., Parton, R.G. and Hancock, J.F. (2003) J. Biol. Chem. **278**, 23738–23746
20. Sabourin, T., Bastien, L., Bachvarov, D.R. and Marceau, F. (2002) Mol. Pharmacol. **61**, 546–553
21. Lamb, M.E., Zhang, C., Shea, T., Kyle, D.J. and Leeb-Lundberg, L.M. (2002) Biochemistry **41**, 14340–14347
22. de Weerd, W.F. and Leeb-Lundberg, L.M. (1997) J. Biol. Chem. **272**, 17858–17866
23. Kifor, O., Diaz, R., Butters, R., Kifor, I. and Brown, E.M. (1998) J. Biol. Chem. **278**, 31167–31176
24. Manes, S., Mira, E., Goemez-Mouton, C., Lacalle, R.A., Keller, P., Labrador, J.P. and Martinez, A.C. (1999) EMBO J. **18**, 6211–6220
25. Nguyen, D.H. and Taub, D. (2002) Blood **99**, 4298–4306
26. Percherancier, Y., Planchenault, T., Valenzuela-Fernandez, A., Virelizier, J.L., Arenzana-Seisdedos, F. and Bachelerie, F. (2001) J. Biol. Chem. **276**, 31936–31944
27. Chun, M., Liyanage, U.K., Lisanti, M.P. and Lodish, H.F. (1994) Proc. Natl. Acad. Sci. U.S.A. **91**, 11728–11732
28. Okamoto, Y., Ninomiya, H., Miwa, S. and Masaki, T. (2000) J. Biol. Chem. **275**, 6439–6446
29. Yamaguchi, T., Murata, Y., Fujiyoshi, Y. and Doi T. (2003) Eur. J. Biochem. **270**, 1816–1827
30. Teixeira, A., Chaverot, N., Schroder, C., Strosberg, A.D., Couraud, P.O. and Cazaubon S.J. (1999) J. Neurochem. **72**, 120–128
31. Becher, A., White, J.H. and McIlhinney, R.A.J. (2001) J. Neurochem. **79**, 787–795
32. Becher, A., Green, A., Ige, A.O., Wise, A., White, J.H. and McIlhinney, R.A.J. (2004) Biochem. Biophys. Res. Commun., **321**, 981–987
33. Navratil, A.M., Bliss, S.P., Berghorn, K.A., Haughian, J.M., Farmerie, T.A., Graham, J.K., Clay, C.M. and Roberson, M.S. (2003) J. Biol. Chem. **278**, 31593–31602
34. Feron, O., Smith, T.W., Michel, T. and Kelly, R.A. (1997) J. Biol. Chem. **272** 17744–17748
35. Dessy, C., Kelly, R.A., Balligand, J.L. and Feron, O. (2000) EMBO J. **19**, 4272–4280
36. Gimpl, G. and Fahrenholz, F. (2000) Eur. J. Biochem. **267**, 2483–2497
37. Guzzi, F., Zanchetta, D., Cassoni, P., Guzzi, V., Francolini, M., Parenti, M. and Chini, B. (2002) Oncogene **21**, 1658–1667
38. Dreja, K., Voldstedlund, M., Vinten, J., Tranum-Jensen, J., Hellstrand, P. and Sward, K. (2002) Arterioscler. Thromb. Vasc. Biol. **22**, 1267–1272

39. Igarashi, J. and Michel, T.J. (2000) J. Biol. Chem. **275**, 32363–32370
40. Latif, R., Ando, T., Daniel, S. and Davies, T.F. (2003) Endocrinology **144**, 4725–4728
41. van Deurs, B., Roepstorff, K., Hommelgaard, A.M. and Sandvig, K. (2003) Trends Cell Biol. **13**, 92–100
42. Masserini, M., Palestini, P. and Pitto, M. (1999) J. Neurochem. **73**, 1–11
43. Moffett, S., Brown, D.A. and Linder, M.E. (2000) J. Biol. Chem. **275**, 2191–2198
44. Coffin, 3rd, W.F., Geiger, T.R. and Martin, J.M. (2003) J. Virol. **77**, 3749–3758
45. Barman, S., Adhikary, L., Chakrabarti, A.K., Bernas, C., Kawakoa, Y. and Nayak, D.P. (2004) J. Virol. **10**, 5258–5269
46. Eckert, G.P., Igbavboa, U., Muller, W.E. and Wood, W.G. (2003) Brain Res. **962**, 144–150
47. Matko, J. and Edidin, M. (1997) Methods Enzymol. **278**, 444–462
48. Chamberlain, L.H. (2004) FEBS Lett. **559**, 1–5
49. Simons, K. and Toomre, D. (2000) Nat. Rev. Mol. Cell Biol. **1**, 31–39
50. Munro, S. (2003) Cell **115**, 377–388
51. Venkatesan, S., Rose, J.J., Lodge, R., Murphy, P.M. and Foley, J.F. (2003) Mol. Biol. Cell **14**, 3305–3324
52. Blanpain, C., Wittamer, V., Vanderwinden, J.M., Boom, A., Renneboog, B., Lee, B., Le Poul, E., El Asmar, L., Govaerts, C., Vassart, G. et al. (2001) J. Biol. Chem. **276**, 23795–23804
53. Percherancier, Y., Lagane, B., Planchenault, T., Staropoli, I., Altmeyer, R., Virelizier, J.L., Arenzana-Seisdedos, F., Hoessli, D.C. and Bachelerie, F. (2003) J. Biol. Chem. **278**, 3153–3161
54. Xiang, Y., Rybin, V.O., Steinberg, S.F. and Kobilka, B. (2002) J. Biol. Chem. **277**, 34280–34286
55. Ostrom, R.S., Violin, J.D., Coleman, S. and Insel, P.A. (2000) Mol. Pharmacol. **57**, 1075–1079
56. Ostrom, R.S. (2002) Mol. Pharmacol. **61**, 473–476
57. Odley, A., Hahn, H.S., Lynch, R.A., Marreez, Y., Osinska, H., Robbins, J. and Dorn, 2nd, G.W. (2004) Proc. Natl. Acad. Sci. U.S.A. **101**, 7082–7087
58. Ushio-Fukai, M., Hilenski, L., Santanam, N., Becker, P.L., Ma, Y., Griendling, K.K. and Alexander, R.W. (2001) J. Biol. Chem. **276**, 48269–48275
59. Leeb-Lundberg, L.M., Cotecchia, S., DeBlasi, A., Caron, M.G. and Lefkowitz, R.J. (1987) J. Biol. Chem. **262**, 3098–3105
60. Pawson, A.J., Maudsley, S.R., Lopes, J., Katz, A.A., Sun, Y.M., Davidson, J.S. and Millar, R.P. (2003) Endocrinology **144**, 3860–3871

Biochem. Soc. Symp. **72**, 165–175
(Printed in Great Britain)

16

Functional role of lipid rafts in CD20 activity?

Eva Janas[1], Richard Priest and Rajneesh Malhotra

Rheumatoid Arthritis Biology, RI-CEDD, Medicines Research Centre, GlaxoSmithKline, Gunnel's Wood Road, Stevenage, Hertfordshire SG1 2NY, U.K.

Abstract

CD20 is a B-lymphocyte-specific integral membrane protein, implicated in the regulation of transmembrane calcium conductance, cell-cycle progression and B-lymphocyte proliferation. CD20 is proposed to function as a SOCC (store-operated calcium channel). SOCCs are activated by receptor-stimulated calcium depletion of intracellular stores. Sustained calcium conductivity across the plasma membrane mediated by SOCC activity is required for long-term calcium-dependent processes, such as transcriptional control and gene expression. Cross-linking of CD20 by antibodies (e.g. Rituxan) has been reported to induce a rapid redistribution of CD20 into specialized microdomains at the plasma membrane, known as lipid rafts. Recruitment of CD20 into lipid rafts and its homo-oligomerization are suggested to be crucial for CD20 activity and regulation. This review outlines recent biochemical studies characterizing the role of CD20 in calcium signalling in B-lymphocytes and evaluates an engagement of lipid rafts in the regulation of CD20-mediated calcium conductivity.

CD20: a B-lymphocyte-specific antigen and its clinical relevance

CD20 belongs to the MS4A (membrane-spanning 4A) gene family, which is expressed in haematopoietic cells, and consists of at least 25 members clustered at human chromosome locus 11q12–13 [1,2]. The MS4A family members [e.g. CD20, FcεRIβ (β-subunit of the high-affinity IgE receptor) and HTm4 (haematopoietic transmembrane-4 protein)] have a predicted tetraspanning membrane topology with an N- and C-terminal cytoplasmic domain (Figure 1). Despite their shared chromosomal localization and conservation in protein domain organization, the function of the members is not well characterized and is likely to be diverse [3,4]. CD20 is the best studied member of this family.

[1]To whom correspondence should be addressed (email eva.x.janas@gsk.com).

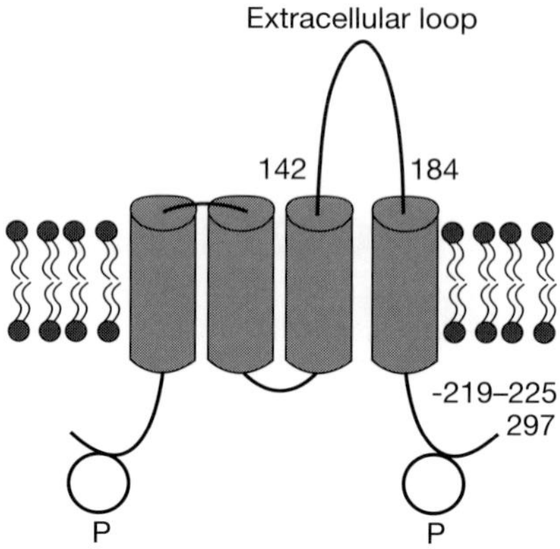

Figure 1 Predicted topology model of CD20. CD20 is a 297-amino-acid phosphoprotein that is predicted to have four membrane-spanning domains with cytoplasmic N- and C-termini. The cytoplasmic sequence (219–225: TCSRPKS in single letter code) required for CD20 raft localization is indicated [60].

CD20 is specifically expressed on the surface of B-cells and cells from most B-cell lymphoproliferative disorders [5]. Different isoforms of CD20 (33, 35 and 37 kDa) result from multiple phosphorylation of serine and threonine residues in the cytoplasmic domains, implying that CD20 activity is highly regulated by phosphorylation [6,7]. At present, the exact stoichiometry of the cell-surface-expressed CD20 complex has not been determined; however, after chemical cross-linking, CD20 has been observed to be separated as dimers or tetramers by non-reducing SDS/PAGE [8]. Binding of a subset of anti-CD20 antibodies to CD20 is reported to increase calcium conductivity across the plasma membrane, to trigger cell-cycle block at the G_1 phase and to inhibit *in vitro* B-cell differentiation and mitogen-induced immunoglobulin secretion [8,9]. Cross-linking of CD20 by these antibodies is also observed to induce activation of protein kinases and to increase expression levels of several surface proteins [10–14]. Rituxan, a chimaeric anti-CD20 antibody (Rituximab, Mabthera, C2B8), is in clinical use for non-Hodgkin's B-cell lymphoma, and has also shown excellent efficacy in inducing clinical improvement and remission in rheumatoid arthritis [15–19]. Binding of Rituxan occurs at the extracellular loop between the predicted third and fourth transmembrane domain of CD20, and does not induce internalization of the CD20 after antibody ligation [20]. The antibody-mediated therapy achieves its effectiveness through B-cell depletion. Several mechanisms of anti-CD20-mediated depletion of B-cells have been postulated, including ADCC (antibody-dependent cell cytotoxicity), activation of the complement system, and CD20-mediated regulation of the cell cycle and apoptosis [21–25]. The induction of apoptosis after anti-CD20-antibody treatment is blocked by chelating either extracellular or intracellular calcium, suggesting that variation in calcium levels contribute to the onset of apoptosis. To evaluate which mechanism contributes mainly to the therapeutic outcome is complicated by the fact that the mode of anti-CD20 antibodies action on tumour cells and cell lines may differ from that on non-malignant peripheral blood cells and lymphoid organ B-cells [26].

Store-operated calcium entry in B-lymphocytes

Intracellular calcium serves as a ubiquitous signal for cellular activation and regulates a broad array of cellular functions ranging from induction of exocytosis, cell proliferation, cell differentiation and cell death [27,28]. In most non-excitable cells, such as lymphocytes, the generation of the receptor-induced increase of intracellular calcium signal involves two phases: ligand binding to a receptor at the plasma membrane triggers a rapid, but transient, mobilization of calcium from intracellular stores such as ER (endoplasmic reticulum) or Golgi apparatus, followed by a slow, but sustained, entry of calcium across the plasma membrane (Figure 2). In B-lymphocytes, the first phase of the calcium response can be initiated by binding of antigen to the BCR (B-cell receptor) resulting in an activation of PLCγ (phospholipase Cγ), which then cleaves PtdIns(4,5)P_2 to diacylglycerol, a membrane-integrated messenger, and Ins(1,4,5)P_3 [29]. Ins(1,4,5)P_3 is released from the plasma membrane and activates ligand-dependent channels [Ins(1,4,5)P_3 receptors] in the membrane of the ER, generating an initial calcium signal. Depletion of intracellular stores mediates a signal that activates SOCCs (store-operated calcium channels) at the plasma membrane. The sustained calcium entry phase is required both to refill intracellular calcium pools and to ensure long-term calcium-dependent processes, e.g. transcriptional control, gene expression, developmental transition, maturation and function of cells. The store-operated calcium current is also termed capacitative calcium entry [30]. Although calcium entry across the plasma membrane has been observed experimentally, the molecular identity of the SOCC has not been determined. The coupling mechanism by which the decrease in luminal calcium concentration is transmitted to SOCCs at the plasma membrane is also poorly understood. Despite intensive efforts to understand the coupling mechanism(s), no defined pathways have been identified [31]. One hypothesis includes a low-molecular-mass diffusible factor (calcium influx factor), which is released from the ER after store depletion and activates SOCCs at the plasma membrane [32]. An alternative model favours a direct physical interaction between proteins localized in the plasma membrane with proteins embedded in the membrane of the ER [33,34]. This interaction can be dynamic and induced by store depletion. A further development of the second model proposes an inducible vesicular transfer of proteins from the ER membrane to the plasma membrane after store depletion. This model has been supported by the observation that communication between SOCCs and store depletion is sensitive to the function of a membrane fusion protein SNAP-25 (soluble *N*-ethylmaleimide-sensitive-factor-attachment protein 25) [35].

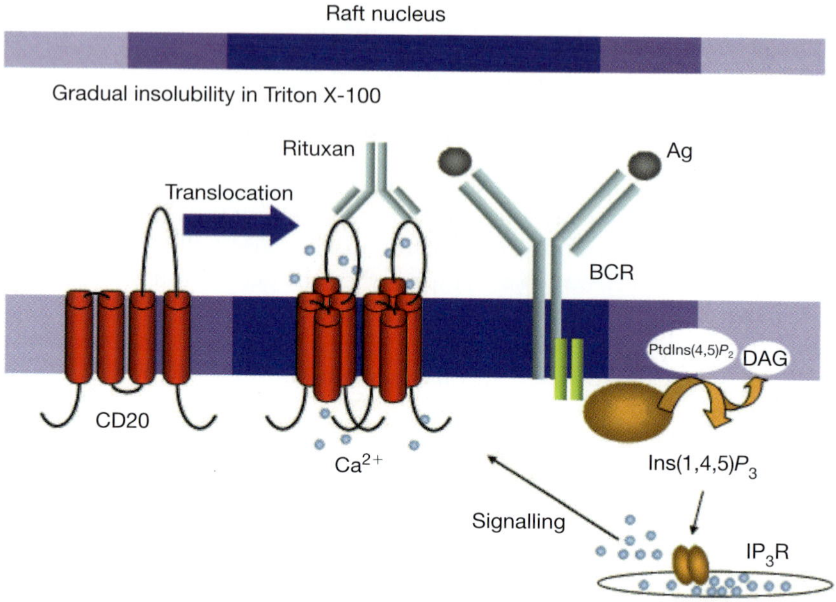

Figure 2 Model for intracellular calcium signalling in B-lymphocytes. Antigen (Ag) presentation to the BCR triggers its recruitment into lipid rafts inducing activation of PLCγ which then cleaves PtdIns(4,5)P_2 to diacylglycerol (DAG) and Ins(1,4,5)P_3. Ins(1,4,5)P_3 binds to the Ins(1,4,5)P_3-dependent receptor (IP_3R) in the ER, generating a rapid increase in cytosolic calcium. Depletion of intracellular stores mediates a signal that activates calcium channels such as CD20 at the plasma membrane which are regulated by the filling state of intracellular stores. Alternatively, direct cross-linking of CD20 by Rituxan has been shown to stimulate calcium influx without preceding depletion of intracellular calcium stores (E. Janas, R. Priest and R. Malhotra, unpublished work). Binding of Rituxan was also observed to induce a high-affinity association of CD20 with Triton-X-100-resistant rafts. Based on the raft structure model of Pizzo and Viola [57], we propose that binding of Rituxan induces the migration of CD20 from the edge of rafts (Brij-58-insoluble, Triton-X-100-sensitive area) to the nucleus raft area (Triton-X-100-insoluble). In the centre of rafts, CD20 acquires an active-channel conformation, possibly supported by the liquid-ordered phase of rafts and/or by close proximity to other regulatory proteins enriched in rafts. The blue colour-coded bar above plasma membrane symbolizes the gradual insolubility of the rafts towards Triton X-100.

Is CD20 operating as a SOCC?

CD20 is proposed to be one of the possible candidate channels mediating capacitative calcium entry in B-lymphocytes. Early electrophysiological studies with cell lines transfected with CD20 cDNA showed that CD20 contributes to transmembrane calcium conductance [8]. The conductive activity of CD20 is proposed to be regulated by a G-protein-dependent mechanism [36]. Recent data demonstrate that CD20, when heterologously expressed in CHO (Chinese-hamster ovary) cells, significantly enhanced calcium entry across the plasma membrane [37]. In general, activation of store-operated channels can be achieved experimentally by reducing the calcium content in the lumen of intracellular stores. Inhibitors of SERCAs (sarcoplasmic/endoplasmic-reticulum Ca^{2+}-ATPases), such as thapsigargin or cyclopiazonic acid, induce passive release of calcium from intracellular stores by blocking calcium re-uptake into the ER. Treatment of B-cells with thapsigargin resulted in a sustained influx of calcium [37]. Thapsigargin did not increase transmembrane calcium conductivity when CD20 expression was down-modulated using siRNA (short interfering RNA), indicating that CD20 is responsible for calcium influx [37]. Furthermore, the sustained calcium influx downstream of the BCR was blocked by use of SKF96365, a known inhibitor of SOCCs. Further evidence supporting CD20 function in calcium entry came from studies with CD20-deficient mice [38]. Murine and human CD20 are well conserved in amino acid sequence (73%). The CD20-deficient mouse did not show an abnormal phenotype in B-cell development and maturation. The most notable observation in these CD20-deficient B-cells was a significant reduction in transmembrane calcium influx following CD19 or BCR ligation. Since receptor-induced calcium response was significantly affected by CD20 deficiency or chelation of extracellular calcium, the authors concluded that CD20 predominantly contributes to conductive calcium responses. The incomplete blockage of calcium transmembrane conductance in CD20-deficient cells is likely to be due to the expression of other MS4A family members which may contribute to transmembrane calcium transport in addition to CD20.

Association of CD20 with lipid rafts

Lipid rafts are specialized microdomains of the plasma membrane, which are highly enriched in sphingolipids intercalated by cholesterol. These microdomains have been proposed to provide a mechanism for lateral compartmentalization of the plasma membrane and to be implicated in the organization of numerous membrane-associated signalling pathways [39,40]. The tight packing of sphingolipids with a high degree of saturated fatty acids impart to these microdomains a liquid-ordered phase which is distinguishable from the rather loosely packed, liquid-disordered membrane [41,42]. Owing to their characteristic lipid composition, lipid rafts resist solubilization in non-ionic detergents (classically Triton X-100) at low temperature, and are isolated as DRMs (detergent-resistant membranes) floating on the top of sucrose gradients [43]. This biochemical approach has been extended with the use of new types of detergents, such as Lubrol WX, or Brij 35 or 98, which preserve the structure of rafts that are solubilized by Triton X-100. This allows the isolation of a subset of rafts with an increased population of proteins embedded compared with classical Triton-X-100-resistant rafts [44].

The densely packed environment of lipid rafts excludes most integral membrane proteins from rafts. Raft-associated proteins can be divided into two groups. First, proteins that are tightly anchored in rafts through covalently bound fatty acyl chains [e.g. GPI (glycosylphosphatidylinositol)-linked proteins and dually acylated Src-family kinases] have been shown to be accommodated permanently in lipid rafts. A second class of proteins [e g. BCR, TCR (T-cell receptor), CD20, CD19, CD40, CD24, FcεRI and PLCγ] associate with lipid rafts in a transient and activation-dependent manner [45–52]. Ligand binding or antibody cross-linking has been shown to trigger lipid raft association of these proteins, accrediting rafts a key role in the initiation and regulation of receptor-mediated signalling. It is attractive to speculate that the ability of certain proteins to associate with lipid rafts in a dynamic manner may effect and modulate proteins' function [44]. For instance, after antigen presentation, BCR is translocated rapidly into rafts; however, its residence inside the micro-environment is only temporary, and the loss of BCR from rafts is complete within 15 min of stimulation [45]. The duration of BCR raft association can be modulated by the presence of co-receptors, such as CD19, whose raft association is also inducible [53,54]. Interestingly, BCR and CD20 were shown to co-localize on the cell surface of stimulated B-cells for a short period of time before they dissociated into different raft populations. CD20 remains surface-exposed, whereas the antigen-decorated BCR is internalized [55]. Binding of anti-CD20 antibodies, such as 2H7 or Rituxan, to CD20 induces a redistribution of CD20 molecules to lipid rafts, which resist Triton X-100 solubilization [47]. However, recent data provide evidence that, in the absence of cross-linking anti-CD20 antibody, CD20 molecules are constitutively associated with a low affinity to lipid rafts. This association is sensitive to solubilization by Triton X-100, but not by Brij 58 (Figure 2) [56]. Thus the binding of anti-CD20 antibodies increases the affinity of CD20 to lipid rafts as observed by a required insolubility of CD20 to Triton X-100. Furthermore, binding of univalent Fab fragments of 2H7 or Rituxan is sufficient to induce migration of CD20 into Triton-X-100-resistant rafts [56]. This observation suggests that cross-linking of two adjacent epitopes by anti-CD20 antibodies is not essential for clustering of CD20 molecules inside lipid rafts, but binding of the monovalent Fab fragment stabilizes a CD20 conformation, which has an intrinsic high affinity to lipid rafts. Pizzo and Viola [57] proposed a lipid raft model, in which a Triton-X-100-resistant raft nucleus is surrounded by lipid layers with gradually increased solubility towards Triton X-100. We assume that, after antibody binding, CD20 molecules preferentially migrate from the edge of a raft (which is resistant to Brij 58 solubilization) to a nucleus raft area (which is resistant to Triton X-100 solubilization), increasing the local concentration of CD20 molecules in a restricted membrane area (Figure 2). Alternatively, Rituxan binding to CD20 could result in the merger of smaller lipid rafts with larger ones. Thus rafts that coalesce may display a platform for new protein–protein interactions among proteins originally localized to different rafts. CD20 was shown to interact with known raft proteins, such as Src-family kinases and adaptor proteins, CD40, and G-coupled-protein subunits, indicating the formation of a raft-dependent functional complex [10,12,36,58,59]. However, the nature of a physiological intra- or extra-cellular stimulus, which increases CD20 raft association, remains to be identified.

Is recruitment of CD20 into lipid rafts crucial for CD20 activity?

Although antibody-induced CD20 lipid raft association is well documented, only a few studies are available investigating the role of lipid rafts in CD20 activity. Intriguing questions remain whether and why CD20 association with lipid rafts might be critical for CD20 function. Raft association might affect function of a protein in two possible ways: first, accumulation of proteins in rafts might facilitate interactions, and, secondly, the liquid-ordered environment of a raft might affect protein function by altering protein conformation. Polyak et al. [60] generated a truncated CD20 mutant in which seven residues in the C-terminal cytoplasmic sequence were removed (Δ219–225). Deletion of this sequence resulted in an inability of CD20 to associate with lipid rafts. This mutant was shown to be inefficient in mediating sustained calcium entry, demonstrating an involvement of lipid rafts in regulating CD20 channel activity [37]. This deletion did not influence oligomerization of CD20.

A valuable tool in investigating raft-associated signalling is the disruption of lipid raft integrity by reduction of cholesterol concentration in the plasma membrane. Cholesterol can be extracted from the plasma membrane by drugs such as methyl-β-cyclodextrin. This cyclic hydrophilic oligomer of glucose has been demonstrated to extract cholesterol from the plasma membrane without its entry into the membrane [61,62]. Therefore the effect of methyl-β-cyclodextrin differs from those of detergents that first incorporate themselves into membranes and then extract membrane components. Li et al. [37] observed that treatment of B-lymphocytes with methyl-β-cyclodextrin disturbs CD20 raft association, and results in a reduction of capacitative calcium entry after stimulation by cross-linking BCR or by depleting intracellular stores by thapsigargin. In line with these data, we monitored calcium influx in B-lymphocytes stimulated by hypercross-linked Rituxan. We observed in the presence of an increasing concentration of methyl-β-cyclodextrin, a dose-dependent reduction of sustained calcium influx, which correlated with a loss of association of CD20 with lipid rafts (E. Janas, R. Priest and R. Malhotra, unpublished work). Replenishing the pool of plasma membrane cholesterol with a methyl-β-cyclodextrin/cholesterol mixture re-established both calcium influx and association of CD20 with lipid rafts. These data provide evidence that both raft integrity and CD20 raft recruitment are required for CD20-mediated calcium conductivity.

A functional importance of raft association has also been shown for regulation of other channels. For instance, treatment of platelet cells with methyl-β-cyclodextrin inhibited calcium conductance in response to a number of agonists [63]. In these cells, methyl-β-cyclodextrin diminished calcium conductance mediated by TRPC1 (transient receptor potential canonical 1), a putative SOCC, implying a correlation between lipid raft recruitment of TRPC1 and its channel activity [64]. The integrity of lipid rafts is also required for Kv1.3 channel activity, since a change in the composition of rafts by cholesterol extraction was shown to inhibit channel function [65,66].

At present, it is not known how the specific environment of lipid rafts might support channel activity. It is conceivable that the integrity of rafts is required for correct assembly of the channel forming units or for regulation of CD20 activity by raft resident proteins. The targeting mechanism for proteins into lipid rafts is not completely understood. Phosphorylation of CD20 was shown not to be critical for CD20 raft association [56]. BCR raft association is also independent of phosphorylation, suggesting that translocation may rather be the result of alteration in conformation and/or oligomerization of BCR. The oligomer could assume a conformation in the membrane that, unlike the monomer, prefers the environment of liquid-ordered phase of lipid rafts. The formation of CD20 oligomeric complexes in the plasma membrane is regarded as a prerequisite for channel activity, and lipid rafts could provide a convenient environment to stabilize these complexes [8,47]. These conclusions are supported by studies with a unique anti-CD20 antibody (FMC7), which exclusively detects multimeric CD20 structures, which are strongly dependent on cholesterol presence in the plasma membrane. Disturbing raft integrity by methyl-β-cyclodextrin completely abolishes binding of FMC7 to CD20 [67]. Thus the oligomerization of CD20 seems to be a cholesterol-dependent process.

Cragg et al. [68] investigated homo-oligomerization of CD20 at the cell-surface by FRET (fluorescence resonance energy transfer). Binding of Rituxan to CD20 resulted in a high FRET signal, suggesting an antibody-induced oligomerization of CD20 molecules. However, no energy transfer was observed between CD20 molecules after treating the cells with another class of anti-CD20 antibodies, such as B1, demonstrating that binding of B1 did not induce a close proximity of CD20 molecules at the cell surface. In line with this observation, the authors showed by detergent-mediated raft isolation that B1 was inefficient in inducing CD20 raft localization [68]. Electrophysiological studies showed a very low calcium conductance even after 24 h incubation with the B1 antibody [8]. Both Rituxan and , although documented to mediate their therapeutic effect by B-cell depletion, obviously operate through a different mechanism. Treatment by Rituxan and B1 have different impacts on oligomerization of CD20 molecules, CD20 lipid raft association, induction of calcium influx, effects on apoptosis and complement activation [8,24,37]. We conclude that both antibodies stabilize a different conformation of CD20, and that binding of Rituxan to CD20 results in lipid raft association and subsequent calcium influx.

Critical evaluation of the functional role of lipid rafts in CD20 channel activity

At present, it is attractive to speculate that the specific environment of lipid rafts might support CD20 channel function. However, there are conflicting data on the functional role of lipid rafts for protein activity. For instance, Aman and Ravichandran [69], as well as Guo et al. [70], reported that BCR-mediated calcium flux is severely diminished after cholesterol extraction. In line with these observations, disturbance of raft formation by deletion of Raftlin, a raft protein, which was recently demonstrated to play a pivotal role in maintenance of lipid rafts in B-lymphocytes, resulted in a reduced BCR-mediated calcium mobilization, implying a role for rafts in BCR activation [71]. However, Petrie et al. [72] observed an increase in BCR-induced intracellular calcium mobilization after lipid raft disruption by methyl-β-cyclodextrin, but described a reduced capacitative calcium influx. These observations allow two conclusions: first, lipid rafts are negatively regulating BCR-mediated calcium influx from intracellular stores, and, secondly (and more importantly for this review), the sustained calcium influx across the plasma membrane (partially dependent on CD20) is dependent on lipid raft integrity. Pizzo and Viola [57] re-evaluated studies with methyl-β-cyclodextrin which were used to demonstrate functional role of lipid rafts in T-cell-mediated calcium signalling. The authors observed that treatment with methyl-β-cyclodextrin not only disturbs lipid raft integrity (an effect which most authors tend to obtain using this drug), but also induces depolarization of the plasma membrane [57]. As calcium influx is sensitive to the membrane potential, fluctuations in electrical gradient across the plasma membrane might have considerable effects on calcium influx. In summary, studies using methyl-β-cyclodextrin must be tightly controlled, since depletion of cholesterol by methyl-β-cyclodextrin might cause an intervention to cell integrity. The strongest evidence for the involvement of rafts in protein function is provided when several approaches point to the same conclusion. Mutagenesis of proteins, which resulted in the loss of raft association without altering protein function, is a valuable tool in investigating the functional role of rafts for the protein in focus [60].

Conclusions

The plasma membrane appears to be structurally compartmentalized to allow separation or integration of signalling events. A variety of signalling molecules can associate either constitutively or inducibly to lipid rafts. In this review, we have outlined the functional role of lipid rafts in calcium signalling in B-lymphocytes and we have focused on studies of the B-cell-specific antigen CD20. CD20 is reported to exist in a dynamic association with Triton-X-100-resistant lipid rafts, and the antibody-induced relocalization of CD20 to these microdomains is suggested to be crucial for capacitative calcium entry in B-lymphocytes. Capacitative calcium entry was reduced either by cholesterol depletion or by deletion of a seven-residue cytoplasmic CD20 sequence that controls its localization to lipid rafts. Raft association of CD20 appears to provide a micro-environment for CD20 homo-oligomerization, as well as for protein–protein interaction with regulatory molecules that modulate CD20 activity.

We thank Jonathan Wilde for critical reading of the manuscript, and the Marie Curie Foundation for financial support.

References

1. Liang, Y. and Tedder, T.F. (2001) Genomics **72**, 119–127
2. Ishibashi, K., Suzuki, M., Sasaki, S. and Imai, M. (2001) Gene **264**, 87–93

3. Kinet, J.P. (1999) Annu. Rev. Immunol. **17**, 931–972
4. Donato, J.L., Ko, J., Kutok, J.L., Cheng, T., Shirakawa, T., Mao, X.Q., Beach, D., Scadden, D.T., Sayegh, M.H. and Adra, C.N. (2002) J. Clin. Invest. **109**, 51–58
5. Tedder, T.F., Streuli, M., Schlossman, S.F. and Saito, H. (1988) Proc. Natl. Acad. Sci. U.S.A. **85**, 208–212
6. Tedder, T.F. and Schlossman, S.F. (1988) J. Biol. Chem. **263**, 10009–10015
7. Valentine, M.A., Meier, K.E., Rossie, S. and Clark, E.A. (1989) J. Biol. Chem. **264**, 11282–11287
8. Bubien, J.K., Zhou, L.J., Bell, P.D., Frizzell, R.A. and Tedder, T.F. (1993) J. Cell Biol. **121**, 1121–1132
9. Gazitt, Y., Gangavalli, A. and Thurman, G.B. (1989) Mol. Biother. **1**, 297–304
10. Kansas, G.S. and Tedder, T.F. (1991) J. Immunol. **147**, 4094–4102
11. Deans, J.P., Schieven, G.L., Shu, G.L., Valentine, M.A., Gilliland, L.A., Aruffo, A., Clark, E.A. and Ledbetter, J.A. (1993) J. Immunol. **151**, 4494–4504
12. Deans, J.P., Kalt, L., Ledbetter, J.A., Schieven, G.L., Bolen, J.B. and Johnson, P. (1995) J. Biol. Chem. **270**, 22632–22638
13. Clark, E.A. and Shu, G. (1987) J. Immunol. **138**, 720–725
14. Semac, I., Palomba, C., Kulangara, K., Klages, N., Echten-Deckert, G., Borisch, B. and Hoessli, D.C. (2003) Cancer Res. **63**, 534–540
15. Sacchi, S., Federico, M., Dastoli, G., Fiorani, C., Vinci, G., Clo, V. and Casolari, B. (2001) Crit. Rev. Oncol. Hematol. **37**, 13–25
16. Goronzy, J.J. and Weyand, C.M. (2003) Arthritis Res. Ther. **5**, 131–135
17. Edwards, J.C., Leandro, M.J. and Cambridge, G. (2002) Biochem. Soc. Trans. **30**, 824–828
18. Shaw, T., Quan, J. and Totoritis, M.C. (2003) Ann. Rheum. Dis. **62**, Suppl. 2, II55–II59
19. Edwards, J.C., Szczepanski, L., Szechinski, J., Filipowicz-Sosnowska, A., Emery, P., Close, D.R., Stevens, R.M. and Shaw, T. (2004) N. Engl. J. Med. **350**, 2572–2581
20. Polyak, M.J. and Deans, J.P. (2002) Blood **99**, 3256–3262
21. Manches, O., Lui, G., Chaperot, L., Gressin, R., Molens, J.P., Jacob, M.C., Sotto, J.J., Leroux, D., Bensa, J.C. and Plumas, J. (2003) Blood **101**, 949–954
22. Cragg, M.S. and Glennie, M.J. (2004) Blood **103**, 2738–2743
23. Byrd, J.C., Kitada, S., Flinn, I.W., Aron, J.L., Pearson, M., Lucas, D. and Reed, J.C. (2002) Blood **99**, 1038–1043
24. Deans, J.P., Li, H. and Polyak, M.J. (2002) Immunology **107**, 176–182
25. Szodoray, P., Alex, P., Dandapani, V., Nakken, B., Pesina, J., Kim, X., Wallis, G.L., Wilson, P.C., Jonsson, R. and Centola, M. (2004) Scand. J. Immunol. **60**, 209–218
26. Vugmeyster, Y. and Howell, K. (2004) Int. Immunopharmacol. **4**, 1117–1124
27. Carafoli, E. (2004) Trends Biochem. Sci. **29**, 371–379
28. Pike, L.J. (2003) J. Lipid Res. **44**, 655–667
29. Winslow, M.M., Neilson, J.R. and Crabtree, G.R. (2003) Curr. Opin. Immunol. **15**, 299–307
30. Putney, Jr, J.W. and McKay, R.R. (1999) BioEssays **21**, 38–46
31. Venkatachalam, K., van Rossum, D.B., Patterson, R.L., Ma, H.T. and Gill, D.L. (2002) Nat. Cell Biol. **4**, E263–E272
32. Randriamampita, C. and Tsien, R.Y. (1993) Nature (London) **364**, 809–814
33. Rosado, J.A., Brownlow, S.L. and Sage, S.O. (2002) J. Biol. Chem. **277**, 42157–42163
34. Boulay, G., Brown, D.M., Qin, N., Jiang, M., Dietrich, A., Zhu, M.X., Chen, Z., Birnbaumer, M., Mikoshiba, K. and Birnbaumer, L. (1999) Proc. Natl. Acad. Sci. U.S.A. **96**, 14955–14960
35. Yao, Y., Ferrer-Montiel, A.V., Montal, M. and Tsien, R.Y. (1999) Cell **98**, 475–485
36. Kanzaki, M., Lindorfer, M.A., Garrison, J.C. and Kojima, I. (1997) J. Biol. Chem. **272**, 14733–14739
37. Li, H., Ayer, L.M., Lytton, J. and Deans, J.P. (2003) J. Biol. Chem. **278**, 42427–42434

38. Uchida, J., Lee, Y., Hasegawa, M., Liang, Y., Bradney, A., Oliver, J.A., Bowen, K., Steeber, D.A., Haas, K.M., Poe, J.C. and Tedder, T.F. (2004) Int. Immunol. **16**, 119–129
39. Simons, K. and Toomre, D. (2000) Nat. Rev. Mol. Cell Biol. **1**, 31–39
40. Simons, K. and Ikonen, E. (1997) Nature (London) **387**, 569–572
41. Brown, D.A. and London, E. (1998) Annu. Rev. Cell Dev. Biol. **14**, 111–136
42. Brown, D.A. and London, E. (1998) J. Membr. Biol. **164**, 103–114
43. London, E. and Brown, D.A. (2000) Biochim. Biophys. Acta **1508**, 182–195
44. Lucero, H.A. and Robbins, P.W. (2004) Arch. Biochem. Biophys. **426**, 208–224
45. Cheng, P.C., Dykstra, M.L., Mitchell, R.N. and Pierce, S.K. (1999) J. Exp. Med. **190**, 1549–1560
46. Hostager, B.S., Catlett, I.M. and Bishop, G.A. (2000) J. Biol. Chem. **275**, 15392–15398
47. Deans, J.P., Robbins, S.M., Polyak, M.J. and Savage, J.A. (1998) J. Biol. Chem. **273**, 344–348
48. Suzuki, T., Kiyokawa, N., Taguchi, T., Sekino, T., Katagiri, Y.U. and Fujimoto, J. (2001) J. Immunol. **166**, 5567–5577
49. Horejsi, V. (2003) Immunol. Rev. **191**, 148–164
50. Cherukuri, A., Cheng, P.C., Sohn, H.W. and Pierce, S.K. (2001) Immunity **14**, 169–179
51. Young, R.M., Holowka, D. and Baird, B. (2003) J. Biol. Chem. **278**, 20746–20752
52. Hur, E.M., Park, Y.S., Lee, B.D., Jang, I.H., Kim, H.S., Kim, T.D., Suh, P.G., Ryu, S.H. and Kim, K.T. (2004) J. Biol. Chem. **279**, 5852–5860
53. Tedder, T.F., Haas, K.M. and Poe, J.C. (2002) Biochem. Soc. Trans. **30**, 807–811
54. Cherukuri, A., Dykstra, M. and Pierce, S.K. (2001) Immunity **14**, 657–660
55. Petrie, R.J. and Deans, J.P. (2002) J. Immunol. **169**, 2886–2891
56. Li, H., Ayer, L.M., Polyak, M.J., Mutch, C.M., Petrie, R.J., Gauthier, L., Shariat, N., Hendzel, M.J., Shaw, A.R., Patel, K.D. and Deans, J.P. (2004) J. Biol. Chem. **279**, 19893–19901
57. Pizzo, P. and Viola, A. (2003) Curr. Opin. Immunol. **15**, 255–260
58. Popoff, I.J., Savage, J.A., Blake, J., Johnson, P. and Deans, J.P. (1998) Mol. Immunol. **35**, 207–214
59. Anolik, J., Looney, R.J., Bottaro, A., Sanz, I. and Young, F. (2003) Eur. J. Immunol. **33**, 2398–2409
60. Polyak, M.J., Tailor, S.H. and Deans, J.P. (1998) J. Immunol. **161**, 3242–3248
61. Ohtani, Y., Irie, T., Uekama, K., Fukunaga, K. and Pitha, J. (1989) Eur. J. Biochem. **186**, 17–22
62. Pitha, J., Irie, T., Sklar, P.B. and Nye, J.S. (1988) Life Sci. **43**, 493–502
63. Wonerow, P., Obergfell, A., Wilde, J.I., Bobe, R., Asazuma, N., Brdicka, T., Leo, A., Schraven, B., Horejsi, V., Shattil, S.J. and Watson, S.P. (2002) Biochem. J. **364**, 755–765
64. Brownlow, S.L., Harper, A.G., Harper, M.T. and Sage, S.O. (2004) Cell Calcium **35**, 107–113
65. Szabo, I., Adams, C. and Gulbins, E. (2004) Pflugers Arch. **448**, 304–312
66. Bock, J., Szabo, I., Gamper, N., Adams, C. and Gulbins, E. (2003) Biochem. Biophys. Res. Commun. **305**, 890–897
67. Polyak, M.J., Ayer, L.M., Szczepek, A.J. and Deans, J.P. (2003) Leukemia **17**, 1384–1389
68. Cragg, M.S., Morgan, S.M., Chan, H.T., Morgan, B.P., Filatov, A.V., Johnson, P.W., French, R.R. and Glennie, M.J. (2003) Blood **101**, 1045–1052
69. Aman, M.J. and Ravichandran, K.S. (2000) Curr. Biol. **10**, 393–396
70. Guo, B., Kato, R.M., Garcia-Lloret, M., Wahl, M.I. and Rawlings, D.J. (2000) Immunity **13**, 243–253
71. Saeki, K., Miura, Y., Aki, D., Kurosaki, T. and Yoshimura, A. (2003) EMBO J. **22**, 3015–3026
72. Petrie, R.J., Schnetkamp, P.P., Patel, K.D., Awasthi-Kalia, M. and Deans, J.P. (2000) J. Immunol. **165**, 1220–1227

Biochem. Soc. Symp. **72**, 177–188
(Printed in Great Britain)

17

Biophysics (and sociology) of ceramides

Félix M. Goñi[1], F.-Xabier Contreras, L.-Ruth Montes, Jesús Sot and Alicia Alonso

Unidad de Biofísica (Centro Mixto CSIC-UPV/EHU), and Departamento de Bioquímica, Universidad del País Vasco, P.O. Box 644, 48080 Bilbao, Spain

Abstract

In the past decade, the long-neglected ceramides (N-acylsphingosines) have become one of the most attractive lipid molecules in molecular cell biology, because of their involvement in essential structures (stratum corneum) and processes (cell signalling). Most natural ceramides have a long (16–24 C atoms) N-acyl chain, but short N-acyl chain ceramides (two to six C atoms) also exist in Nature, apart from being extensively used in experimentation, because they can be dispersed easily in water. Long-chain ceramides are among the most hydrophobic molecules in Nature, they are totally insoluble in water and they hardly mix with phospholipids in membranes, giving rise to ceramide-enriched domains. *In situ* enzymic generation, or external addition, of long-chain ceramides in membranes has at least three important effects: (i) the lipid monolayer tendency to adopt a negative curvature, e.g. through a transition to an inverted hexagonal structure, is increased, (ii) bilayer permeability to aqueous solutes is notoriously enhanced, and (iii) transbilayer (flip-flop) lipid motion is promoted. Short-chain ceramides mix much better with phospholipids, promote a positive curvature in lipid monolayers, and their capacities to increase bilayer permeability or transbilayer motion are very low or non-existent.

> *"A society can be described only as a set of particular organisms, and even then it is difficult to extrapolate the joint activity of this ensemble from the instant of specification, that is, to predict social behaviour".*
> E.O. Wilson, *Sociobiology* (1975)

[1]To whom correspondence should be addressed (email gbpgourf@lg.ehu.es).

Introduction

Ceramides (N-acylsphingosines) consist of a fatty acid bound to the amino group of C-2 of sphingosine or another related long-chain base. Free ceramides are only found in large amounts in the skin stratum corneum, where their hydrophobic properties may be put to use in preventing body water evaporation by uncontrolled perspiration [1]. In addition, ceramides exist in minute proportions in many cell membranes, where they are now considered to play an important role in cell signalling [2]. The outcome of ceramide signalling is cell-type-dependent. However, the most commonly reported result of ceramide signalling is induction of apoptosis. Furthermore, it has become clear that tumour cells have developed strategies to reduce ceramide elevations and attenuate the pro-apoptotic effects of ceramide.

Ceramide was originally conceived of as a classic second messenger, directly and stoichiometrically activating a set of target proteins [3,4]. Over the past decade, a number of direct targets for ceramide have been described as mediators of ceramide signalling. Ceramide reportedly interacts directly with, among other proteins, kinase suppressor of Ras (or ceramide-activated protein kinase) [5], ceramide-activated protein phosphatase [6], certain isoforms of protein kinase C [7] or c-Raf-1 [8]. One problem with these target proteins is that they are usually soluble and intracellular, while ceramide, as discussed below, is probably generated in the outer leaflet of the plasma membrane, thus it is not clear at present how these ceramide molecules may reach their cytoplasmic targets.

In addition to its action through direct binding to target proteins, it is becoming increasingly clear that ceramide exerts part of its physiological effects through changes in the physical properties of the membrane. This idea was proposed in the review by Kolesnick et al. [9]. More recently, experimental evidence has been published suggesting that ceramide generated in membrane rafts induces their coalescence into larger clusters or platforms, and that these macrodomains serve as a site for oligomerization of cell-surface receptors, such as Fas and CD40 [10,11], and the internalization of bacteria, such as *Pseudomonas aeruginosa* and *Neisseria gonorrhoeae* [12]. In this paradigm, acid sphingomyelinase is initially translocated on to the outer leaflet of the plasma membrane, where it catalyses the hydrolysis of raft-associated sphingomyelin.

Ceramides have been found in Nature with fatty acids containing from two to 28 C atoms, being saturated, unsaturated or hydroxylated. However, the most commonly found, by far, are fatty acids of 16–24 C atoms. These 'long-chain' (i.e. long N-acyl-chain) ceramides are among the least polar, most hydrophobic lipids in membranes. Their solubility in water is negligible. Thus free long-chain ceramides cannot exist in solution in biological fluids or in the cytosol, they belong to the category of 'insoluble, non-swelling amphiphiles' [13], implying that they cannot even give rise to micelles or other aggregates in aqueous suspensions as do, for example, common phospholipids, sphingolipids or surfactants. Short-chain ceramides (e.g. N-acetylceramide, or Cer2) appear to 'swell' in water, giving rise to homogeneous dispersions that are treated, in practice, as solutions. Since most naturally occurring ceramides cannot be dispersed in water, because of their hydrophobicity, short-chain ceramides, and particularly Cer2, have been used extensively in cell physiology studies when ceramide effects had to be elicited, both in intact cell and in cell-free systems. The advantages and disadvantages of the use of short-chain ceramides in cell studies have been discussed in the review by Kolesnick et al. [9]. In summary, these authors consider the short-chain analogues as useful experimental tools, although they introduce the caveat that some physical effects of the short- and long-chain molecules may be significantly different.

The fact is, however, that the physical properties of the short-chain ceramides have received little attention up to now, apart from the studies by Simon and co-workers [14,15], and that comparative studies of short- and long-chain ceramides are virtually non-existent. In this chapter, some data will be shown on the behaviour of long- and short-chain ceramides in water and in lipid bilayer environments, and some fundamental differences in their interaction with water and phospholipid molecules (i.e. their 'molecular sociology') will be discussed.

Ceramides in aqueous media

The properties of ceramides in aqueous media were tested in a Langmuir balance. Ceramides (in DMSO solution) were either deposited on top of the interface or injected into the subphase, and the change in surface pressure (π) was recorded. DMSO alone did not cause any significant change in π. Representative curves of ceramide effects are shown in Figure 1. For Cer16 (Figure 1A), addition on top of the buffer surface causes an immediate increase in π that remains constant afterwards. However, injection of Cer16 into the subphase does not cause any change in π. A different situation is found for Cer2 (Figure 1B). Whatever the mode of dispensation, this ceramide partitions between the subphase and the interface, and the same equilibrium value of π (approx. 6 $mN \cdot m^{-1}$ for a 1 μM bulk concentration of Cer2) is measured. When different concentrations of Cer2 or Cer6 are injected into the subphase, a dose-dependent increase in π is observed (results not shown).

The results demonstrate that ceramides containing an N-fatty acyl residue of at least C_{16} ('long-chain ceramides') should be classified as insoluble non-swelling amphiphiles. This group of substances, of which triacylglycerols are the most common example, can orient themselves at the air/water interface, but do not partition into it, i.e. water surface pressure is unchanged when they are introduced in the bulk of the solvent. They do not undergo any kind of water interactions leading to their dispersibility in water either. As members of this group of non-swelling amphiphiles, long-chain ceramides are among the most hydrophobic molecules in Nature. This explains, for example, why Cer16 is virtually unable to exchange between membranes, at least without the aid of an exchange protein [15]. In contrast, our results with two 'short-chain ceramides' Cer2 and Cer6, in agreement with some previous observations, reveal a very different behaviour, because they can be easily dispersed in water, giving rise to optically clear or slightly opalescent dispersions. Thus short-chain ceramides should be classified as

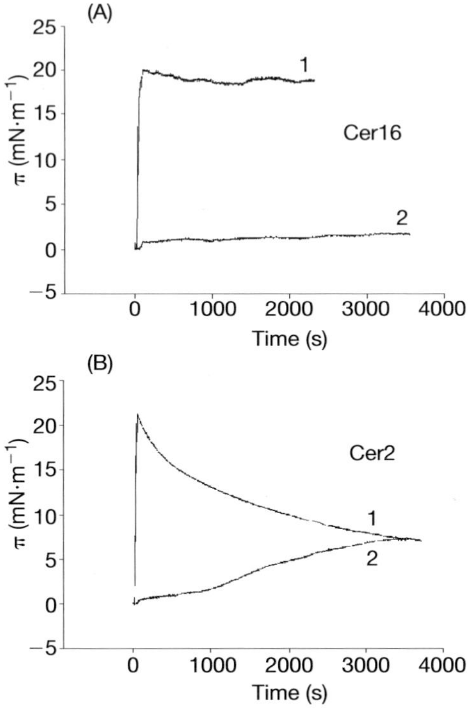

Figure 1 Time course of ceramide-induced changes in surface pressure (π) at an air/water interface. Subphase composition was 20 mM Pipes, 150 mM NaCl and 1 mM EDTA, pH 7.4. Ceramides in DMSO were either deposited on to the water surface (curves 1) or injected into the subphase (curves 2). (**A**) Cer16. (**B**) Cer2. The amounts of added ceramide correspond to bulk concentrations of 2 μM (Cer16, curve 1), 10 μM (Cer16, curve 2) and 1 μM (Cer2, both curves).

'soluble amphiphiles', together with other biomolecules, such as lysophospholipids or fatty acylcarnitines, and with the common detergents.

Ceramide solubilization of phospholipids in monolayers

Soluble amphiphiles, such as Cer2 or Cer6, should be considered as detergents, according to Helenius and Simons [16]. Detergents are able to insert into, and interact with, phospholipid monolayers/bilayers, giving rise to detergent-phospholipid mixed micelles. In order to test the ability of short- and long-chain ceramides to solubilize phospholipids, a fluorescent derivative of PE (phosphatidylethanolamine) namely rhodamine–PE (Rh–PE), was spread as a monolayer at the air/water interface. Then ceramides (in DMSO solution) were injected into the subphase at 10 μM bulk concentration. After equilibrium, aliquots of the subphase were withdrawn, and Rh fluorescence was assayed. As shown in Table 1, short-chain ceramides, and Cer2 in particular,

Table 1 Solubilization of Rh–PE monolayers by ceramides. Ceramides in 50 μl of DMSO were injected into the subphase of monolayers of fluorescent Rh–PE. After equilibration, aliquots of the subphase were removed and Rh fluorescence was assayed. The final ceramide concentration was 10 μM. Results are average values±S.E.M. (n=3).

Ceramide added	Rh in buffer (fluorescence intensity, arbitrary units)
None (50 μl of DMSO)	15±2.5
Cer16	22±2.0
Cer6	47±2.0
Cer2	79±0.75

brought about the solubilization of Rh–PE, while Cer16 was inactive in this respect. Note that Rh–PE by itself does not, to any significant extent, form micelles in water.

Previous studies by Simon and Gear [14] described properties of short-chain ceramides that were not shared by their long-chain homologues, in particular the ability of the former to perturb cell membranes. These observations, together with the behaviour of ceramides in aqueous systems as seen in Figure 1, prompted us to examine these molecules as putative detergents. The results in Table 1 confirm the view that short-chain ceramides do act as soluble amphiphiles, i.e. detergents, towards phospholipid monolayers, thus their effect on cell membranes is likely to be very different from that of Cer16 or longer-chain members of the family.

Ceramides and the gel–fluid phase-transition of phospholipids

Synthetic membranes consisting of aqueous dispersions of DEPE (dielaidoyl PE), in common with many other synthetic phospholipids, exhibit a thermotropic transition from a gel phase to a fluid (or liquid-crystalline) phase. In the case of DEPE, this transition takes place at a T_m (melting temperature) of approx. 37.5°C. The gel–fluid transition is conveniently monitored by DSC (differential scanning calorimetry), and is very sensitive to the presence of other lipids in the bilayer. In fact, modifications of the thermodynamic parameters of the transition in the presence of foreign lipids may provide information on the interaction between host and foreign lipids.

Thermograms corresponding to the gel–fluid transition of pure DEPE and DEPE/ceramide mixtures are shown in Figure 2. The corresponding thermodynamic parameters are given in Table 2. Cer16 has the effect of spreading the phase transition over a wide range of temperatures, approx. 12°C at 20 molar percent, while increasing the $\Delta T_{1/2}$ (endotherm width at mid-height) by approx. 4°C. The latter value is difficult to establish with accuracy, because the endotherms are clearly asymmetric, and reveal the existence of several components. The thermograms are thus indicating an immiscibility of Cer16 with DEPE, and the coexistence of mixtures of somewhat different compositions. These thermograms are similar to those described, for mixtures of synthetic

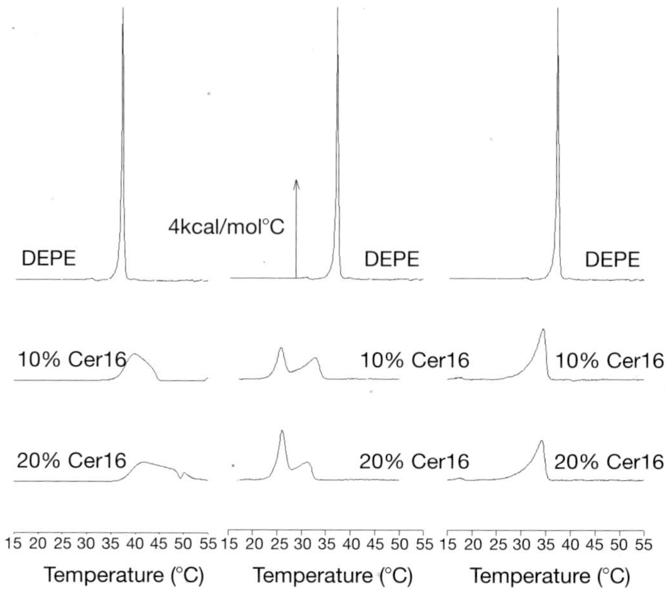

Figure 2 DSC thermograms of the gel–fluid transition of pure DEPE and DEPE/ceramide mixtures in excess water. The molar proportions of ceramide are indicated at the side of each curve. The endotherms shown correspond to the third heating scan of each experiment.

phospholipids and natural ceramides, by Veiga et al. [17] and by Carrer and Maggio [18]. These authors noted that some degree of immiscibility was already apparent at ceramide concentrations in the bilayer below 5 molar percent [17,18]. The new endotherm components appearing at the high-temperature side of the calorimetric peak as more ceramide was added were interpreted as resulting from ceramide-enriched domains (note that pure hydrated ceramide has an order–disorder transition well above the T_m of DEPE).

Again Cer6 and Cer2 behaved differently from Cer16. The most obvious difference in Figure 2 is that both Cer6 and Cer2 shifted the DEPE transition temperature downwards, i.e. they made the DEPE bilayer more fluid, in contrast with the effect of Cer16. This is particularly surprising, because both Cer6 and Cer2 have T_ms above the T_m of DEPE (although lower than the T_m of Cer16). The effect is more notable with Cer6, which gives rise to a low-temperature separate endotherm, than with Cer2, which induces only an asymmetry, at the low-temperature side of the DEPE endotherm. Thus, in contrast with Cer16, both Cer2 and Cer6 mix and interact with DEPE, giving rise to domains whose co-operative transition temperature occurs below that of the pure phospholipid. Of the two effects that are not seen with Cer16, formation of low-temperature melting domains and mixing with DEPE, low-temperature domain formation appears to predominate with Cer6, while mixing is better with Cer2. Thus the capacity of ceramide to mix with DEPE increases in the order

Table 2 Thermodynamic parameters of the gel–fluid transition of DEPE, either pure or in mixtures with ceramides. Data taken from thermograms such as those shown in Figure 2. Results are average values±S.E.M. (n=3). 1 kcal=4.184 kJ.

Mixture (molar ratios) in excess water	T_m (°C)	ΔH (kcal/mol)	$\Delta T^{1/2}$ (°C)
DEPE (pure)	37.6±0.00	6.5±0.38	0.43±0.039
DEPE/Cer16 (9:1)	39.8±0.02	5.8±0.03	5.67±0.136
DEPE/Cer16 (8:2)	41.4±0.24	7.7±0.10	8.68±0.957
DEPE/Cer6 (9:1)	34.9±0.26	7.5±0.49	2.93±0.123
DEPE/Cer6 (8:2)	26.0±0.09	6.7±0.36	3.90±0.167
DEPE/Cer2 (9:1)	36.2±0.10	6.0±0.34	1.44±0.009
DEPE/Cer2 (8:2)	34.7±0.02	6.0±0.30	2.40±0.073

Cer16<Cer6<Cer2. Carrer et al. [19] have studied mixtures of Cer8 with DMPC [dimyristoyl PC (phosphatidylcholine)] or DPPC (dipalmitoyl PC), and have found a behaviour that is qualitatively more similar to that of Cer16 than to that of Cer6. Thus the limit between 'short' and 'long' N-acyl chains, for the purpose of ceramide behaviour, appears to be between six and eight C atoms.

The DSC studies on the effect of ceramides on the gel–fluid transition may be summarized as follows: (i) long-chain ceramides do not mix well with phospholipids, and tend to make the bilayer less fluid, (ii) short-chain ceramides mix better with phospholipids, and have the effect of increasing bilayer fluidity, and (iii) the limit between short- and long-chain ceramides is at six to eight C atoms of the N-acyl chain.

Ceramide-induced solute efflux through lipid membranes

One striking property of ceramides that may be linked to their physiological effect is their capacity to restructure the permeability barrier of model and cell membranes, thus giving rise to vesicle or cell efflux. Ceramide-induced release of aqueous contents from liposomes and resealed erythrocyte ghosts was first observed by Ruiz-Argüello et al. [20], who induced *in situ* generation of ceramide by treating the sphingomyelin-containing model or cell membranes with bacterial sphingomyelinase. Ceramide-induced membrane efflux may be important physiologically, e.g. in generating local ion fluxes, or even in the release of large molecules, such as cytochrome *c*, whose efflux from mitochondria is crucial for the activation of apoptosis.

In more recent years, a number of studies have dealt with the issue of membrane permeabilization by ceramides. Simon and Gear [14] found that Cer2 caused release of [^{3}H]adenine from platelets at a ceramide/lipid ratio of 0.2. Ghafourifar et al. [21] showed that Cer2 and Cer6 induced cytochrome *c* release from isolated mitochondria. Di Paola et al. [22] found that Cer2, but not Cer16, perturbed the permeability barrier of the inner mitochondrial membrane, and dissipated the electrochemical H^+ potential. Also, Siskind and Colombini [23] used electrophysiological methods to demonstrate channel formation by short-

and long-chain ceramides in planar lipid bilayers at total ceramide/lipid ratios of 0.05. Siskind and co-workers have proposed that ceramides form large channels, through which efflux takes place [23–25].

With the aim of clarifying certain aspects of ceramide-induced efflux, Montes et al. [26] prepared LUVs (large unilamellar vesicles) consisting of egg sphingomyelin/egg phosphatidylethanolamine/cholesterol (2:1:1, by vol.), and loaded them with water-soluble fluorescent molecules. Ceramides were either generated *in situ* by sphingomyelinase, or added to the pre-formed vesicles. These authors showed that the presence of long-chain (egg) ceramides could indeed allow the efflux of large molecules (approx. 20 kDa) through membranes, and that the same proportion of long-chain ceramides generated by sphingomyelinase induced faster and more extensive efflux than when added in organic solution to the pre-formed vesicles.

A comparison of the efflux of small molecules [ANTS (8-aminonaphthalene-1,3,6-trisulphonic acid)/DPX (p-xylenebispyridinium bromide)] induced by long- and short-chain ceramides from large unilamellar vesicles can be seen in Figure 3. Egg ceramide, containing essentially Cer16, induced release of entrapped molecules, while Cer6 was less active in this respect, and Cer2 was virtually inactive. Ceramides were added in ethanol, the solvent effect is shown in the control curve (Figure 3). The final molar fraction of ceramide with respect to total lipid was 0.1. It should be noted that because of the high lipid/water partition coefficient of Cer2 [14], at least 85% of the total amount added is expected to be incorporated into the bilayer. These results appear to contradict the observations by other authors [14,22] that Cer2 was particularly effective in causing membrane efflux, but, as will be discussed below, the conditions used by those authors probably led to partial membrane solubilization, whereas the measurements shown in Figure 3 occur in the absence of membrane lysis.

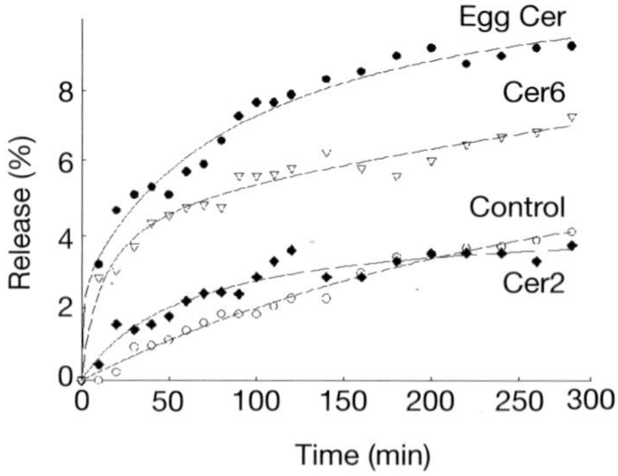

Figure 3 Time course of ceramide-induced efflux of ANTS/DPX. Ceramide dissolved in dodecane/ethanol (2:98, by vol.) was added to pre-formed vesicles composed of sphingomyelin/PE/cholesterol (2:1:1, by vol.) containing entrapped ANTS/DPX. Total lipid concentration was 0.3 mM in 1 ml. Ceramide was 10 molar percent of total lipid. Additions: (○) 4 μl of dodecane/ethanol (control), (●) Cer16, (▽) Cer6, (◆) Cer2. Additional experimental details are given in [26].

Ceramide-induced transbilayer (flip-flop) lipid motion in membranes

Although ceramide was known to induce structural rearrangements in membrane bilayers [9], including the formation of ceramide-rich and -poor domains and the efflux of aqueous solutes, only recently, Contreras et al. [27] described a novel effect of ceramide, namely the induction of transbilayer lipid movements (flip-flop motion). This effect was demonstrated in both model (LUVs) and cell (erythrocyte ghost) membranes in which ceramide generation took place *in situ* through the action of an externally added sphingomyelinase. Two different novel assays were developed to detect transbilayer lipid movement. One of the assays required the preparation of vesicles containing a ganglioside only in the outer monolayer. The same vesicles contained entrapped neuraminidase. Sphingomyelinase activity induced ganglioside hydrolysis under conditions in which no neuraminidase was released from the vesicles. The second assay involved the preparation of liposomes or erythocyte ghosts labelled with a fluorescent energy donor in their inner leaflets. Sphingomyelin hydrolysis was accompanied by fluorescence energy transfer to an impermeable acceptor in the outer aqueous medium. This was interpreted as ceramide generated in one monolayer undergoing flip-flop motion and inducing thereby the motion of other lipid molecules between leaflets. The need to conserve a certain proportion of lipid mass between both monolayers may explain this ceramide-induced motion of various lipids.

Müller et al. [28] have described yet another procedure to assay transbilayer motions of lipids, this one based on the excimer-forming capacity of the fluorescent probe pyrene, incorporated into phospholipid analogues (e.g. pyrene-PC). We have used this technique to assay flip-flop motion induced by ceramides. In this case, ceramides were added to the pre-formed LUV suspension, rather than being generated *in situ* through enzyme action. The results are shown in Figure 4. The vesicles were prepared containing pyrenyl PC in the outer monolayer only, at a concentration such that formation of pyrene excimers was favoured. If pyrenyl PC molecules flip to the inner monolayer, the excimer/monomer ratio of pyrene molecules decreases, and the fluorescence spectrum changes accordingly. The q parameter on the ordinate axis corresponds to the fraction of pyrenyl PC molecules in the inner monolayer, computed from changes in the excimer and monomer fluorescence peaks. For a completely symmetrical membrane, $q=0.5$. The data in Figure 4 indicate that all three ceramides tested induce transbilayer lipid motion, but egg ceramide, containing mainly Cer16, is more effective in this respect.

The possible physiological significance of the above observations should not be ignored. As mentioned in the Introduction, a large number of direct targets for ceramide are intracellular. In turn, most of the sphingomyelin in the plasma membrane exists in the outer leaflet, perhaps in the form of rafts [29], and acid sphingomyelinase appears to be translocated on to the outer leaflet to perform the hydrolytic cleavage of sphingomyelin [11,12]. The data in this section support the idea that ceramide, generated in the outer leaflet, will distribute spontaneously between both monolayers, thus becoming accessible

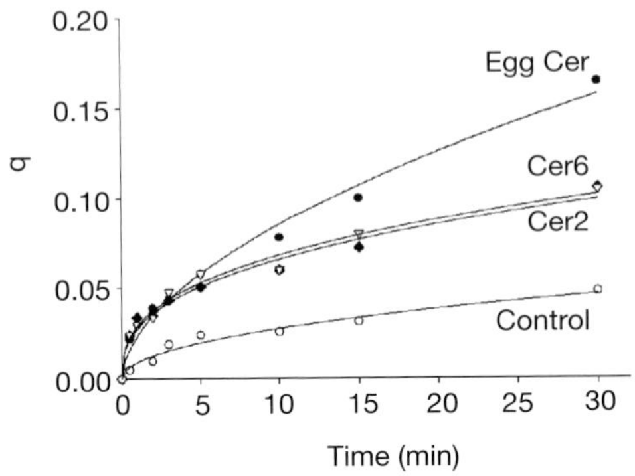

Figure 4 Transbilayer lipid motion induced by ceramides. Vesicles as in Figure 3, except these contained 5 molar percent pyrenyl PC asymmetrically located in the outer monolayer. The following additions were made to 1 ml of the LUV suspension: (●) 30 μM Cer16 in 2 μl of ethanol, (▽) 30 μM Cer6 in buffer, (◆) 30 μM Cer2 in buffer, (○) 2 μl of ethanol. Additional experimental details are given in [28].

to cytosolic target proteins that may transiently dock to the plasma membrane inner monolayer.

Discussion

The available experimental evidence on the physical changes induced by ceramides in membranes can be understood on the basis of three main principles, namely (i) the tendency of long-chain (but not of short-chain) ceramides to form separate domains in bilayers, (ii) the capacity of short-chain (but not of long-chain) ceramides to act as detergents and solubilize lipid bilayers, and (iii) the propensity of long-chain (but not of short-chain) ceramides to favour formation of inverted non-lamellar phases.

Ceramide-rich domain formation involving Cer16 is demonstrated by the calorimetric data [17,18] (Figure 2) and also by fluorescence [30] and ^{2}H-NMR data [31]. Some of these observations indicate as well that short-chain ceramides, and Cer2 in particular, mix well with phospholipids, so that no in-plane domain separation is observed. This property of ceramides explains also their effect on membrane permeability. In our view, solute efflux occurs through the structural defects occurring at the interfaces between gel (ceramide-rich) and fluid (ceramide-poor) domains. Such structural defects do not arise with Cer2, hence the chain-length-dependent differences observed in Figure 3.

A somewhat opposing property of ceramides is that short-chain, but not long-chain, ones will act as detergents and solubilize membranes (Table 1). This is due to the fact that Cer2 and Cer6, but not Cer16, are soluble amphiphiles [13,16] (Figure 1). Solubilization by Cer2 is observed at ceramide/phospholipid ratios higher than 0.2 (J. Sot, unpublished work); in addition, bilayers containing sphingomyelin and cholesterol are particularly resistant to detergent solubilization [29,32–34]. For these reasons, efflux observed in Figure 3 and in [26], with ceramide/phospholipid ratios of 0.1, is not related to solubilization. However, di Paola et al. [22] did find that Cer2, but not Cer16, impaired the mitochondrial proton gradient. The inner mitochondrial membrane is virtually devoid of sphingomyelin and cholesterol, and is thus easily solubilized. In addition, the Cer2 concentrations used by these authors were significantly higher than those used in Figure 3. Thus it is very probable that Cer2 in the work by di Paola et al. [22] was acting as a detergent. Cer16, in the same experimental work, was giving rise to lateral phase separation in the outer membrane, and to cytochrome *c* release.

The propensity of Cer16 to favour inverted non-lamellar phases has been shown repeatedly [17,20] using different techniques. Ceramide-induced formation in the bilayers of (perhaps transient) non-lamellar structures, whose architecture may remind the inverted hexagonal or inverted cubic lipid phases, is probably the origin of the flip-flop movements induced by this lipid. It has been found that lipids that promote inverted-phase formation, e.g. PE or cholesterol, increase the rate of ceramide-promoted transbilayer lipid motion, and the opposite is found with lyso-PC, that inhibits the formation of inverted lipid phases (F.-X. Contreras, unpublished work). Cer2 does not exhibit such a tendency to promote lamellar to non-lamellar transitions in lipids (J. Sot, unpublished work), and, correspondingly, it is a poor inducer of flip-flop movements (Figure 4).

The data discussed in the present paper concur in presenting ceramides as lipids with a high capacity to modify the physical properties of cell membranes. This is in agreement with the views put forward by van Blitterswijk et al. [35]. It is thus possible that, apart from those effects mediated by specific ceramide binding to target proteins, ceramides may act through changes in the bilayer properties. One good example is the finding that ceramide, because of its capacity to segregate laterally, giving rise to ceramide-enriched domains, is able to convert rafts into macroscopic clusters or platforms, that play important roles in cell signalling and bacterial infection [11,12]. With respect to the different physical properties of short- and long-chain ceramides that have been extensively discussed above, the possibility exists that they are at the origin of the different effects that, in some cases, have been found with long- and short-chain ceramides in cell studies [36–39].

This work was supported in part by grants from MCYT (Ministerio de Ciencia y Tecnología) (BMC 2002-00784) and the University of the Basque Country (13552/2001). F.-X.C., L.-R.M. and J.S. were graduate students supported by the Basque Government.

References

1. Hill, J.R. and Wertz, P.W. (2003) Biochim. Biophys. Acta **1616**, 121–126
2. Cremesti, A.E., Goñi, F.M. and Kolesnick, R. (2002) FEBS Lett. **531**, 47–53
3. Hannun, Y.A., Luberto, C. and Argraves, K.M. (2001) Biochemistry **40**, 4893–4903
4. Mathias, S., Pena, L.A. and Kolesnick, R.N. (1998) Biochem. J. **335**, 465–480

5. Yan, F. and Polk, D.B. (2001) Cancer Res. **61**, 963–969
6. Chalfant, C.E., Rathman, K., Pinkerman, R.L., Wood, R.E., Obeid, L.M., Ogretmen, B. and Hannun, Y.A. (2002) J. Biol. Chem. **277**, 12587–12595
7. Kashiwagi, K., Shirai, Y., Kuriyama, M., Sakai, N. and Saito, N. (2002) J. Biol. Chem. **277**, 18037–18045
8. Zhou, M., Horita, D.A., Waugh, D.S., Byrd, R.A. and Morrison, D.K. (2002) J. Mol. Biol. **315**, 435–446
9. Kolesnick, R.N., Goñi, F.M. and Alonso, A. (2000) J. Cell. Physiol. **184**, 285–300
10. Cremesti, A., Paris, F., Grassme, H., Holler, N., Tschopp, J., Fuks, Z., Gulbins, E. and Kolesnick, R. (2001) J. Biol. Chem. **276**, 23954–23961
11. Grassme, H., Jendrossek, V., Bock, J., Riehle, A. and Gulbins, E. (2002) J. Immunol. **168**, 298–307
12. Grassme, H., Jendrossek, V., Riehle, A., von Kurthy, G., Berger, J., Schwarz, H., Weller, M., Kolesnick, R. and Gulbins, E. (2003) Nat. Med. **9**, 322–330
13. Small, D.M. (1970) Fed. Proc. **29**, 1320–1326
14. Simon, Jr, C.G. and Gear, A.R. (1998) Biochemistry **37**, 2059–2069
15. Simon, Jr, C.G., Holloway, P.W. and Gear, A.R. (1999) Biochemistry **38**, 14676–14682
16. Helenius, A. and Simons, K. (1975) Biochim. Biophys. Acta **415**, 29–79
17. Veiga, M.P., Arrondo, J.L., Goñi, F.M. and Alonso, A. (1999) Biophys. J. **76**, 342–350
18. Carrer, D.C. and Maggio, B. (1999) J. Lipid Res. **40**, 1978–1989
19. Carrer, D.C., Härtel, S., Mónaco, H.L. and Maggio, B. (2003) Chem. Phys. Lipids **122**, 147–152
20. Ruiz-Argüello, M.B., Basanez, G., Goñi, F.M. and Alonso, A. (1996) J. Biol. Chem. **271**, 26616–26621
21. Ghafourifar, P., Klein, S.D., Schucht, O., Schenk, U., Pruschy, M., Rocha, S. and Richter, C. (1999) J. Biol. Chem. **274**, 6080–6084
22. Di Paola, M., Cocco, T. and Lorusso, M. (2000) Biochemistry **39**, 6660–6668
23. Siskind, L.J. and Colombini, M. (2000) J. Biol. Chem. **275**, 38640–38644
24. Siskind, L.J., Kolesnick, R.N. and Colombini, M. (2002) J. Biol. Chem. **277**, 26796–26803
25. Siskind, L.J., Davoody, A., Lewin, N., Marshall, S. and Colombini, M. (2003) Biophys. J. **85**, 1560–1575
26. Montes, L.R., Ruiz-Argüello, M.B., Goñi, F.M. and Alonso, A. (2002) J. Biol. Chem. **277**, 11788–11794
27. Contreras, F.X., Villar, A.V., Alonso, A., Kolesnick, R.N. and Goñi, F.M. (2003) J. Biol. Chem. **278**, 37169–37174
28. Müller, P., Schiller, S., Wieprecht, T., Dathe, M. and Herrmann, A. (2000) Chem. Phys. Lipids **106**, 89–99
29. Simons, K. and Vaz, W.L. (2004) Annu. Rev. Biophys. Biomol. Struct. **33**, 269–295
30. London, M. and London, E. (2004) J. Biol. Chem. **279**, 9997–10004
31. Huang, H.W., Goldberg, E.M. and Zidovetzki, R. (1998) Eur. Biophys. J. **27**, 361–366
32. London, E. and Brown, D.A. (2000) Biochim. Biophys. Acta **1508**, 182–195
33. Patra, S.K., Alonso, A., Arrondo, J.L.R. and Goñi, F.M. (1999) J. Liposome Res. **9**, 247–260
34. Sot, J., Collado, M.I., Arrondo, J.L.R., Alonso, A. and Goñi, F.M. (2002) Langmuir **18**, 2828–2835
35. Van Blitterswijk, W.J., van der Luit, A.H., Veldman, R.J., Verheij, M. and Borst, J. (2003) Biochem. J. **369**, 199–211
36. Olivera, A., Romanowski, A., Rani, C.S. and Spiegel, S. (1997) Biochim. Biophys. Acta **1348**, 311–323
37. Bosel, A. and Pfeuffer, T. (1998) FEBS Lett. **422**, 209–212
38. Allan-Yorke, J., Record, M., de Preval, C., Davrinche, C. and Davignon, J.L. (1998) J. Virol. **72**, 2316–2322
39. Fukunaga, T., Nagahama, M., Hatsuzawa, K., Tani, K., Yamamoto, A. and Tagaya, M. (2000) J. Cell Sci. **113**, 3299–3307

Biochem. Soc. Symp. **72**, 189–198
(Printed in Great Britain)

18

Reversible — through calmodulin — electrostatic interactions between basic residues on proteins and acidic lipids in the plasma membrane

Stuart McLaughlin[1], Gyöngyi Hangyás-Mihályné, Irina Zaitseva and Urszula Golebiewska

Department of Physiology and Biophysics, Health Science Center, SUNY Stony Brook, New York, NY 11794, U.S.A.

Abstract

The inner leaflet of a typical mammalian plasma membrane contains 20–30% univalent PS (phosphatidylserine) and ~1% multivalent PtdIns(4,5)P_2. Numerous proteins have clusters of basic (or basic/hydrophobic) residues that bind to these acidic lipids. The intracellular effector CaM (calmodulin) can reverse this binding on a wide variety of proteins, including MARCKS (myristoylated alanine-rich C kinase substrate), GAP43 (growth-associated protein 43, also known as neuromodulin), gravin, GRK5 (G-protein-coupled receptor kinase 5), the NMDA (*N*-methyl-D-aspartate) receptor and the ErbB family. We used the first principles of physics, incorporating atomic models and the Poisson–Boltzmann equation, to describe how the basic effector domain of MARCKS binds electrostatically to acidic lipids on the plasma membrane. The theoretical calculations show the basic cluster produces a local positive electrostatic potential that should laterally sequester PtdIns(4,5)P_2, even when univalent acidic lipids are present at a physiologically relevant 100-fold excess; four independent experimental measurements confirm this prediction. Ca^{2+}/CaM binds with high affinity (K_d ~10 nM) to this domain and releases the PtdIns(4,5)P_2. MARCKS, a major PKC (protein kinase C) substrate, is present at concentrations comparable with those of PtdIns(4,5)P_2 (approx. 10 μM) in many cell types. Thus MARCKS can act as a reversible PtdIns(4,5)P_2 buffer, binding PtdIns(4,5)P_2 in a quiescent cell, and releasing it locally when the intra-

[1]To whom correspondence should be addressed (email smcl@epo.som.sunysb.edu).

cellular Ca^{2+} concentration increases. This reversible sequestration is important because PtdIns(4,5)P_2 plays many roles in cell biology. Less is known about the role of CaM-mediated reversible membrane binding of basic/hydrophobic clusters for the other proteins.

The electrostatic potential adjacent to the inner leaflet of a plasma membrane

The 19th Century physicist Helmholtz recognized that fixed charges at a surface attract counter-ions, ions of opposite sign, from the aqueous phase to form an electrical 'double layer' that has a 'small, but not disappearing' thickness [1]. The counter-ions do not stick to the surface for the same reason that the earth's atmosphere does not fall to the ground: the electrical attraction and the statistical tendency of counter-ions to diffuse away from a region of high concentration balance each other. In the early part of the 20th century, Gouy [2] and Chapman [3] used the Poisson equation to describe the electrostatic attraction of the counter-ions to the surface and the Boltzmann relation to describe the statistical distribution of the counter-ions in the ion atmosphere. They made three key assumptions: (i) the fixed charges on the surface (the acidic phospholipids in the inner leaflet of a plasma membrane, in the case that we will consider) are smeared uniformly over the surface, (ii) water may be considered a structureless medium with a uniform dielectric constant, and (iii) ions in the aqueous phase are point charges. Their theory predicts that, when the surface potential is small, it is proportional to the surface-charge density (mole fraction of acidic lipids in the membrane) and the Debye length. (An identical equation describes the potential difference between the plates of a capacitor separated by a distance equal to the Debye length.) The magnitude of the potential decreases exponentially with the distance from the surface with a space constant equal to the Debye length, which is approx. 1 nm for physiological (~100 mM) salt concentration. The surprising feature of this simple theory is that it works remarkably well for phospholipid membranes, as reviewed in detail elsewhere [4]. Over a dozen different laboratories have used numerous techniques to demonstrate that this simple theory provides an adequate description of the potential adjacent to a bilayer, and by inference, the bilayer component of a biological membrane. The cytoplasmic leaflet of mammalian plasma membrane contains approx. 20–30% univalent acidic phospholipid, mainly PS (phosphatidylserine); the potential adjacent to such a membrane is −30 to −40 mV in 100 mM salt (see Figure 3 in [4]).

About a decade ago, researchers developed methods for applying the Poisson–Boltzmann equation to atomic models of phospholipid bilayers; these models assign the appropriate partial charge to each atom of the lipids in the membrane. The Poisson–Boltzmann equation can then be used to calculate the electrostatic potential in the aqueous phase adjacent to the bilayer. The theoretical calculations predict the same flat equipotential profile described by the Gouy–Chapman (smeared charged) theory for mole fractions of acidic lipid >15% [5]. If the mole fraction of charged lipid is very low, however, the equipotential profile is not flat. Rather, the equipotential profile around each acidic lipid is approximately a hemisphere, with a value twice that predicted by Debye–Huckel theory: the factor of two is due to the image charge effect.

If we want to calculate the electrostatic binding energy of a basic peptide to a membrane from first principles, we need atomic level descriptions of not only the bilayer, but also the peptide. We place the peptide distance x from the bilayer, and calculate the potential everywhere in space. Once we know the potential and the location of each charge, we can calculate the energy of the peptide at that distance x. Repeating the calculation for different distances and orientations of the peptide allows us to calculate the average energy, and thus the concentration of the peptide as a function of distance. As discussed in detail elsewhere [6–8], we then use the integral of the excess peptide concentration adjacent to the membrane, i.e. the 'Gibbs surface excess', to deduce net number of peptides 'bound' electrostatically.

Binding of peptides that mimic clusters of basic residues on proteins

The theoretical treatment makes very simple predictions for interaction of a basic peptide with a membrane containing univalent acidic lipid: the binding energy will increase (i) linearly with the number of basic residues in the peptide, (ii) linearly with the mole fraction of acidic lipid in the membrane, and (iii) as the salt concentration decreases. The experimental results agree satisfactorily with these predictions. For example, simple peptides with three, five and seven basic residues (e.g. Lys_3, Lys_5 and Lys_7) bind with energies of 3, 5 and 7 kcal/mol (1 kcal=4.184 kJ) to 2:1 PC (phosphatidylcholine)/PS vesicles in 100 mM KCl [6]. Both centrifugation and FCS (fluorescence correlation spectroscopy) measurements show that the theory also predicts the membrane-binding behaviour of a more complicated basic peptide corresponding to the MARCKS (myristoylated alanine-rich C kinase substrate) effector region, MARCKS-(151–175), which has 13 basic and five hydrophobic residues: its molar partition coefficient increases exponentially from 10^2 to 10^8 (the binding energy increases linearly) as the mole fraction of PS in the membrane increases from 0 to 25% [9]. Finally, the electrostatic binding energy of simple basic peptides and small proteins increases as the salt concentration decreases, as predicted theoretically [6,7].

MARCKS as a paradigm for protein–membrane electrostatic interactions

Our studies with MARCKS and basic/hydrophobic peptides corresponding to its effector domain illustrate several general features of electrostatic membrane–protein interactions. The major PKC (protein kinase C) substrate in many different cell types, MARCKS is a 'natively unfolded' protein that binds to the plasma membrane through insertion of a myristate into the hydrocarbon core of the bilayer, and electrostatic interactions between a cluster of basic amino acids in its effector domain and acidic lipids in the membrane [9,10]. The basic cluster binds Ca^{2+}/CaM (calmodulin) strongly, which produces translocation of the protein to the cytoplasm [10]. The structure of the effector domain–Ca^{2+}/CaM complex is unusual because the complexed effector domain is not an amphipathic α-helix, but rather retains most of its elongated structure [11]. PKC phosphorylation of three serine residues in the effector domain also produces translocation of MARCKS to the cytoplasm, because it reduces the electrostatic interaction with the membrane, and the binding energy of myristate alone is not strong enough to anchor a protein to the plasma membrane.

What does MARCKS do in a cell? Our working hypothesis is that MARCKS acts as a reversible PtdIns(4,5)P_2 buffer: the basic effector domain laterally sequesters PtdIns(4,5)P_2 through non-specific electrostatic interactions, then releases it when Ca^{2+}/CaM binds to or PKC phosphorylates this domain [12–15].

Electrostatic calculations show that when a cluster of basic residues, such as the MARCKS-(151–175) peptide, binds to a PC/PS bilayer, it produces a local positive electrostatic potential (see Figure 4d of [12], Figure 9F of [14], Figure 12 of [15], or [16]) that acts as a basin of attraction for multivalent acidic lipids such as PtdIns(4,5)P_2, which has a valence of approx. −4 at pH 7 [12]. Specifically, the potential experienced by PtdIns(4,5)P_2 far (>2 nm) from the peptide is approx. kT/e=−25 mV and the potential experienced by PtdIns(4,5)P_2 close to the adsorbed cluster is +25 mV. If we oversimplify and treat PtdIns(4,5)P_2 as a point charge with a valence of −4 that does not perturb the potential when it approaches the bound peptide, the Boltzmann relation predicts the potential of the positively charged adsorbed cluster will concentrate PtdIns(4,5)P_2 in the adjacent membrane region by approx. three orders of magnitude. Of course PtdIns(4,5)P_2 does have a finite size and does perturb the potential in its neighbourhood, but a more rigorous theoretical treatment based on the Poisson–Boltzmann relation and atomic models makes roughly the same prediction: the peptide should sequester PtdIns(4,5)P_2 even when PS is present at the 100-fold excess characteristic of biological membranes [16].

Experimental evidence that MARCKS sequesters PtdIns(4,5)P_2 via electrostatic interactions

Our first indication that the MARCKS protein could sequester PtdIns(4,5)P_2 was admittedly indirect: we observed that it inhibited the activity of the enzyme PLC (phospholipase C), which catalyses hydrolysis of PtdIns(4,5)P_2, on phospholipid membranes containing PC, PS and PtdIns(4,5)P_2 [8,12]. The MARCKS-(151–175) peptide also inhibits PLC, and PKC phosphorylation or binding of Ca^{2+}/CaM to the peptide reverses the inhibition [8,12]. These data led us to explore the interactions between MARCKS and PtdIns(4,5)P_2 more deeply, and perform experiments that have provided more direct evidence that the cluster of basic residues on MARCKS does indeed sequester PtdIns(4,5)P_2 [15].

We began by measuring binding of MARCKS-(151–175) to electrically neutral PC membranes. The effector domain peptide binds, albeit not very strongly, to PC bilayers, because its five phenylalanine residues penetrate into the hydrocarbon interior of the bilayer, as shown by magic angle spinning NMR and other experiments (see Figure 1 of [15] and references therein). When we incorporate 1% PtdIns(4,5)P_2 into the PC membrane, the binding of MARCKS-(151–175) increases by four orders of magnitude [17]; specifically the molar partition coefficient, *K*, is ~10^2 M^{-1} for a PC vesicle and ~10^6 M^{-1} for a 99:1 PC/PtdIns(4,5)P_2 vesicle. (*K* is the reciprocal of the lipid concentration required to bind 50% of the peptide. 1/*K* may be regarded as an apparent dissociation constant of the peptide with a single lipid.) As expected for simple electrostatic interactions, the binding is independent of both the chemical nature of the basic residues [Lys_{13} and Arg_{13} peptides also bind with K=10^6 M^{-1} to 99:1 PC/PtdIns(4,5)P_2 vesicles] and the chemical nature of the PtdIns(4,5)P_2 {the peptide binds equally well to membranes formed from mixtures of PC with either PtdIns(3,4)P_2 or PtdIns(4,5)P_2 [13,14]}.

When the peptide binds to a PC/PtdIns(4,5)P_2 membrane, three PtdIns(4,5)P_2 molecules diffuse towards the peptide and form an electrostatic binding site; this is why the forward rate constant depends on the mole fraction of PtdIns(4,5)P_2 in the membrane [14]. In contrast, peptide binding to a membrane formed from PC and a univalent acidic lipid, such as PS, does not perturb the lateral distribution of lipids in the membrane significantly; the forward rate constant remains diffusion limited as the mole fraction of PS in the membrane changes [18]. Biological membranes, of course, comprise numerous acidic lipids, and PtdIns(4,5)P_2 is present at much lower concentrations than univalent acidic lipids, such as PS. Can a basic peptide, or cluster of basic residues on a protein, laterally sequester PtdIns(4,5)P_2 via electrostatic interactions in membranes where PS is present at a 100-fold excess? Theoretical calculations predict that sequestration should occur [8,16]. We measured the ability of MARCKS-(151–175) to sequester PtdIns(4,5)P_2 in PC/PS/PtdIns(4,5)P_2 membranes using four independent techniques — FRET (fluorescence resonance energy transfer), EPR, fluorescence self quenching, and PLC activity — and found significant sequestration using each technique [15].

Binding due to non-specific electrostatic interactions differs from conventional stereospecific binding

We have focused on the role of non-specific electrostatic protein–membrane interactions, but it is important to recognize that many proteins have domains that bind specifically to phosphoinositides [19,20]. The PH (pleckstrin homology) domain of PLCδ1, for example, forms hydrogen bonds with both the 4- and 5-phosphates in PtdIns(4,5)P_2, holding it in a classical 'lock and key' complex. Comparison of the PtdIns(4,5)P_2 binding characteristics of this domain and the MARCKS basic effector region peptide show four important differences. First, stereospecific interactions require structure: the PLCδ1 PH domain has a known structure [19], while MARCKS-(151–175) is unstructured, both in solution and when bound to a membrane [15]. Secondly, the ratio of protein/ligand in the complex is different: the PLCδ1 PH domain (and other similar phosphoinositide binding domains of known structure) binds one

PtdIns(4,5)P_2, while the MARCKS basic peptide, which has 13 positive residues, binds three PtdIns(4,5)P_2 molecules. Thirdly, the strength of interaction differs significantly: the PLCδ1 PH domain binds to 99:1 PC/PtdIns(4,5)P_2 vesicles with an affinity that is 100-fold weaker than that of MARCKS-(151–175). The high affinity of the unstructured peptide for the vesicle follows from its ability to combine with three PtdIns(4,5)P_2 molecules, whereas the PH domain binds to just one. Fourthly, the domains function differently: the well-established role of the PLCδ1 PH domain is to anchor the protein to the plasma membrane, enhancing the local concentration of PtdIns(4,5)P_2 substrate experienced by the catalytic domain [19]. We postulate that the function of the basic effector domain of MARCKS is to help control the free PtdIns(4,5)P_2 concentration in the membrane, laterally sequestering this signalling lipid PtdIns(4,5)P_2, then releasing it locally in response to either PKC phosphorylation or binding of Ca^{2+}/CaM [12–15].

Why controlling the free PtdIns(4,5)P_2 concentration in membranes is important

One of the enigmas of PtdIns(4,5)P_2 is that it plays an important role in such a wide variety of cell signalling processes; Figure 1 illustrates six of these functions.

Moving clockwise from the upper left of Figure 1, we first note that PtdIns(4,5)P_2 produces three important second messengers. PLC-catalysed hydrolysis of PtdIns(4,5)P_2 yields the second messengers Ins(1,4,5)P_3 and DAG (diacylglycerol), and phosphorylation by PI3K (phosphoinositide 3-kinase) produces PtdIns(1,4,5)P_3, which functions as a membrane anchor for several important proteins. Secondly, binding to PtdIns(4,5)P_2 targets some proteins with PH or other phosphoinositide-binding domains to the plasma membrane [19,20]. As noted above, the PLCδ1 PH domain binds to PtdIns(4,5)P_2 very selectively. As this was the first PH domain to be understood in atomic detail, it was initially thought that most PH domains might also exhibit high selectivity. More recent detailed studies of several of the ~200 other PH domains reveals that the majority are much less selective for phosphoinositides [21]. Thirdly, several enzymes, e.g. PLD (phospholipase D),

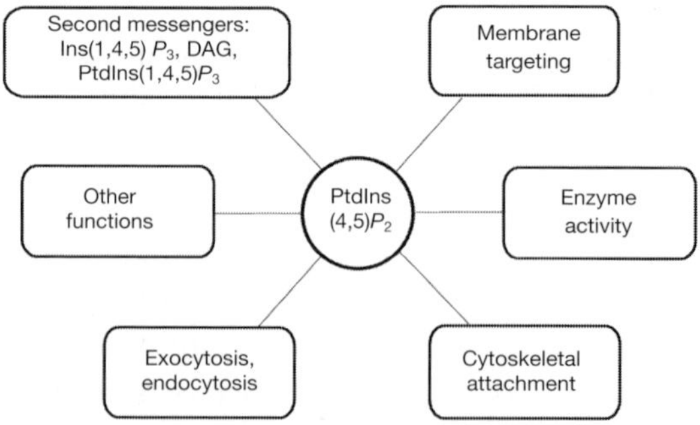

Figure 1 Functions of PtdIns(4,5)P_2. See text for a description of the Figure.

require PtdIns(4,5)P_2 for activation [22]. Fourthly, PtdIns(4,5)P_2 plays a role in maintaining the cytoskeleton, as shown by elegant laser tweezer experiments that demonstrate a dramatic release of this vital cellular structure in response to decreasing the level of PtdIns(4,5)P_2 in a cell [23]. Fifthly, there is good evidence that PtdIns(4,5)P_2 is involved in endocytosis [24], and mounting evidence that suggests it may also be important in exocytosis [24]. Remarkably, Figure 1 does not cover all known PtdIns(4,5)P_2 functions; e.g. it activates several different classes of ion channels [25–27].

The plethora of functions for this simple lipid molecule implies that its availability must be controlled very stringently. But how? One possibility is that kinases could produce PtdIns(4,5)P_2 locally on demand, as discussed by Anderson and colleagues [28]. Another possibility is that MARCKS, and other plasma membrane proteins, act as reversible PtdIns(4,5)P_2 buffers. The cell probably uses both mechanisms: both MARCKS and the kinases that synthesize PtdIns(4,5)P_2 are concentrated in the nascent phagosomes of macrophages and the ruffles of fibroblasts, two regions with high concentrations of PtdIns(4,5)P_2.

MARCKS and other molecules may act as reversible PtdIns(4,5)P_2 buffers

We begin by defining the characteristics necessary for a putative PtdIns(4,5)P_2 buffer. Our biophysical studies suggest any protein with a cluster of four or more proximal basic residues that binds to the inner leaflet of the plasma membrane should be capable of laterally sequestering PtdIns(4,5)P_2. The putative buffer must also be very abundant, since it must be present at a concentration comparable with that of PtdIns(4,5)P_2 (approx. 10 μM) if it is to sequester a significant fraction of the PtdIns(4,5)P_2 in the plasma membrane. Only a few proteins are present at this concentration in cells. Finally, the sequestration must be reversible if the putative buffer functions to help control the level of PtdIns(4,5)P_2. Caroni and colleagues discussed the cell biological studies suggesting that MARCKS, GAP43 (growth-associated protein 43, also known as neuromodulin) and CAP23 (cortical cytoskeleton-associated protein of 23 kDa) could function as buffers, or pipmodulins [29]. We recently began investigating GAP43, a peripheral membrane protein that is concentrated in the growth cones of developing neurons and has two palmitates (rather than a single myristate for MARCKS or CAP23) that anchor it strongly to the membrane. Our experiments with a peptide corresponding to the basic cluster on GAP43, GAP43-(30–57), indicate that the peptide binds both Ca^{2+}/CaM and apoCaM with roughly the same affinity in 150 mM salt (U. Golebiewska, unpublished work), in agreement with an earlier report that the GAP43 protein binds apoCaM roughly as strongly as Ca^{2+}/CaM [30]. Moreover, the GAP43-(30–57) peptide binds both Ca^{2+}/CaM and apoCaM with sufficiently high affinity to compete with physiological (millimolar or greater) concentrations of 2:1 PC/PS membranes. These data suggest that both forms of CaM may bind to this region of GAP43 in a cell, blocking its interaction with PtdIns(4,5)P_2. The basic cluster also contains a serine residue that is a target for PKC phosphorylation; phosphorylation reduces the affinity of this region for CaM, allowing it to bind to PtdIns(4,5)P_2 in the membrane. If the basic region does bind to the membrane, it can laterally sequester PtdIns(4,5)P_2: FRET experiments with GAP43-(30–57) produce results similar to those reported for MARCKS (U. Golebiewska, unpublished work). This leads us to speculate that GAP43 could be a PtdIns(4,5)P_2 buffer that functions in a complementary way to MARCKS; i.e. GAP43 sequesters PtdIns(4,5)P_2 in response to PKC phosphorylation (i.e. upon dissociation from CaM), acting as a reversible PtdIns(4,5)P_2 'sink', whereas MARCKS releases PtdIns(4,5)P_2 upon phosphorylation of its three serine residues (or binding to Ca^{2+}/CaM), acting as a reversible PtdIns(4,5)P_2 'source'.

Other important proteins may also reversibly sequester PtdIns(4,5)P_2

Most plasma membrane proteins are not present at a sufficiently high concentration in the cell (approx. 10 μM) to be capable of acting as reversible PtdIns(4,5)P_2 buffers. Nevertheless, the kinase GRK5 (G-protein-coupled receptor kinase 5) [31], the scaffolding protein Gravin [32], the NMDA (*N*-methyl-D-aspartate) receptor [14] and the ErbB family of receptor tyrosine kinases all have clusters of basic/hydrophobic residues that probably bind to the inner leaflet of a plasma membrane. These basic clusters are capable of binding PtdIns(4,5)P_2, and the available evidence suggests this binding will be reversed by Ca^{2+}/CaM. For most of these proteins, the biological role of their basic/hydrophobic cluster has not yet been determined.

We have initiated a study of the ErbB family members, which all have a basic/hydrophobic region in the cytoplasmic juxtamembrane domain immediately adjacent to the transmembrane helix [33]. Peptides corresponding to these regions bind strongly to membranes containing acidic lipids; in each case, Ca^{2+}/CaM can reverse this binding (S. McLaughlin, S.O. Smith, M. Hayman and D. Murray, unpublished work). Villalobo and co-workers have also shown that ErbB1 and ErbB2 are CaM-binding proteins [34,35]. We have preliminary data that these basic clusters can laterally sequester PtdIns(4,5)P_2 based on measuring FRET between labels on PtdIns(4,5)P_2 and on a peptide corresponding to the basic/hydrophobic juxtamembrane cluster in ErbB1, ErbB1-(645–660). Activation of ErbB1 results in trans-autophosphorylation that produces a binding site for PLCγ on the C-terminal portion of the ErbB molecule, suggesting that ErbB1 reversibly binds both PLC and its PtdIns(4,5)P_2 substrate. Moreover, the reversible (through Ca^{2+}/CaM) binding of the juxtamembrane region to a bilayer leads to a new model for the activation of this important class of receptor tyrosine kinases (S. McLaughlin, S.O. Smith, M. Hayman and D. Murray, unpublished work). A new model is necessary because the ErbB family, unlike other receptor tyrosine kinases, does not require phosphorylation of tyrosine residues in the activation loop within the kinase domain for activation [36]. Thus we argue that biophysical studies of how clusters of basic/hydrophobic residues bind to membranes, laterally sequester PtdIns(4,5)P_2, and how Ca^{2+}/CaM may reverse this membrane binding and PtdIns(4,5)P_2 sequestration, may lead to important new insights into the mechanism of action of important membrane signalling proteins.

References

1. Helmholtz, H.L. (1879) Ann. Phys. (Leipzig) **7**, 337 [translated by Bocquet, P. (1951) Two Monographs on Electrokinetics, Engineering Research Institute, University of Michigan, Ann Arbor]
2. Gouy, M. (1910) J. Phys. (Paris) **9**, 457
3. Chapman, D.L. (1913) Philos. Mag. **25**, 475
4. McLaughlin, S. (1989) Annu. Rev. Biophys. Biophys. Chem. **18**, 113–136
5. Peitzsch, R.M., Eisenberg, M., Sharp, K.A. and McLaughlin, S. (1995) Biophys. J. **68**, 729–738
6. Ben-Tal, N., Honig, B., Peitzsch, R.M., Denisov, G. and McLaughlin, S. (1996) Biophys. J. **71**, 561–575
7. Ben-Tal, N., Honig, B., Miller, C. and McLaughlin, S. (1997) Biophys. J. **73**, 1717–1727
8. Murray, D., Arbuzova, A., Honig, B. and McLaughlin, S. (2002) Curr. Top. Membr. **52**, 271–302
9. Rusu, L., Gambhir, A., McLaughlin, S. and Rädler, J. (2004) Biophys. J. **87**, 1044–1053
10. McLaughlin, S. and Aderem, A. (1995) Trends Biochem. Sci. **20**, 272–276
11. Yamauchi, E., Nakatsu, T., Matsubara, M., Kato, H. and Taniguchi, H. (2003) Nat. Struct. Biol. **10**, 226–231
12. McLaughlin, S., Wang, J., Gambhir, A. and Murray, D. (2002) Annu. Rev. Biophys. Biomol. Struct. **31**, 151–175
13. Wang, J., Arbuzova, A., Hangyas-Mihalyne, G. and McLaughlin, S. (2001) J. Biol. Chem. **276**, 5012–5019
14. Wang, J., Gambhir, A., Hangyas-Mihalyne, G., Murray, D., Golebiewska, U. and McLaughlin, S. (2002) J. Biol. Chem. **277**, 34401–34412
15. Gambhir, A., Hangyas-Mihalyne, G., Cafiso, D.S., Wang, J., Murray, D., Pentyala, S.N., Smith, S.O. and McLaughlin, S. (2004) Biophys. J. **86**, 2188–2207
16. Wang, J., Gambhir, A., McLaughlin, S. and Murray, D. (2004) Biophys. J. **86**, 1969–1986
17. Arbuzova, A., Wang, L., Wang, J., Hangyas-Mihalyne, G., Murray, D., Honig, B. and McLaughlin, S. (2000) Biochemistry. **39**, 10330–10339
18. Arbuzova, A., Murray, D. and McLaughlin, S. (1998) Biochim. Biophys. Acta **1376**, 369–379
19. Lemmon, M.A. (2003) Traffic **4**, 201–213
20. DiNitto, J.P., Cronin, T.C. and Lambright, D.G. (2003) Sci. STKE 2003, re16
21. Yu, J.W., Mendrola, J.M., Audhya, A., Singh, S., Keleti, D., DeWald, D.B., Murray, D., Emr, S.D. and Lemmon, M.A. (2004) Mol. Cell **13**, 677–688
22. Liscovitch, M., Chalifa, V., Pertile, P., Chen, C.S. and Cantley, L.C. (1994) J. Biol. Chem. **269**, 21403–21406
23. Raucher, D., Stauffer, T., Chen, W., Shen, K., Guo, S., York, J.D., Sheetz, M.P. and Meyer, T. (2000) Cell **100**, 221–228
24. Stenmark, H. (2003) in Phosphoinositides in Subcellular Targeting and Enzyme Activation (Stenmark, H. ed.), Springer-Verlag, Berlin
25. Hilgemann, D.W., Feng, S. and Nasuhoglu, C. (2001) Sci. STKE 2001, re19
26. Suh, B.C., Horowitz, L.F., Hirdes, W., Mackie, K. and Hille, B. (2004) J. Gen. Physiol. **123**, 662–683
27. Prescot, E.D. and Julius, D. (2003) Science **300**, 1284–1288
28. Doughman, R.L., Firestone, A.J. and Anderson, R.A. (2003) J. Membr. Biol. **194**, 77–89
29. Laux, T., Fukami, K., Thelen, M., Golub, T., Fray, D. and Caroni, P. (2000) J. Cell Biol. **149**, 1455–1472

30. Alexander, K.A., Cimler, B.M., Meier, K.E. and Storm D.R. (1987) J. Biol. Chem. **262**, 6108–6113
31. Pronin, A.N., Satpaev, D.K., Slepak, V.Z. and Benovic, J.L. (1997) J. Biol. Chem. **272**, 18273–18280
32. Nauert, J.B., Klauck, T.M., Langeberg, L.K. and Scott, J.D. (1996) Curr. Biol. **7**, 52–62
33. Ullrich, A., Coussens, L., Hayflick, J.S., Dull, T.J., Gray, A., Tam, A.W., Lee, J., Yarden, Y., Libermann, T.A., Schlessinger, J. et al. (1984) Nature (London) **309**, 418–425
34. Martin-Nieto, J. and Villalobo, A. (1998) Biochemistry **37**, 227–236
35. Li, H., Sanhez-Torres J., Del Carpio, A., Salas, V. and Villalobo, A. (2004) Biochem. J. **381**, 257–266
36. Hubbard, S.R. (2004) Nat. Rev. Mol. Cell Biol. **5**, 464–471

Biochem. Soc. Symp. **72**, 199–209
(Printed in Great Britain)

19

Translocation of the cell-penetrating Tat peptide across artificial bilayers and into living cells

Paul Curnow[1,3], Harry Mellor, David J. Stephens, Mark Lorch[2] and Paula J. Booth[3]

Department of Biochemistry, University of Bristol, University Walk, Bristol BS8 1TD, U.K.

Abstract

The ability of a short, charged peptide to penetrate synthetic DOPC (1,2-dioleoyl-*sn*-3-glycerophosphocholine) liposomes was investigated by fluorescence confocal microscopy. The peptide, termed Tat (*trans*-activating transcription factor), was a 14-mer derived from the region of the HIV-1 Tat protein responsible for cellular internalization. This Tat peptide was labelled at a C-terminal cysteine residue with the fluorescent probes IAF (5-iodo-acetamidofluorescein) or A568 (Alexa Fluor 568). The Tat–IAF conjugate was directly observed entering liposomes at room temperature (approx. 25°C) in the absence of pH gradient, ATP or other energy source. The uptake of the Tat–A568 conjugate in unfixed, live HeLa cells was found to be via endocytosis, as expected. In contrast, when the peptide was attached to an IAF-labelled 25 kDa protein corresponding to the catalytic domain of *Clostridium botulinum* C3 exotoxin, this larger, Tat–C3–IAF construct was not able to enter liposomes, although it localized similarly to Tat–A568 in live cells. The data suggest that Tat peptide can cross synthetic bilayers spontaneously *in vitro*, but that size and type of cargo may limit this behaviour.

[1]Present address: Marine Biotechnology Centre, University of California Santa Barbara, Santa Barbara, CA 93106, U.S.A.
[1]Present address: Institute for Biophysical Chemistry, Goethe University, Frankfurt, 9 Marie Curie Street, 60439 Frankfurt am Main, Germany
[3]To whom correspondence should be addressed (email Curnow@lifesci.ucsb.edu or paula.booth@bristol.ac.uk).

Introduction

The intriguing discovery that conjugation to small, positively charged peptides can enhance cellular uptake of a range of biologically active macromolecules has stimulated a great deal of research interest. Many of these so-called CPPs (cell-penetrating peptides) are currently being investigated as potential intracellular delivery agents for a range of therapeutic moieties (e.g. [1–3]).

The mechanism of CPP uptake by cells is still unclear, and has been closely scrutinized. Microscopy and FACS experiments seem to show that the internalization of these peptides and their cargos takes place independently of any receptor proteins [4]. Furthermore, internalization also seems to occur at 4°C in the presence of inhibitors of endocytosis [5,6], as well as in the absence of ATP [7]. These results were initially interpreted as evidence for a spontaneous, non-endocytic membrane translocation mechanism for these peptides (for a recent review, see [8]). However, more recently, it has been shown that these earlier experiments were compromised by artifactual data [9,10], and it now appears that a major route for the cellular internalization of CPPs is in fact by caveolar endocytosis [11,12]. Negatively charged membranes and the presence of sulphated heparan sulphate proteoglycans also seem to enhance the rate of uptake as a result of electrostatic attraction between the peptides and the cell [13,14].

One widely used CPP of particular interest is derived from a short region of the HIV-1 Tat (*trans*-activating transcription factor) protein. It was shown previously that the Tat protein was able to enter cells and localize to the nucleus when added exogenously [15]. Mutagenesis studies later revealed that a 13-amino-acid region of the protein (residues 48–60) contained both the NLS (nuclear localization signal) and PTD (protein transduction domain). This 13-amino-acid sequence featured a disproportionately high number of lysine and arginine residues. Further work showed that a peptide composed of these amino acids alone, termed Tat-(48–60), exhibited the bilayer-translocating properties common to the CPP family [5,16], and is able to deliver various cargos to the cell (see, for example, [17,18]).

The mechanism by which this entry takes place remains contentious, and some of the latest experiments on the Tat peptide in live cells have suggested a role for both endocytosis and the more mysterious, non-endocytic, spontaneous transduction pathway [19,20]. *In vitro* studies on a CPP derived from the *Antennapedia* homoeodomain and with some sequence homology with Tat-(48–60) has shown that the peptide can enter giant unilamellar liposomes (5–10 μm in diameter) composed of a crude PC (L-α-phosphatidylcholine) mixture [21]. Spontaneous bilayer translocation was observed by microscopy both into and out of the liposomes, depending on the concentration gradient. The same investigation also used small unilamellar vesicles to demonstrate that incubation with the peptide did not cause liposome leakage in the manner of the pore-forming, bilayer-associated antibiotic peptide melittin. In contrast with these results, however, a recent study did not observe a fluorescence enhancement consistent with translocation using fluorescein-labelled Tat peptide and smaller (200 nm diameter) liposomes of egg-PC containing Tb^{3+} [22].

Few experiments in synthetic systems have been performed, and yet a simple, single-component liposome sample may provide a good means of investigating the controversial peptide-translocation phenomena [23]. Certainly such *in vitro* experiments will indicate whether peptides can indeed spontaneously translocate in a lipid bilayer. In the present review, we investigate whether the Tat peptide is capable of such spontaneous translocation and how a peptide cargo affects this ability. The Tat peptide was synthesized such that an additional cysteine residue was present at the C-terminus, allowing the specific covalent attachment of the relatively small fluorescent probes IAF (5-iodoacetamidofluorescein) and A568 (Alexa Fluor 568). The behaviour of the peptide was investigated in live cells, as well as in giant (5–10 μm diameter) liposomes of DOPC (1,2-dioleoyl-*sn*-3-glycerophosphocholine). A Tat–protein fusion construct was also studied, which consisted of the peptide linked to the 25 kDa catalytic domain of *Clostridium botulinum* toxin (C3) [24]. The Tat–protein construct was also fluorescently labelled.

In agreement with previous models, we show here the ability of the Tat peptide to translocate single-component synthetic bilayers, while the dominant means of internalization in live cells remains conventional endocytosis. Moreover, the Tat–protein construct cannot translocate into synthetic liposomes, suggesting a cargo size limitation.

Materials and methods

Peptide synthesis and labelling

The 14-mer NH_2-GRKKRRQRRRPPQC-CO_2H was synthesized by standard Fmoc (fluoren-9-ylmethoxycarbonyl) chemistry by Graham Bloomberg, University of Bristol Peptide Synthesis Facility (Bristol, U.K.). The peptide is identical with that Tat region 48–60, except for the introduction of an additional cysteine group at the C-terminus. The thiol-reactive labels 5′-IAF and A568 were purchased from Molecular Probes, and were conjugated to the reactive C-terminal thiol by incubation at room temperature for 4 h in 20 mM Tris/HCl, pH 8.0, with an equimolar peptide/dye ratio. The conjugate was purified further by HPLC, dried under vacuum and re-suspended to 1 mM in 50 mM potassium phosphate buffer, pH 7.8. Conjugation efficiency was >95%, as determined by the lack of an individual Tat peptide peak in the HPLC eluate.

Preparation of Tat–C3

The bacterial expression vector encoding the GST (glutathione S-transferase)–Tat–C3 fusion protein has been described previously [25], and was a gift from Dr E. Sahai (Cancer Research UK, London, U.K.). Recombinant Tat–C3 was purified from *Escherichia coli* by standard methods. Briefly, *E. coli* BL21 cells were transformed with the vector encoding the GST–Tat–C3 fusion protein and grown to an A_{600} of 0.6. Protein expression was induced by the addition of 0.1 mM IPTG (isopropyl β-D-thiogalactoside) for 3 h, and the bacteria were collected by centrifugation. The cells were resuspended in extraction buffer

(50 mM Tris/HCl, pH 7.5, 100 mM NaCl, 1 mM sodium dithionite, 1% Triton X-100 and 0.1 mM PMSF), frozen in liquid nitrogen and broken by sonication. The protein extract was clarified by centrifugation and the GST–Tat–C3 was purified on glutathione–Sepharose. Tat–C3 was cleaved from the column as described previously [25] by incubation with 30 mM thrombin overnight at 4°C. The protease was then removed by incubation with *p*-amino-benzamidine–agarose beads.

This Tat–C3 polypeptide was found to contain a single cysteine residue, and so conjugation with thiol-reactive probes was again appropriate. Tat–C3 (400 μM) was reacted with 4 mM 5′-IAF in 20 mM Tris/HCl, pH 8.0. Conjugated protein was separated from free dye by passage through a Pierce Minispin gel-filtration column.

Preparation of large unilamellar liposomes

Liposomes for confocal microscopy were prepared following the method of Thoren et al. [21]. DOPC (20 mg) was dissolved in 2 ml of cyclohexane, freeze-dried under vacuum in round-bottom flasks and stored at −20°C until required. Lipid samples were then resuspended in 50 mM potassium phosphate buffer, pH 7.8, with 1% (v/v) glycerol, and 5 μl aliquots were dropped on to 3 mm coverslips. Samples were then dried by vacuum desiccation and transported to the microscope. Once mounted, 5 μl of 50 mM potassium phosphate buffer, pH 7.8, with 1% (v/v) glycerol was added, and the dried lipid film was allowed to rehydrate for 5 min before use.

Confocal microscopy

Confocal microscopy was performed essentially identically for both liposome and live cell samples, using a Leica SP2 AOBS confocal microscope. In the liposome experiment, 20 μl of a 20 μM solution of the bilayer-staining fluorescent dye FM 4-64 (*N*-(3-triethylammoniumpropyl)-4-{6-[4-(diethylamino) phenyl]hexatrienyl} pyridinium dibromide) (Molecular Probes) was first added to the large unilamellar vesicle solution on the coverslip to demonstrate liposome integrity. A 5 μl volume of either 1 μM IAF, for Tat-less control, or 1 μM Tat–IAF/Tat–C3–IAF was then added. These concentrations remained identical for the live-cell experiment, but 100 μl of 1 μM Tat–A568 and 1 μM Tat–C3–IAF was used. Images of fluorescein and A568 were acquired using 488 nm or 543 nm excitation wavelengths respectively.

Live-cell imaging

Data were collected using a widefield fluorescence imaging system from TILL Photonics essentially as described in [26]. Human cervical carcinoma (HeLa) cells were incubated with 5 μl of 1 μM Tat conjugates. The temperature was maintained at 37°C, and images were acquired 30 min after addition of Tat peptides. Time-lapse images were acquired using sequential acquisition of individual channels.

Results

Tat–IAF, but not Tat–C3–IAF, penetrates synthetic DOPC liposomes *in vitro*

Confocal microscopy was used to investigate the behaviour of Tat–IAF and Tat–C3–IAF in large unilamellar liposomes (~5 μm diameter) of DOPC. Figure 1 shows the results of four individual experiments. The non-specific membrane stain FM 4-64 was used to demonstrate vesicle integrity (Figure 1A). FM 4-64 is believed to insert into bilayers, whereupon it becomes intensely red fluorescent, with an emission maximum at 750 nm. In a separate control experiment, the cell-impermeant green fluorophore 5′-IAF (emission maximum 515 nm) was introduced (Figure 1B). The images show that the 5′-IAF dye alone is not able to accumulate in the internal space of the liposomes.

However, when Tat–IAF was added to similar liposomes (Figure 1C), the peptide-conjugated fluorophore was seen to accumulate rapidly (<1 min) within the liposomes. Staining was noted within the membrane as well as inside the liposomes themselves. After washing the liposomes, the fluorescence within them was seen to reduce slowly, although the membrane staining remained the same. In contrast, Figure 1(D) demonstrates that the introduction

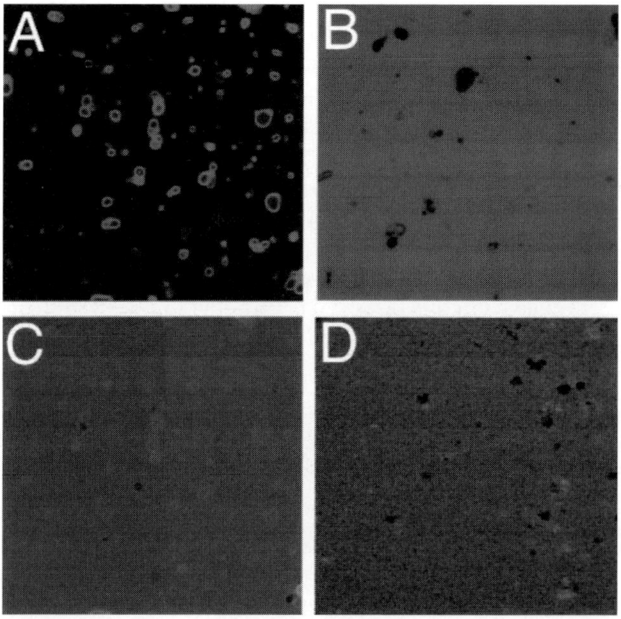

Figure 1 Confocal microscope images of FM4-64 (A), IAF (B), Tat–IAF (C) and Tat–C3–IAF (D) introduced to large unilamellar liposomes of DOPC. The control experiments in (**A**) and (**B**) demonstrate liposome integrity and the inability of IAF to permeate liposomes. (**C**) shows the ability of Tat–IAF to penetrate liposomal bilayers and to accumulate within the bilayers. However, when Tat-(48–60) is fused to a larger (25 kDa) protein cargo (**D**), no accumulation is observed.

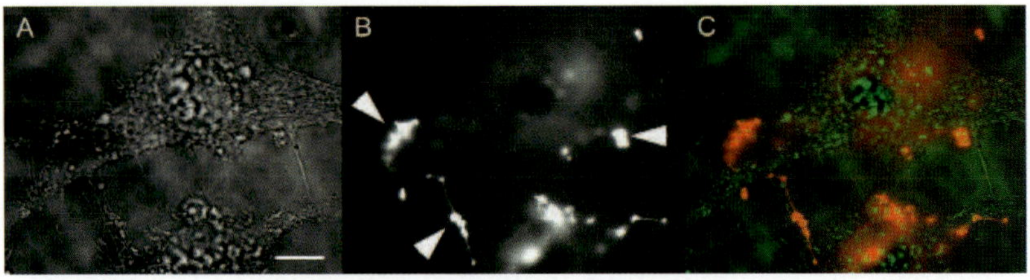

Figure 2 Localization of TAT–A568 to membrane ruffles and extensions in the cell periphery of live HeLa cells. (**A**) Brightfield image (green in merged image, **C**), (**B**) A568 fluorescence (red in merged image, **C**), (**C**) merged image of (**A**) and (**B**). Note the accumulation of A568 in membrane ruffles and extensions (arrowheads). TAT–C3–IAF localizes to these same structures (not shown). Scale bar, 10 μm.

of the larger (<25 kDa), protein-conjugated, Tat–C3–IAF construct at approximately equivalent concentrations did not result in the transfer of fluorescence to the inside of the liposome. There was also no association with the bilayer.

Tat–A568 and Tat–C3 co-localize in unfixed HeLa cells

Widefield microscopy was used to characterize the distribution of the fluorescent constructs in live HeLa cells (Figure 2). Equivalent amounts of Tat–C3–IAF and Tat–A568 were added simultaneously to unfixed HeLa cells. Both of the constructs were seen to localize to similar regions, and to associate with the cell membrane, although only Tat–A568 is shown here. In particular, Tat constructs appear to aggregate in clusters around regions of the cell with significant membrane ruffling (Figure 2B). This might be due to previously predicted electrostatic interaction between the positively charged Tat peptide and the overall negative charge of the cell membrane.

Tat is internalized by endocytosis in HeLa cells

Time-lapse imaging showed that the fluorescent Tat constructs entered cells via fluid-phase endocytosis. Figure 3 shows that punctate structures corre-

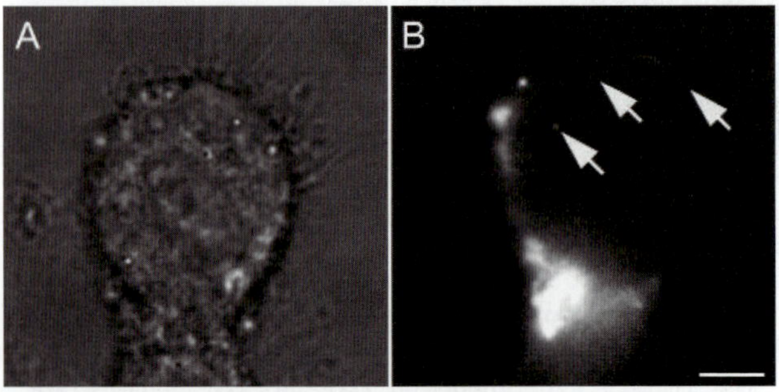

Figure 3 Internalization of TAT–A568 in live HeLa cells. (**A**) Brightfield image, (**B**) A568 fluorescence. Note localization to internal vesicular structures. Time-lapse imaging shows that these undergo rapid directed movements within the cell (not shown). Scale bar, 10 μm.

sponding to intracellular transport intermediates were clearly visible translocating through the cytoplasm. Confocal microscopy was then used to determine the cellular-uptake mechanism of Tat–IAF and images were taken every 20 s after the introduction of Tat–IAF. Figure 4 shows a post-introduction time-course with images 2 min apart, progressing sequentially from Figure 4(A) to Figure 4(D). No significant internalization of the construct was seen, although the expected enhanced binding in membrane regions was observed. It remains possible that internalization of the fluorescently labelled peptide occurred, but was below the limit of detection.

Discussion

The supposed ability of small, basic peptides (CPPs) to cross lipid bilayers without receptor recognition or energetic input is a field of continuing research interest (for review see [23]). Covalent attachment of a variety of macromolecules, including proteins [18], liposomes [17] and phages [6] to these peptides has led to the efficient delivery of these moieties *in vitro* and *in vivo*. This unusual translocation phenomenon could be usefully exploited in the design of a range of potential therapies.

Recent work has suggested that the positively charged CPPs display enhanced uptake behaviour through electrostatic interactions with cell surfaces [13,14], which enhances standard endocytic uptake rates [9,10,12]. However, other studies have continued to suggest the concurrent existence of a hitherto

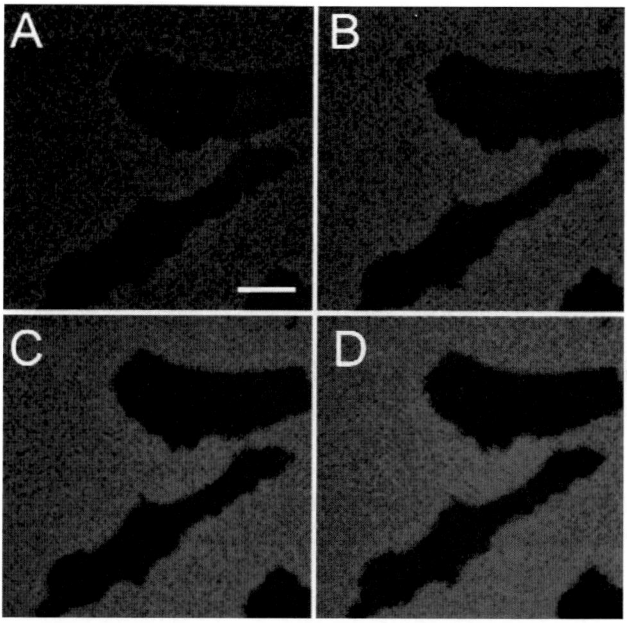

Figure 4 TAT–IAF introduced to live HeLa cells does not show significant translocation to the inside of the cells by scanning confocal microscopy. Frames are 2 min apart. Scale bar, 10 μm.

uncharacterized, direct transmembrane-translocation mechanism that may be used by these peptides [19]. Simple synthetic bilayer systems offer a useful opportunity to analyse any alternative bilayer-translocating phenomena in the absence of other cellular events. Such experiments are essential to confirm some of the proposed translocation mechanisms, but are currently underrepresented in the CPP literature [21,22]. The system presented here, using large unilamellar liposomes of the zwitterionic lipid DOPC, removes any potential contribution from lipid headgroup charge, and indeed any other possible energetic input, such as pH gradient or ATP, and obviously excludes any receptor-recognition event. We used covalent fusions of a C-terminal thiol-modified Tat peptide to enable the direct visualization of the *in vitro* translocation process by confocal and widefield microscopy.

An observed translocation behaviour *in vitro* may be affected by cargo size

A thioreactive fluorescein derivative, 5′-IAF, was seen to be internalized into liposomes only when covalently attached to the peptide (Figures 1B and 1C). There is also significant accumulation within the bilayer region, yet there does not appear to be any disruption of the bilayer, in agreement with previous studies [19–21]. However, when this small IAF fluorophore was replaced with a large (25 kDa) polypeptide corresponding to the active site of botulinum toxin (C3), no internalization was seen (Figure 1D). These results imply that the Tat peptide can cross synthetic liposomal bilayers *in vitro*, in the total absence of other cellular components. Moreover, penetration has a rate of influx much higher than that seen in live cells [7,27], being in the order of seconds rather than minutes. However, there appears to be a size limit on translocation *in vitro*. Further work must be performed to validate this observation, in light of the recent publications questioning the mechanism of Tat peptide uptake [9–12]. However, if penetration is indeed mediated by the physical properties of the peptide, these properties could be influenced by the attachment of a large peptide group.

Two mechanisms for CPP translocation have now been suggested. The first predicts the formation of inverted micelles within the membrane due to electrostatic interactions between the positively charged peptide and negatively charged bilayer lipids [4]. These structures have a hydrophilic centre that can accommodate the peptide charge, and are able to traverse the bilayer. The second postulates an 'electroporation-like' mechanism [28], where the asymmetric distribution of peptide across the bilayer generates a transmembrane electrical field. This alters the lateral pressure and curvature stress within the bilayer, which, at certain values, causes the bilayer to become permeable and so permit the passage of the peptide across it. Implicit in both of these ideas is that peptide translocation occurs down a concentration gradient, and that the phenomenon is potentially a reversible equilibrium process [27].

Our data agree with both models. With regard to the first model, there will be a limit to the size of the moiety that can be accumulated within the inverted micelles. Potentially, the presence of the charged peptides at the membrane surface leads to transient dipole formation at the zwitterionic DOPC headgroups, which allows translocation via the formation of small inverted micelles. One might also postulate that the charge density at a membrane site is an important factor in instigating this process, and that peptides tethered to large proteins cannot accumulate at the bilayer surface at sufficient concentrations to induce micelle formation. The alteration of bilayer properties in the second electroporation model for spontaneous translocation could result in the bilayer becoming permeable only to cargos of a certain size or physical property (e.g. overall charge).

Accumulation of Tat–IAF within the DOPC bilayer

Tat–IAF also appears to associate with the membrane itself, as evidenced by the concentration of fluorescence at the liposome bilayer in samples in which translocation is taking place (Figure 1C). It might be supposed that if a peptide concentration gradient is the driving force behind penetration to the inside of the liposome, a proportion of the peptide will be trapped within the bilayer at an equilibrium state. This bilayer localization only occurs in samples that demonstrate the translocation phenomena, which suggests that the bilayer-associated peptides are more likely to represent an intermediate state rather than a partitioning event.

Tat–A568 is internalized by endocytosis in live HeLa cells

Imaging of live HeLa cells confirms reports that Tat–peptide internalization is primarily mediated by fluid-phase endocytosis in live cells (Figure 3). Under confocal microscopy (Figure 4), the peptide was not found to accumulate inside cells at observable levels on short timescales. This does not necessarily rule out a role for the alternative translocation mechanism, but suggests that it is at best a secondary mechanism for entry into HeLa cells. Previous studies have shown that Tat–C3 is taken up in a similar manner [25].

Co-localization of Tat–A568 and Tat–C3–IAF

Widefield microscopy was used to demonstrate the equivalent distribution of Tat–A568 and Tat–C3–IAF in HeLa cells (Figure 2). In both cases, the peptide conjugates were localized to equivalent regions of the cell, with particular localization in areas of membrane ruffling (i.e. of significant membrane surface area) as expected.

Comparison with previous *in vitro* studies

A CPP derived from the *Antennapedia* homoeodomain and with some sequence homology with Tat has been shown to enter giant unilamellar liposomes of 5–10 μm in diameter [21]. The liposomes used in this previous study were made of a crude lipid extract, containing about 40% PC, 33% phosphatidylethanolamine, 14% phosphatidylinositol, 5% lyso-PC and 4% cardiolipin. This should be compared with the pure (>95%) DOPC liposomes used here.

However, in a second and more recent study using egg-PC at 200 nm diameter, no entry of a fluorescently labelled TAT-(44–57) was observed [22]. This work used a novel assay involving Tb^{3+}-containing liposomes, whereby the fluorescence of the label is enhanced by reaction with Tb^{3+} after bilayer translocation, and no translocation event was observed. We have also tried simple fluorescence measurements using quenches for fluorescein, relief of self-quench and even the quenching of the tryptophan fluorescence in a different peptide derivative in liposomes containing the fluorescence quencher iodide with no success. An appreciation of the difficulties of measuring fluorescence signals in liposomal suspension — such as high background scattering, and the internal volume of liposomes being several orders of magnitude smaller than the external solution volume — persuaded us to use a microscopy approach to study Tat behaviour *in vitro*.

Conclusions

We demonstrate the successful *in vitro* bilayer translocation of Tat peptide conjugates into single-component, small DOPC liposomes. Interestingly, we observed a size limitation on the cargo that can be translocated into such liposomes. The conjugates are internalized into unfixed HeLa cells by fluid-phase endocytosis, and are localized to membrane ruffles at the periphery of the cells. This work contributes to the ongoing debate regarding a possible spontaneous bilayer translocation mechanism for CPPs.

This work was supported by the BBSRC (Biotechnology and Biological Sciences Research Council) (studentship to P.J.B. for P.C.).

References

1. Asoh, S., Ohsawa, I., Mori, T., Katsura, K., Hiraide, T., Katayama, Y., Kimura, M., Ozaki, D., Yamagata, K. and Ohta, S. (2002) Proc. Natl. Acad. Sci. U.S.A. **99**, 17107–17112
2. Gustafsson, A.B., Sayen, M.R., Williams, S.D., Crow, M.T. and Gottlieb, R.A. (2002) Circulation **106**, 735–739
3. Gratton, J.-P., Yu, J., Griffith, J.W., Babbitt, R.W., Scotland, R.S., Hickey, R., Giordano, F.J. and Sessa, W.C. (2003) Nat. Med. **9**, 357–362
4. Derossi, D., Calvet, S., Trembleau, A., Brunissen, A., Chassaing, G. and Prochiantz, A. (1996) J. Biol. Chem. **271**, 18188–18193
5. Vives, E., Brodin, P. and Lebleu, B. (1997) J. Biol. Chem. **272**, 16010–16017
6. Eguchi, A., Akuta, T., Okuyama, H., Senda, T., Yokoi, H., Inokuchi, H., Fujita, S., Hayakawa, T., Takeda, K., Hasegawa, M. and Nakanishi, M. (2001) J. Biol. Chem. **276**, 26204–26210
7. Suzuki, T., Futaki, S., Niwa, M., Tanaka, S., Ueda, K. and Suguira, Y. (2002) J. Biol. Chem. **277**, 2437–2443
8. Lindsay, M.A. (2002) Curr. Opin. Pharmacol. **2**, 587–594
9. Richard, J.P., Melikov, K., Vives, E., Ramos, C., Verbeure, B., Gait, M.J., Chernomordik, L.V. and Lebleu, B. (2003) J. Biol. Chem. **278**, 585–590
10. Lundberg, M. and Johansson, M. (2002) Biochem. Biophys. Res. Commun. **291**, 367–371
11. Potocky, T.B., Menon, A.K. and Gellman, S.H. (2003) J. Biol. Chem. **278**, 50188–50194
12. Fittipaldi, A., Ferrari, A., Zoppe, M., Arcangeli, C., Pellegrini, V., Beltram, F. and Giacca, M. (2003) J. Biol. Chem. **278**, 34141–34149
13. Tyagi, M., Rusnati, M., Presta, M. and Giacca, M. (2001) J. Biol. Chem. **276**, 3254–3261
14. Bellet-Amalric, E., Blaudez, D., Desbat, B., Graner, F., Gauthier, F. and Renault, A. (2000) Biochim. Biophys. Acta **1467**, 131–143
15. Mann, D.A. and Frankel, A.D. (1991) EMBO J. **10**, 1733–1739
16. Lindgren, M., Hallbrink, M., Prochiantz, A. and Langel, U. (2000) Trends Pharm. Sci. **21**, 99–103

17. Torchillin, V.P., Rammohan, R., Weissig, V. and Levchenko, T.S. (2001) Proc. Natl. Acad. Sci. U.S.A. **98**, 8786–8791
18. Schwarze, S.R., Ho, A., Vocero-Akbani, A. and Dowdy, S.F. (1999) Science **285**, 1569–1572
19. Thoren, P.E., Persson, D., Isakson, P., Goksor, M., Onfelt, A. and Norden, B. (2003) Biochem. Biophys. Res. Commun. **307**, 100–107
20. Zaro, J.L. and Shen, W.-C. (2003) Biochem. Biophys. Res. Commun. **307**, 241–247
21. Thoren, P.E., Persson, D., Karlsson, M. and Norden, B. (2000) FEBS Lett. **482**, 265–268
22. Kramer, S.D. and Wunderli-Allenspach, H. (2003) Biochim. Biophys. Acta **1609**, 161–169
23. Vives, E. (2003) J. Mol. Recognit. **16**, 265–271
24. Sauzeau, V. (2001) Circulation Res. **88**, 1102–1104
25. Coleman, M.L., Sahai, E.A., Yeo, M., Bosch, M., Dewar, A. and Olson, M.F. (2001) Nat. Cell Biol. **3**, 339–345
26. Stephens, D.J. (2003) EMBO Rep. **4**, 210–217
27. Hallbrink, M., Anders, F., Elmquist, A., Pooga, M., Bartfai, T. and Langel, U. (2001) Biochim. Biophys. Acta **1515**, 101–109
28. Binder, H. and Lindblom, G. (2003) Biophys. J. **85**, 982–995

Biochem. Soc. Symp. **72**, 211–222
(Printed in Great Britain)

20

Aggregation and fibrillization of prions in lipid membranes

Jurate Kazlauskaite and Teresa J.T. Pinheiro[1]

Department of Biological Sciences, University of Warwick, Coventry CV4 7AL, U.K.

Abstract

A key molecular event in prion diseases is the conversion of PrP (prion protein) from its normal cellular form (PrP^c) into the disease-specific form (PrP^{Sc}). The transition from PrP^c to PrP^{Sc} involves a major conformational change, resulting in amorphous aggregates and/or fibrillar amyloid deposits. Here, we review several lines of evidence implicating membranes in the conversion of PrP, and summarize recent results from our own work on the role of lipid membranes in conformational transitions of prion proteins. By establishing new correlations between *in vivo* biological findings with *in vitro* biophysical results, we propose a role for lipid rafts in prion conversion, which takes into account the structural heterogeneity of PrP in different lipid environments.

The hidden role of membranes in prion conversion

The key molecular event in prion diseases is the conversion of the normal PrP^c (cellular prion protein) to the aberrant pathogenic form, generally denoted as the scrapie isoform (PrP^{Sc}) [1,2]. Prion diseases, or TSEs (transmissible spongiform encephalopathies) include CJD (Creutzfeldt–Jakob disease) in humans, BSE (bovine spongiform encephalopathy), scrapie in sheep, and CWD (chronic wasting disease) in deer and elk, among others. In these diseases, as in other neurodegenerative diseases, such as Alzheimer's and Parkinson's, the accumulation of amyloid plaques and/or amorphous aggregates in the brain is associated with neuronal loss and consequent degeneration of brain function, and, invariably, in the death of the afflicted individuals [3]. The molecular details of the conversion of PrP^c into PrP^{Sc} are not fully understood, but appear to involve a refolding of the monomeric α-helical cellular protein into an oligomeric β-sheet-rich form prone to aggregation and/or fibrillization. The present review highlights *in vivo* and *in vitro* experimental evidence implicating membranes in prion conversion.

[1]To whom correspondence should be addressed (email t.pinheiro@warwick.ac.uk).

In normal healthy conditions, prion is a harmless plasma membrane protein predominantly expressed in the brain. The function of PrP^{c} is not fully established, but it is implicated in copper binding and cell signalling [4–6]. Prion is a 209-amino-acid-long protein, with two N-glycosylation sites at asparagine residues, attached to the outside face of the cell membrane via a GPI (glycosylphosphatidylinositol) lipid anchor coupled to the C-terminal Ser^{231} residue [7]. Like other GPI-anchored proteins, PrP^{c} is sequestered into cholesterol- and sphingomyelin-rich membrane domains, known as lipid rafts [7,8].

Experimental observations suggest that an interaction of PrP with the membrane may play a role in the conversion of PrP^{c} into PrP^{Sc}. Early studies revealed that purified PrP^{Sc} and infectivity co-partition into lipid membranes [9]. In addition, PrP^{c} is easily cleaved and released from the cell surface after digestion with PIPLC (phosphatidylinositol-specific phospholipase C), whereas PrP^{Sc} is not, after similar treatment [10,11]. These and other findings [12] support the idea that an altered membrane association of PrP may be an important factor in prion pathogenesis.

PrP^{c} is synthesized by ribosomes on the cytosolic face of the ER (endoplasmic reticulum) membrane, and transits through the Golgi before reaching its destination at the cell surface. From the plasma membrane, PrP^{c} can be recycled, degraded or converted into PrP^{Sc} (Figure 1). The subcellular site for the formation of PrP^{Sc} is unknown; however, conversion occurs after PrP^{c} reaches the plasma membrane [17,19]. Evidence from scrapie-infected cultured cells implicates the plasma membrane and endocytic organelles as relevant sites, but it is unclear which provides a more favourable environment for conversion and whether compartments along the secretory pathway might also be involved. Once PrP^{Sc} is formed, it appears to accumulate in late endosomes, lysosomes, on the cell surface and in extracellular spaces in the form of amorphous deposits, diffuse fibrils or dense amyloid plaques [20–23].

In addition to fully secreted GPI-anchored PrP, anchorless forms of PrP are produced in the ER or released as soluble forms. Two transmembrane forms of PrP, with either the N- or C-terminus in the lumen, are associated with the ER membrane via a common transmembrane segment, comprising a hydrophobic stretch of residues from 113 to 135 [24,25]. Release of soluble forms of PrP (lacking GPI anchor) from cells has been reported some years ago [16,26–28], and, more recently, similar results have been obtained using splenocytes or cerebellar granule neurons from transgenic mice overexpressing PrP [29]. Also, 10–20% of PrP^{Sc} molecules extracted from infected hamsters were found to be truncated at Gly^{228}, and therefore lack GPI anchor [30].

The majority of PrP^{Sc} generated in infected animals contains a GPI anchor, and the relevance of anchorless forms of PrP is not clear. However, it is possible that anchorless forms of PrP may be physiologically relevant in the conversion process of PrP. A recent study has tested the effect of GPI anchor on prion conversion [12]. This work shows that either GPI-anchored (GPI^{+}) PrP or an anchor-deficient (GPI^{-}) form of PrP expressed in fibroblasts is stably associated with raft membranes, and the latter alternative

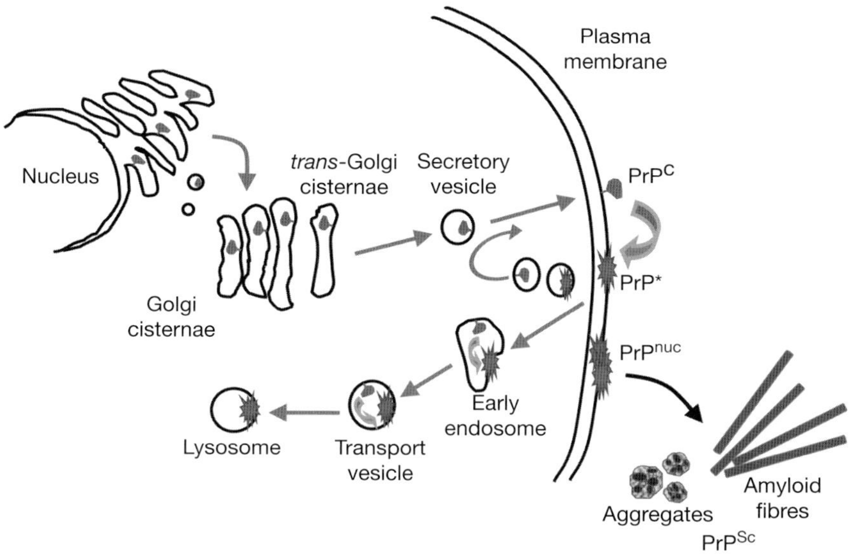

Figure 1 Cellular trafficking of PrP^c and potential subcellular sites for the formation of PrP^{Sc}. PrP^c is synthesized in the rough ER, as an approx. 250-residue-long polypeptide chain, containing an N-terminal signal peptide and a C-terminal hydrophobic sequence that signals for the addition of the GPI anchor. The polypeptide enters the ER maturation route and transits through the Golgi before it reaches its location at the plasma membrane. As it trafficks through the ER and Golgi, the protein is subjected to the cleavage of the N-terminal signal peptide, addition of N-linked oligosaccharide chains at two sites, formation of a single disulphide bond and attachment of the GPI anchor [7,13,14]. At the cell surface, the protein is incorporated into the outer layer of the plasma membranes via its GPI anchor and is sequestered into rafts (not shown) [7,8]. PrP^c is shown as a 'splodge' associated with the various membranes via a tail that represents the GPI anchor. From the plasma membrane, PrP^c can be endocytosed and recycled to the cell surface as rapidly as within 3 min [15], degraded in the endosome and lysosome pathway, with a half-life of 4–6 h [16–18], or converted into PrP^{Sc}, comprising amorphous aggregates and amyloid fibrils. This conversion may occur via a membrane-associated altered conformation (PrP^*) that is prone to aggregation. Lipid rafts within the plasma membrane assist on the formation of a nucleus (PrP^{nuc}), which can either progress into amyloid fibrils or amorphous aggregates.

mode of membrane association (independent of GPI) is not detectably altered by glycosylation and is not markedly reduced by deletion of the N-terminus. Furthermore, using a cell-free assay, it was found that GPI^+ PrP conversion into PrP^{Sc} requires either the action of PIPLC or addition of the membrane-fusion agent PEG [poly(ethylene glycol)], whereas GPI^- PrP is converted without PIPLC or PEG treatment [12]. These findings support the view that additional direct lipid–polypeptide interactions may play a role in the conversion of PrP^c into PrP^{Sc}.

In vitro binding of anchorless prion proteins to lipid membranes

Despite the several lines of experimental evidence that suggest a potential role of lipid in the conversion of PrP, few *in vitro* structural studies have addressed the role of membranes. The main difficulty resides in the fact that PrP is anchored to the plasma membrane via a GPI anchor, and some would argue that studies of PrP and membranes should involve a lipid-anchored prion molecule. However, purification of GPI-anchored PrP from animal sources is not a viable approach, as it cannot yield the required amounts for comprehensive structural and/or biophysical studies. For this reason, the vast majority of structural studies are carried out with recombinant forms of PrP expressed in *E. coli*, which are neither glycosylated nor contain a GPI anchor. We reason that if membranes have a role in prion conversion, this is likely to involve a direct interaction of the polypeptide chain with lipid molecules in the membrane; thus, using prion proteins without the GPI anchor provides a reasonable approach to examine the role of membranes in the conversion of PrP. Furthermore, anchorless forms of PrP are produced *in vivo* (as described above), and these may constitute the initial prion conformation that catalyses the conversion reaction. In the absence of the true identity of the pathogenic form of the prion molecule [31], it seems prudent to keep an open view.

The propensity of PrP to adopt various conformations *in vitro* under different conditions [32–34] suggests that within the cell, as PrP experiences various subcellular environments during trafficking in the ER, at the plasma membrane and recycling through the endocytic pathway, PrP may exist in different conformations. In all these various cellular environments, PrP is associated with lipid membranes, providing numerous chances for an adverse association with lipid molecules. The potential role of membranes in PrP conversion can be envisaged in two ways. There could be a direct involvement of membranes in the conversion mechanism, whereby association of PrP^{c} with lipid produces an altered conformation, referred to as PrP^{*}, that is involved in the formation of PrP^{Sc}. Alternatively, PrP^{*} may form independently of a direct interaction with lipid through changes in cellular environment and/or interactions with other factors. The resulting altered molecule (PrP^{*}) may have a higher affinity for lipid membranes that facilitates protein–protein interactions, leading to the formation of PrP^{Sc} (Figure 1). Studies from our group have examined these alternative PrP–membrane interactions in prion conversion by analysing the binding properties of different forms of the prion protein to lipid membranes and their resulting structural changes [35–37]. In the present review, we summarize *in vitro* structural studies that support a model involving membrane-associated intermediate conformations of PrP (PrP^{*} and PrP^{nuc}) in the fibrillization and aggregation of prions (Figure 1).

Recombinant PrP can be refolded to a state with high content of α-helix, referred as α-PrP, which represents PrP^{c}, and forms rich in β-sheet structure, β-PrP, that have some intermediate properties of PrP^{Sc}, and therefore serve as models for early states in the formation of PrP^{Sc} [36,38]. The interaction of α- and β-isoforms of the truncated protein PrP-(90–231) to model lipid mem-

Table 1 Summary of binding of PrP isoforms to lipid membranes. The binding of PrP-(90–231) isoforms to membranes composed of POPC (1-palmitoyl-2-oleoyl phosphatidylcholine), POPG (1-palmitoyl-2-oleoyl phosphatidylglycerol) or DPPC (dipalmitoyl phosphatidylcholine)/cholesterol/sphingomyelin (rafts) is summarized from fluorescence binding curves presented in Sanghera and Pinheiro [35], and Kazlauskaite et al. [36]. The isoform α-PrP is richer is α-helical structure and represents PrP^{c}, while β-PrP is rich in β-sheet and is a model for intermediate conformations of PrP^{Sc}. Measurements at pH 7 model the environment at the plasma membrane, whereas pH 5 represents the endosomal/lysosomal pH. No binding is indicated by −, strong binding as ++ and less strong binding as +.

	α-PrP		β-PrP	
	pH 7	pH 5	pH 7	pH 5
POPC	−	−	−	−
POPG	++	++	+	++
Rafts	+	−	+	++

branes was investigated, at both pH 7 and 5, to model the pH environment of the plasma membrane and within endosomes/lysosomes respectively [35–37]. These studies show that α- and β-PrP have different binding affinities to lipid membranes and exhibit different pH-dependent binding behaviour. Both α- and β-PrP bind to negatively charged lipid membranes in a pH-dependent manner. No binding is observed with zwitterionic lipid membranes for either isoform, but both bind to raft membranes, with β-PrP binding more strongly than the α-helical form (Table 1). A study by Morillas et al. [39], using mixed lipid vesicles and micellar systems, also shows that binding of recombinant human prion proteins to lipid membranes requires negatively charged lipid.

Surface plasmon resonance studies reveal that, overall, the binding of PrP to lipid membranes is favoured at pH 5 relative to pH 7, and that the association of β-PrP with lipid membranes is stronger than that of α-PrP [37]. The stronger binding of PrP to lipid membranes at pH 5 suggests that endosomes and lysosomes, where the pH is more acidic than at the plasma membrane, reaching pH values of ~5, can provide an environment conducive to a strong association of PrP with the membrane surface. Depending on the starting conformation of PrP, this interaction can be irreversible and can result in drastic consequences. In a membrane-associated state, a prion molecule destined for degradation could become more protected to proteolytic digestion and start to accumulate in the cell. Alternatively, because PrP is constitutively recycled between the endocytic compartment and the plasma membrane [15], a corrupted conformation generated in endosomes could appear at the plasma membrane, where it could serve as a catalyst for the conversion of PrP^{c}.

In vitro aggregation and fibrillization of prions in lipid membranes

Binding of PrP to lipid membranes results in structural changes to PrP [35,36]. The extent and nature of structural changes depends on pH, membrane lipid composition and the starting conformation of PrP. The resulting altered membrane-associated prion conformations can lead to either amorphous protein aggregation or formation of amyloid fibrils. Both α- and β-PrP acquire more β-sheet structure upon binding to negatively charged lipid. Calcein release from loaded vesicles showed that the integrity of these membranes is destabilized by PrP, resulting in total release of vesicle-entrapped dye. Furthermore, interaction with negatively charged lipid results in amorphous PrP aggregation, regardless of pH and the starting conformation of PrP (Table 2). In contrast, binding of α-PrP to raft membranes increases α-helical structure, does not destabilize the lipid membrane and does not lead to PrP aggregation, whereas binding of β-PrP to rafts induces unfolding of PrP without disturbing the integrity of the lipid bilayer, and culminates in fibrillization of PrP (Table 2).

Most cases of prion diseases are sporadic (>80% of the total number of cases) and conversion of PrP^{c} into the aberrant disease-associated form, PrP^{Sc}, is usually described as a rare event during the folding of PrP. However, the association of PrP^{c} with membranes is often neglected. An aberrant association of PrP^{c} with cellular membrane components provides an alternative hypothesis for the spontaneous conversion of PrP in prion diseases. Our studies showed that α-PrP binds to ordered lipid membranes rich in cholesterol and sphingomyelin, characteristic of the lipid composition of rafts. This interaction results in a protein conformation with a higher content of α-helix, compared with the structure of PrP in solution. This gain in α-helix is likely to reflect folding of the unstructured N-terminal domain of PrP. The α-helical state formed with raft membranes does not disturb the integrity of the lipid bilayer. Thus an extrapolation to the *in vivo* situation suggests that a direct interaction of PrP^{c} with rafts in the plasma membrane would stabilize the α-helical cellular form of PrP without destabilizing the plasma membrane.

The precise mechanism of PrP-induced cell death is unclear. *In vitro* studies have demonstrated that PrP^{Sc} is neurotoxic to cultured neurons [40,41], but has no toxic effects when grafted in the brain of Prp-$p^{0/0}$ mice [42,43]. Various hypotheses have been postulated to explain neuronal cell death. These include increased oxidative stress as a result of depleted levels of PrP^{c} [44]. Others have proposed a mechanism for prion neurotoxicity through a direct perturbation of the plasma membrane of nerve cells as a consequence of PrP^{Sc} accumulation [45,46]. Our results show that, in lipid rafts, α-PrP acquires a higher α-helical structure and does not induce membrane destabilization. However, if PrP is segregated into other negatively charged lipid domains in the plasma membrane, this results in a β-sheet-rich conformation, PrP^{*}, which destabilizes the lipid membrane, leading to total release of vesicle contents [35,36]. *In vivo*, such destabilization would lead to leakage of cell contents and eventually to cell death. Thus our findings support the proposal of a mechanism for neurotoxicity through a direct interaction of PrP^{Sc} and/or related precursor forms with neuronal membranes.

At present, the molecular events that facilitate PrP^{Sc} aggregation are still unknown. The plasma membrane of most cells typically contains an outer surface that is rich in cholesterol and zwitterionic phosphocholine-based lipids, and

Table 2 Structural changes, membrane effects and resulting aggregation or fibrillization of PrP isoforms in lipid membranes. The structural effects on PrP-(90–231) isoforms upon binding to lipid membranes composed of POPG (1-palmitoyl-2-oleoyl phosphatidylglycerol) or DPPC/cholesterol/sphingomyelin (rafts) is summarized from FTIR (Fourier transform IR) structural measurements. Membrane destabilization was measured from the release of calcein from loaded vesicles and aggregation and fibrillization of PrP analysed by electron microscopy. Only the main changes are indicated. The full description of all structural effects is presented in Sanghera and Pinheiro [35], and Kazlauskaite et al. [36]. *α-PrP does not bind to raft membranes at pH 5 [35]. Measurements at pH 7 model the environment at the plasma membrane, while those at pH 5 represent the endosomal/lysosomal pH.

	Structural effects		Membrane destabilization		Aggregation/fibrillization	
Lipid (pH)	α-PrP	β-PrP	α-PrP	β-PrP	α-PrP	β-PrP
POPG (5)	↑ β-Sheet	↑ β-Sheet	Yes	Yes	Aggregation	Aggregation
POPG (7)	↑ β-Sheet	↑ β-Sheet	Yes	Yes	Aggregation	Aggregation
Rafts (5)	–*	↑ Coil	–*	No	–*	Fibrillization
Rafts (7)	↑ α-Helix	↑ Coil	No	No	None	Fibrillization

a negatively charged inner surface [47]. Under normal circumstances, the majority of PrP is associated with lipid rafts on the outer leaflet of the plasma membrane. Our findings indicate that these conditions do not induce PrP aggregation, if PrP is in its α-helical cellular conformation. Instead, an interaction of PrP^c with the membrane surface would stabilize the protein in the α-helical structure. However, during internalization, either for degradation or recycling, PrP^c leaves rafts [15] and becomes exposed to negatively charged lipids in cell membranes. This change in lipid environment triggers a profound conformational change in PrP^c, resulting in an altered β-sheet-poised state of PrP that could act as a precursor and/or a template for the formation of PrP^{Sc} [35].

The ambivalent role of rafts in prion conversion

Since the pathogenic prion conversion appears to take place at the cell surface [10,11,20], and PrP^c is predominantly sequestered into raft domains, it would follow that conversion is likely to occur in rafts. Several lines of evidence implicate rafts in the conversion of PrP: (i) both PrP^c and PrP^{Sc} co-localize in rafts [8,20]; (ii) cholesterol depletion redistributes PrP^c outside rafts within the plasma membrane and decreases formation of PrP^{Sc} in prion-infected cell lines [19,48]; (iii) recombinant transmembrane forms of PrP^c, not localized in rafts, do not serve as substrates for the formation of PrP^{Sc} [19,49]; (iv) depletion of sphingolipids promotes prion infection in neuroblastoma cells [50]; (v) treatment with the sterol-binding anti-scrapie drug amphotericin B modifies the flotation properties of extracted rafts and inhibits PrP^{Sc} synthesis [51]; and (vi) conversion of raft-associated GPI-anchored PrP requires insertion of PrP^{Sc} into the host membrane in cell-free conversion reactions [12].

However, the above experimental evidence for a role of rafts in prion conversion is ambivalent. Depletion of sphingolipids destabilizes rafts and the resulting increase in formation of PrP^{Sc} suggests that the raft environment may stabilize the cellular conformation of PrP [50]. *In vitro* structural studies from our group showed that, in model raft membranes, the α-helical conformation of PrP, which represents PrP^c, is protected from aggregation and fibrillization [35], but that a poised conformation with elevated β-sheet structure is converted into amyloid fibrils [36]. In contrast with sphingomyelin, interfering with cholesterol decreases the formation of PrP^{Sc} [19,48]. Our own *in vitro* studies show no specific interactions of PrP with cholesterol-containing membranes [35]. The contrasting effects of sphingomyelin and cholesterol may imply that PrP conversion within rafts requires another cellular factor that accumulates in rafts in a cholesterol-dependent manner. Alternatively, conversion of PrP may occur outside rafts. In favour of this view is the fact that PrP^c appears to leave rafts before being endocytosed [15] and *in vitro* results showing that the α-helical isoform of PrP in non-raft membranes forms amorphous aggregates [36]. However, this latter study also shows that a β-sheet conformation of PrP forms amyloid fibrils in rafts. These results suggest that *in vivo* an altered state could be induced outside rafts during recycling of PrP through the endocytic pathway, where the acidic pH favours unfolding of PrP. Alternatively, this altered conformation may be induced through an interaction(s) with other cellular protein(s). Once formed, this corrupted conformation of PrP would accumulate in rafts and lead to amyloid fibrils and/or amorphous aggregates (Figure 2).

In conclusion, *in vitro* studies of prions and lipids support the view that the membrane environment is likely to play a role in the conversion of PrP^c into PrP^{Sc} *in vivo*. The strong binding of PrP to membranes containing negatively

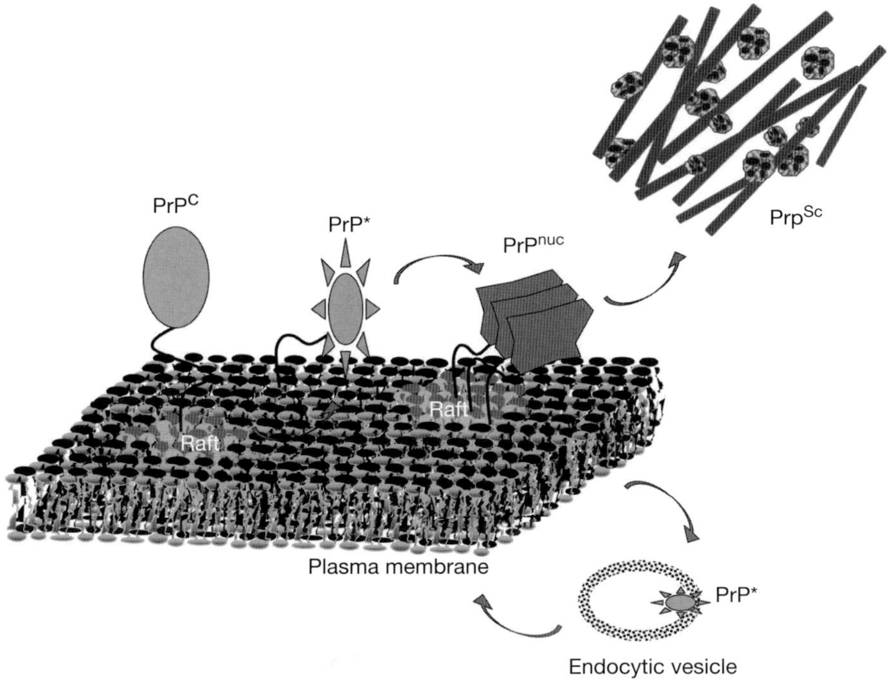

Figure 2 The ambivalent role of rafts in prion conversion. A fragment of the plasma membrane is depicted showing the fluid lipid bilayer composed of unsaturated lipid molecules. PrP^c is sequestered into ordered cholesterol and sphingomyelin-rich lipid domains or rafts via its GPI anchor [7]. *In vitro* biophysical structural studies [35,36] indicate that the α-helical conformation, corresponding to PrP^c, is protected in rafts. In non-raft lipid membranes PrP undergoes a conformational change to an altered state with a high content of β-sheet structure (PrP*). This altered conformation may form *in vivo* either in endocytic vesicles or while PrP transits outside rafts [15]. PrP* is prone to aggregation, and, if accumulated in domains of the plasma membrane with negatively charged lipid molecules, will rapidly form amorphous aggregates [36]. However, *in vivo*, PrP* is likely to re-enter a raft. Conformations of PrP rich in β-sheet structure have a higher affinity for lipid rafts than α-helical forms [37]. Thus PrP* molecules will accumulate in rafts; their confinement within a raft promotes the formation of oligomeric states that constitute a nucleus (PrP^{nuc}) for the fibrillization and/or aggregation of PrP. In this scheme, PrP* is formed outside rafts, while nucleation and fibril growth occur within rafts at the plasma membrane surface. After a critical size, aggregates and fibrils (PrP^{Sc}) fall out of the membrane surface into the extracellular space, carrying with them strongly associated raft lipid components, such as sphingolipids [52].

charged lipid, resulting in vesicle destabilization, suggests that, *in vivo*, PrP could disrupt cell membranes and lead to the release of cell contents and/or secretory and endosomal vesicles. Our studies suggest that interactions of PrP with negatively charged lipid molecules are highly undesirable, as this leads to β-sheet-rich forms of PrP and the rapid formation of amorphous aggregates. As for the involvement of rafts in PrP conversion, they appear to have an ambivalent role, which is reconciled in our studies by considering the different structural forms of PrP. Rafts provide a protective environment for the α-helical cellular conformation of PrP, but a corrupted altered conformation, poised with β-sheet structure, has a higher affinity for raft membranes, and this results in fibrillization of PrP (Figure 2). The β-sheet precursor conformation, PrP*, is likely to form outside rafts, probably through dangerous liaisons with negatively charged lipid molecules during endocytosis and recycling to the cell surface. Alternatively, interactions with other non-lipid cellular factors may induce the altered PrP* conformation. In both cases, the resulting PrP* molecule has a higher affinity for lipid rafts, and this promotes the formation on the membrane surface of a nucleus (PrP^{nuc}) for the aggregation and/or fibrillization of PrP.

Prion research in T.J.T.P.'s laboratory has been funded by the Royal Society, the Wellcome Trust (SHoWCAse 053914/Z/98/Z), the Medical Research Council (G9901445) and the Biotechnology and Biological Sciences Research Council (88/BS516471). We thank Professor John Ellis for helpful discussions and a critical reading of the manuscript.

References

1. Prusiner, S.B. (1998) Proc. Natl. Acad. Sci. U.S.A. **95**, 13363–13383
2. Chesebro, B. (2003) Br. Med. Bull. **66**, 1–20
3. Caughey, B. and Lansbury, P.T. (2003) Annu. Rev. Neurosci. **26**, 267–298
4. Brown, D.R. (2004) Neurobiol. Dis. **15**, 534–543
5. Prado, M.A., Alves-Silva, J., Magalhaes A.C., Prado, V.F., Linden, R., Martins, V.R. and Brentani, R.R. (2004) J. Neurochem. **88**, 769–781
6. Roucou, X., Gains, M. and LeBlanc, A.C. (2004) J. Neuroscience Res. **75**, 153–161
7. Stahl, N., Borchelt, D.R., Hsiao, K. and Prusiner, S.B (1987) Cell **51**, 229–240
8. Vey, M., Pilkuhn, S., Wille, H., Nixon, R., DeArmond, S.J., Smart, E.J., Anderson, R.G.W., Taraboulos, A. and Prusiner, S.B. (1996) Proc. Natl. Acad. Sci. U.S.A. **93**, 14945–14949
9. Gabizon, R., McKinley, M.P. and Prusiner, S.B. (1987) Proc. Natl. Acad. Sci. U.S.A. **84**, 4017–4021
10. Stahl, N., Borchelt, D.R. and Prusiner, S.B. (1990) Biochemistry **29**, 5405–5412
11. Safar, J., Ceroni, M., Gajdusek, D.C. and Gibbs, C.J. (1991) J. Inf. Dis. **163**, 488–494
12. Baron, G.S., Wehrly, K., Dorward, D.W., Cesebro, B. and Caughey, B. (2002) EMBO J. **21**, 1031–1040
13. Turk, E., Teplow, D.B., Hood, L.E. and Prusiner, S.B. (1988) Eur. J. Biochem. **176**, 21–30
14. Haraguchi, T., Fisher, S., Olofsson, S., Endo, T., Groth, D., Tarentino, A., Borchelt, D.R., Teplow, D., Hood, L., Burlingame, A. et al. (1989) Arch. Biochem. Biophys. **274**, 1–13
15. Sunyach, C., Jen, A., Deng, J., Fitzgerald, K.T., Frobert, Y., Grassi, J., McCaffrey, M.W. and Morris, R. (2003). EMBO J. **22**, 3591–3601
16. Borchelt, D.R., Scott, M., Taraboulos, A., Stahl, N. and Prusiner, S.B. (1990) J. Cell Biol. **110**, 743–752
17. Caughey, B. and Raymond, G.J. (1991) J. Biol. Chem. **266**, 18217–18224

18. Borchelt, D.R., Taraboulos, A. and Prusiner, S.B. (1992) J. Biol. Chem. **267**, 16188–16199
19. Taraboulos, A., Scott, M., Semenov, A., Avraham, D., Laszlo, L. and Prusiner, S.B. (1995) J. Cell Biol. **129**, 121–132
20. Naslavsky, N., Stein, R., Yanai, A., Friedlander, G. and Taraboulos, A. (1997) J. Biol. Chem. **272**, 6324–6331
21. Caughey, B. (1991) Semin. Virol. **2**, 189–196
22. Shyng, S.L., Huber, M.T. and Harris, D.A. (1993) J. Biol. Chem. **268**, 15922–15928
23. DeArmond, S.J. and Prusiner, S.B. (1995) Am. J. Pathol. **146**, 785–811
24. Hay, B., Barry, R.A., Lieberburg, I., Prusiner, S.B. and Lingappa, V.R. (1987) Mol. Cell. Biol. **7**, 914–920
25. Lopez, C.D., Yost, C.S., Prusiner, S.B., Myers, R.M. and Lingappa, V.R. (1990) Science **248**, 226–229
26. Caughey, B., Race, R.E., Vogel, M., Buchmeier, M.J. and Chesebro, B. (1988) Proc. Natl. Acad. Sci. U.S.A. **85**, 4657–4661
27. Caughey, B., Race, R.E., Ernst, D., Buchmeier, M.J. and Chesebro, B. (1989) J. Virol. **63**, 175–181
28. Borchelt, D.R., Rogers, M., Stahl, N., Telling, G. and Prusiner, S.B. (1993) Glycobiology **3**, 319–329
29. Parizek, P., Roeckl, C., Weber, J., Flechsig, E., Aguzzi, A. and Raeber, A.J. (2001) J. Biol. Chem. **276**, 44627–44632
30. Stahl, N., Baldwin, M.A., Burlingame, A.L. and Prusiner, S.B. (1990) Biochemistry **29**, 8879–8884
31. Chiesa, R. and Harris, D.A. (2001) Neurol. Dis. **8**, 643–763
32. Zhang, H., Stöckel, J., Mehlhorn, I., Groth, D., Baldwin. M.A., Prusiner, S.B., James, T.L. and Cohen, F.E. (1997) Biochemistry **36**, 3543–3553
33. Swietnicki, W., Morillas, M., Chen, S.G., Gambetti, P. and Surewicz, W.K. (2000) Biochemistry **39**, 424–431
34. Baskakov, I.V., Legname, G., Baldwin, M.A., Prusiner, S.B. and Cohen, F.E. (2002) J. Biol. Chem. **277**, 21140–21148
35. Sanghera, N. and Pinheiro, T.J.T. (2002) J. Mol. Biol. **315**, 1241–1256
36. Kazlauskaite, J., Sanghera, N., Sylvester, I., Vénien-Bryan, C. and Pinheiro, T.J.T. (2003) Biochemistry **42**, 3295–3304
37. Critchley, P., Kazlauskaite, J., Eason, R. and Pinheiro, T.J.T. (2004) Biochem. Biophys. Res. Commun. **313**, 559–567
38. Mehlhorn, I., Groth, D., Stockel, J., Moffat, B., Reilly, D., Yansura, D., Willett, W.C., Baldwin, M., Fletterick, R., Cohen, F.E. et al. (1996) Biochemistry **35**, 5528–5537
39. Morillas, M., Swietnicki, W., Gambetti, P. and Surewicz, W.K. (1999) J. Biol. Chem. **274**, 36859–36865
40. Forloni, G., Angeretti, N., Chiesa, R., Monzani, E., Salmona, M., Bugiani, O. and Tagliavini, F. (1993) Nature (London) **362**, 543–546
41. Brown, D.R., Schmidt, B. and Kretzschmar, H.A. (1996) Nature (London) **380**, 345–347
42. Brandner, S., Isenmann, S., Raeber, A., Fischer, M., Sailer, A., Kobayashi, Y., Marino, S., Weissmann, C. and Aguzzi, A. (1996) Nature (London) **379**, 339–343
43. Brandner, S., Raeber, A., Sailer, A., Blattler, T., Fischer, M., Weissmann, C. and Aguzzi, A. (1996) Proc. Natl. Acad. Sci. U.S.A. **379**, 13148–13151
44. Brown, D.R., Schulz-Schaeffer, W.J., Schmidt, B. and Kretzschmar, H.A. (1997) Exp. Neurol. **146**, 104–112
45. Pillot, T., Lins, L., Goethals, M., Vanloo, B., Baert, J., Vandekerckhove, J., Rosseneu, M. and Brasseur, R. (1997) J. Mol. Biol. **274**, 381–393
46. Lin, M.-C., Mirzabekov, T. and Kagan, B.L. (1997) J. Biol. Chem. **272**, 44–47
47. Van Meer, G. (1989) Annu. Rev. Cell. Biol. **5**, 247–275

48. Bate, C., Salmona, M., Diomede, L. and Williams, A. (2004) J. Biol. Chem. **279**, 14983–14990
49. Kaneko, K., Vey, M., Scott, M., Pilkuhn, S., Cohen, F.E. and Prusiner SB (1997) Proc. Natl. Acad. Sci. U.S.A. **94**, 2333–2338
50. Naslavsky, N., Shmeeda, H., Friedlander, G., Yanai, A., Futerman, A.H., Barenholz, Y. and Taraboulos, A. (1999) J. Biol. Chem. **274**, 20763–20771
51. Mange, A., Nishida, N., Milhavet, O., McMahon, H.E.M., Casanova, D. and Lehmann, S. (2000) J. Virol. **74**, 3135–3140
52. Klein, T.R., Kirsch D., Kaufmann, R. and Riesner, D. (1998) Biol. Chem. **379**, 655–666

Biochem. Soc. Symp. **72**, 223–231
(Printed in Great Britain)

21

BAR domains and membrane curvature: bringing your curves to the BAR

Jennifer L. Gallop[1] and Harvey T. McMahon[1]

MRC Laboratory of Molecular Biology, Hills Road, Cambridge CB2 2QH, U.K.

Abstract

BAR (bin, amphiphysin and Rvs161/167) domains are a unique class of dimerization domains, whose dimerization interface is edged by a membrane-binding surface. In its dimeric form, the membrane-binding interface is concave, and this gives the ability to bind better to curved membranes, i.e. to sense membrane curvature. When present at higher concentrations, the domain can stabilize membrane curvature, generating lipid tubules. This domain is found in many contexts in a wide variety of proteins, where the dimerization and membrane-binding function of this domain is likely to have a profound effect on protein activity. If these proteins function as predicted, then there will be membrane subdomains based on curvature, and thus there is an additional layer of compartmentalization on membranes. These and other possible functions of the BAR domain are discussed.

Introduction

Most cells and organelles are not simply round and therefore membranes do not generally adopt a minimum energy state, but have infoldings, protrusions, compressions and tubulations. These observed larger membrane shapes are effectively differences in local membrane curvature. It is interesting to consider the possibility that membrane curvature itself has a primary role in specifying the location of effector proteins, for example actin filaments, enzymic activities or scaffold proteins. It is already well recognized that distinct membrane morphologies with defined subregions of curvature are generated and maintained in an active fashion by the interplay of cytoskeletal elements and membrane trafficking pathways. Increasingly, membrane-curva-

[1]Correspondence can be addressed to either author (email jennyg@mrc-lmb.cam.ac.uk or hmm@mrc-lmb.cam.ac.uk).

ture sensing, generating and stabilizing molecules are also being recognized to have important functions in membrane morphology [1–3]. These all work together to allow, among other functions, the active movement of cells in response to external cues, the budding and fusion of membrane compartments and the dividing of membrane areas into different subdomains.

There are many reviews on the role of the cytoskeleton in cell dynamics [4–6], but, in the present review, we concentrate on the role of BAR (bin, amphiphysin and Rvs161/167) domains [7]. These are membrane-binding domains that may prove to play a central role in binding and stabilizing areas of high membrane curvature. This protein module is found in many trafficking proteins (Table 1), and also proteins of unknown function.

The BAR domain — a curvature sensor

The BAR domain got its name from the conserved N-terminal region of bin, amphiphysin and Rvs161/167 proteins (all amphiphysin family members). This domain in amphiphysin is known to be responsible for membrane binding and dimerization of the protein [8,9]. In addition, both *in vitro* and *in vivo*, the domain leads to tubulation of membranes [3,9]. This tubulation is due, at least in part, to the banana shape of the domain (Figure 1A), which binds to membranes via its concave surface [7]. Each monomer of the BAR domain has three kinked α-helices (Figure 1A, arrows), which gives rise to a six-helix bundle on dimerization. The membrane interaction is purely electrostatic and thus amphiphysin can bind well to negatively charged membranes, given the high intensity of charge at the ends of the domain (Figure 1B). Moreover, if it is provided with membranes of higher curvature, it binds better, as it makes use of the positive charges on the concave surface [7]. Thus this domain is a dimerization and membrane-curvature-sensing module. Some of the other domains found in proteins with BAR modules are listed in Table 1. This leads to the discussion below on how the presence of BAR domains will affect protein function.

BAR domains as dimerization domains on membrane binding

In the BAR structure, the dimerization interface buries a large hydrophobic surface, and the edge of this dimerization surface forms part of the membrane-binding face. The dissociation constant for the dimer in solution is relatively weak. The K_d for the amphiphysin dimer is 6 μM [7], and, for comparison, the Jun/Fos leucine zipper heterodimer has a K_d of 110 nM [10]. From the structure, we see that for curvature sensing to occur via the concave face of the banana-shaped domain, dimerization is necessary. Given the positive-charge distribution on this surface, we predict that membrane binding will promote dimerization. Inhibiting or promoting dimer formation could therefore be a regulatory mechanism for fine-tuning membrane binding.

Table 1. Example BAR domain proteins, indicating common themes in their domain structure. PTB, phosphotyrosine-binding domain; Snx1, sorting nexin 1.

Protein	BAR membrane binding shown?	Other lipid-binding domains	GEF/GAP activity or small GTPase binding?	SH3 domain?	Other domains
Amphiphysins	Yes	N-terminal amphipathic helix	No	Yes	Endocytic motif domain
Endophilins	Yes	N-terminal amphipathic helix	No	Yes	
Nadrin/RICH	Yes	N-terminal amphipathic helix	RhoGAP activity	No	Proline repeats
BRAP/Bin2	No	N-terminal amphipathic helix	No	No	
APPLs	No	PH and PTB domains	Binds Rab5 GTP	No	
Arfaptins	Yes	N-terminal region	Binds Arf, Rac, Arl	No	
Snx1	Yes	PX	No	No	Sorting nexin domain
Oligophrenins	Yes	PH	RhoGAP activity	Yes	
Centaurin βs	Yes	PH	ArfGAP activity	No	Ankyrin repeats
Tuba	No	None	RhoGEF activity	Yes (several)	

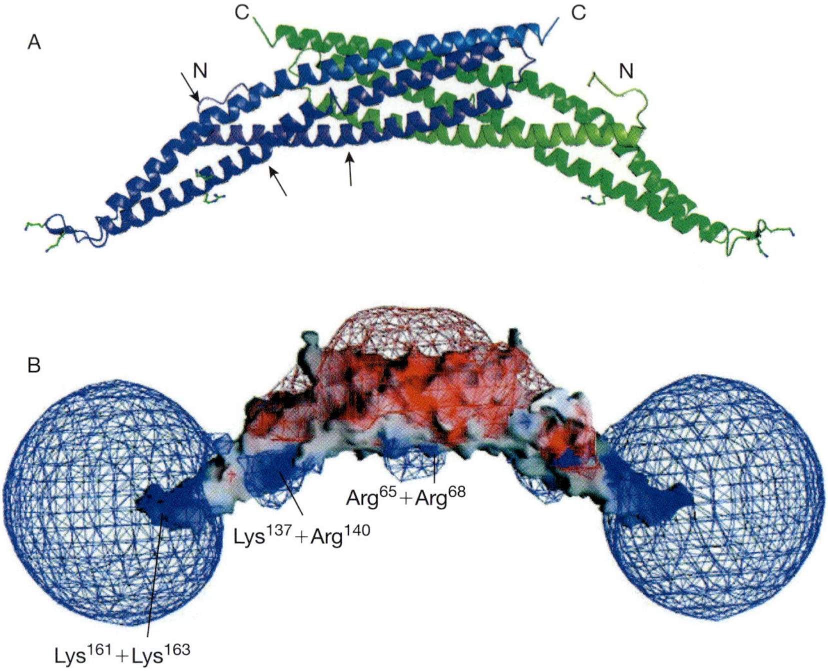

Figure 1 Structure of amphiphysin BAR domain. (**A**) Ribbon representation of the *Drosophila* amphiphysin BAR domain structure, which spans amino acids 25–240 (Protein Data Bank code 1URU). The banana shape is generated by dimerization and kinks in the α-helices. The overall structure is the same as an arfaptin C-terminal fragment (Protein Data Bank code 1I4D), although these proteins are only weakly homologous at the sequence level. (**B**) Surface-charge representation of the BAR domain (red, $-10\,kTe^{-1}$, blue $+kTe^{-1}$) showing electrostatic equipotential surfaces in mesh at $2\,kTe^{-1}$ (+ blue, − red). The positively charged residues on the concave face are labelled. Kinks in α-helices are indicated by arrows.

The amphipathic helix of N-BAR domains may lead to stabilization or driving of membrane curvature

A frequent modification of BAR domains is an N-terminal sequence that is predicted to form an amphipathic helix (Figure 2), and, in all cases where this helix is found, these are called N-BAR domains [7]. In the case of amphiphysin, the N-terminal sequence was shown to form a helix on membrane binding [7]. This N-terminal amphipathic helix is important for the membrane interaction of amphiphysin, endophilin and RICH/nadrin [7,11,12]. The N-terminal amphipathic helix is also present in BRAP (breast-cancer-associated protein)/Bin2, one of the first BAR domains to be identified [13,14]. Based on our modelling of these sequences as α-helices, we see that the polar face of the helix is largely positive (Figure 2A), which suggests an interaction with either a

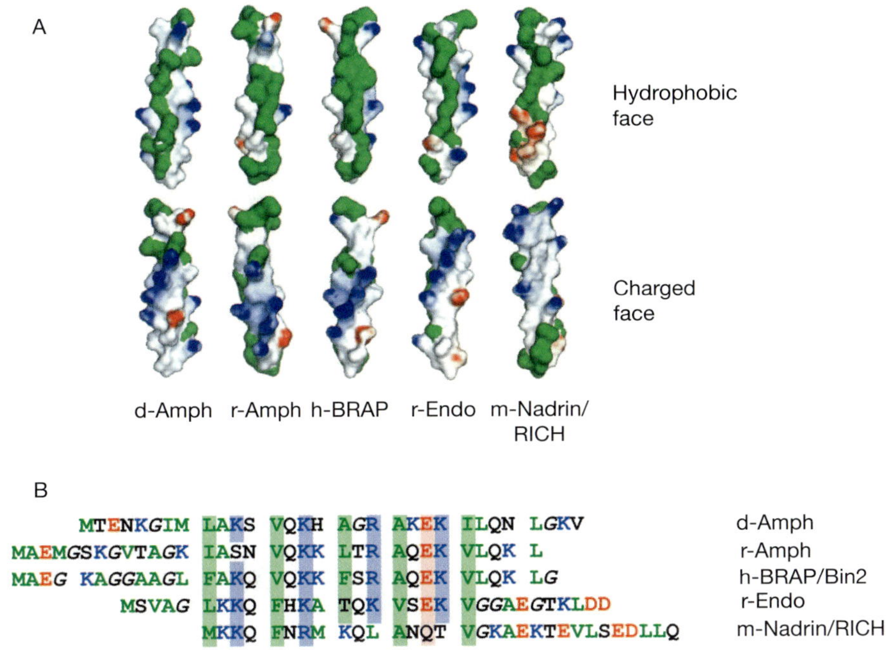

Figure 2 Amphipathic helix predictions for N-BAR domains. (A) α-Helical models of a range of N-terminal sequences from BAR domain proteins show the polar (red negatively charged, blue positively charged) and non-polar (green) faces. (**B**) Alignments of the N-terminal sequences based on the regions of hydrophobicity and charge in the modelled helices. The glycine residues indicated in italics will give rise to flexibility in the helix, and it can seen that where glycine residues occur, there is a misalignment of polar and non-polar residues when modelled on a straight α-helix. This is most striking in RICH/nadrin, where there is a negatively charged patch on the hydrophobic face towards the C-terminal end of the sequence. Amph, amphiphysin; Endo, endophilin. d, *Drosophila*; r, rat; h, human; m, mouse.

negative surface of another protein (potentially the convex face of the BAR domain) or with negatively charged lipid head-groups.

One explanation for the observed superior binding of N-BAR domains is the membrane insertion of the hydrophobic face of this helix in the hydrophobic phase of the membrane. Membrane insertion of the amphipathic helix could have a number of consequences. If the helix does insert into the bilayer, then it will increase the residency time of the domain on the membrane (as seen with epsin [15]), and this increased residency will allow formation of a more effective framework to drive or stabilize membrane-curvature changes. Insertion of an amphipathic helix also has the consequence of displacing lipids leading to the generation of local curvature (Figure 3A). With a high concentration of BAR proteins, helix insertion may contribute significantly to membrane curvature. If the helix inserts into the bilayer perpendicular to the BAR domain, then

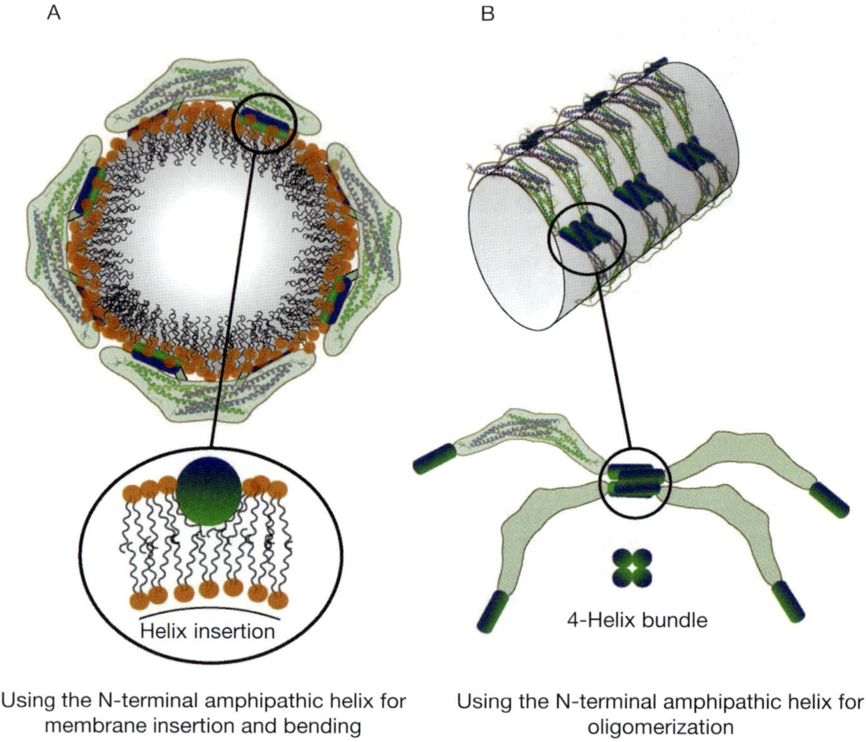

Figure 3 Role of amphipathic helix in membrane binding. (**A**) Helix insertion into the outer leaflet could cause increased membrane binding of N-BAR domains. (**B**) Alternatively, the amphipathic helices could form helical bundles and mediate oligomerization on the surface of the tubule.

this will act to drive one-dimensional curvature, i.e. lipid tubules. If it inserts in parallel to the BAR domain, then the N-BAR domain will generate two-dimensional curvature, i.e. lipid vesicles. In liposome tubulation experiments, there is often a fine transition between complete tubulation via an N-BAR domain and the formation of vesicles, and thus the generation of two-dimensional curvature has been observed.

An alternative model for the action of the N-terminal helix (which needs to be tested) may be to promote higher-order oligomerization on membranes by formation of a three or four-helix bundle which packs the hydrophobic residues in the centre, thus providing a better framework for driving tubulation (Figure 3B). Indeed, striations are sometimes observed on N-BAR domain tubules and cross-linking on liposomes gives rise to higher-order species [11,12]. Of course, these striations may simply be due to side-by-side packing of this banana-shaped molecule, and this conformation will readily allow cross-linking.

For amphiphysin and endophilin, the amphipathic helix is not essential for *in vitro* liposome tubulation ([7], and J. L. Gallop, unpublished work). Even if it is tempting to assume that insertion of the amphipathic helix is solely responsible for driving tubulation, mutation of a key hydrophobic residue, (Phe^{10} in r-endophilin 1, Phe^{9} in r-amphiphysin 1) and indeed deletion of the amphipathic helix still results in tubule formation at high protein concentrations, and, at lower concentrations, the proteins bind better to smaller liposomes. This suggests an amphipathic helix is an important alteration to the BAR domain that helps promote membrane curvature.

Coincidence detection of membrane curvature and composition

BAR domains are frequently found in combination with other lipid binding domains (see Table 1). These domains include PH (pleckstrin homology) domains (which bind phosphoinositides) and PX (phox) domains (which bind 3-phosphorylated phosphoinositides). This conjugation of the BAR domain to other domains with defined lipid specificity means that the protein not only selects a particular membrane compartment, but also selects areas of membrane curvature on that compartment. If both domains are required for membrane binding then the protein acts as a logical AND gate, or coincidence detector. This mechanism is used frequently in signalling pathways, where, for instance, localization and a binding partner are required for activation [16]. This gives rise to precise localization of binding partners and enzymic activities. Centaurin β2 requires both functional BAR and PH domains in order to tubulate intracellular membranes in COS cells [7]. In the case of sorting nexin1, a BAR and a PX domain combination leads to the localization of the protein to tubular extensions on endosomes [17].

BAR domains in creating or stabilizing membrane subdomains

One of the most important functions of the BAR domain may well be in the generation of membrane subcompartments. One such subdomain is the T-tubule network in muscle, which is an extension of the plasma membrane into the muscle cell. This network is dependent on the presence of amphiphysin, and, in its absence, there is a defect in excitation–contraction coupling [3]. The tubular extension on endosomes also require the presence of another BAR protein, sorting nexin1, and failure to form this subdomain on the endosomes leads to defects in intracellular membrane trafficking [17].

BAR domains as regulated recruiters of downstream effectors

We have already referred to the probable increased dimerization potential of BAR domains on membranes. SH3 (Src homology 3) domains are found in many BAR domain proteins, and a BAR protein on a membrane will present two SH3 domains for interactions. This will favour the binding of dimeric SH3-binding proteins or monomeric proteins with multiple SH3-binding sites. Amphiphysin and endophilin both have C-terminal SH3 domains that recruit the mechanochemical GTPase dynamin and the lipid phosphatase synaptojanin to sites of endocytosis. Dynamin is an oligomer and so will bind better to oligomeric BAR proteins, while synaptojanin is a monomer, but has multiple SH3-interaction sequences. SH3-domain interactions can be very weak (in the order of 1 mM), and thus the initial dimerization of the SH3 protein via the BAR domain–membrane interaction may be important for achieving the downstream effect mediated by the SH3 interaction.

BAR domain proteins and small G-proteins

A striking proportion of the BAR domain proteins so far identified either activate the hydrolysis or exchange of nucleotides on small G-proteins [GAP (GTPase-activating protein) and GEF (guanine nucleotide-exchange factor) activities]. This highlights that membrane-curvature sensing is coupled with enzymic activities, both directly, as with GAPs and GEFs, and indirectly, as in the case of synaptojanin bound to endophilin. The regulation of Rho GTPase activity is particularly interesting because of its role in the control of actin polymerization and cytoskeletal remodelling [18], which is associated with changes in membrane morphology. Knockdown of oligophrenin 1 (a BAR protein; see Table 1) in neurons leads to defects in dendrite spine formation [19]. We can hypothesize that the role of the BAR domain, in conjunction with the PH domain, is localization of the GAP activity to the correct membrane position, for instance where spines protrude from dendrites. Tuba [20], another BAR protein, has Rho GEF activity and multiple SH3 domains that bind dynamin, N-WASP (neural Wiskott–Aldrich syndrome protein) and other actin-regulating proteins. Tuba may therefore be a link between the membrane-curvature events that accompany dynamin-mediated fission and cortical actin rearrangements.

There are also two examples of BAR domains that bind small G-proteins directly, and, based on homology, there are likely to be more. Arfaptin binds to a variety of G-proteins, including Rac1 and Arf1 [21]. Rac binds to the membrane-binding surface, while Arf competes with this interaction, but might not bind to the same site. It is known that Arf promotes the interaction of many proteins with membranes, and thus the exact function of this G-protein in arfaptin membrane interaction will be an important paradigm to solve. APPL is also proposed to bind to Rab5 GTP via its BAR+PH domain [22]. The exact interaction mode has not been solved, but by analogy with arfaptin, it is likely to modulate BAR domain membrane binding.

Conclusions

Sensing membrane morphology via protein domains and using this to influence cellular processes is a novel concept in biology. We have presented some ideas of how this can work at a molecular level, but further experiments are needed to verify or disprove these ideas. Given the large number of proteins that have potential BAR domains, the importance of membrane morphology in cell structure and function, and the conservation of BAR modules from yeast to man, we can predict that these modules will have a broad and significant impact on diverse biological processes.

We thank Marijn Ford and Phil Evans for help preparing the α-helix models and surface charge diagrams. J.L.G. is funded by an MRC Pre-doctoral Fellowship.

References

1. Bigay, J., Gounon, P., Robineau, S. and Antonny, B. (2003) Nature (London) **426**, 563–566
2. Ford, M., Mills, I.G., Peter, B.J., Vallis, Y., Praefcke, G., Evans, P.R. and McMahon, H.T. (2002) Nature (London) **419**, 361–366
3. Razzaq, A., Robinson, I., McMahon, H.T., Skepper, J., Su, Y., Zelhof, A., Jackson, A., Gay, N. and O'Kane, C. (2001) Genes Dev. **15**, 2967–2979
4. Qualmann, B. and Kessels, M.M. (2002) Int. Rev. Cytol. **220**, 93–144
5. Lippincott-Schwartz, J. (1998) Curr. Opin. Cell Biol. **10**, 52–59
6. Raftopoulou, M. and Hall, A. (2004) Dev. Biol. **265**, 23–32
7. Peter, B.J., Kent, H.M., Mills, I.G., Vallis, Y., Butler, P.J., Evans, P.R. and McMahon, H.T. (2004) Science **303**, 495–499
8. Wigge, P., Kohler, K., Vallis, Y., Doyle, C.A., Owen, D., Hunt, S.P. and McMahon, H.T. (1997) Mol. Biol. Cell **8**, 2003–2015
9. Takei, K., Slepnev, V.I., Haucke, V. and De Camilli, P. (1999) Nat. Cell Biol. **1**, 33–39
10. Pernelle, C., Clerc, F., Dureuil, C., Bracco, L. and Tocque, B. (1993) Biochemistry **32**, 11682–11687
11. Farsad, K., Ringstad, N., Takei, K., Floyd, S.R., Rose, K. and De Camilli, P. (2001) J. Cell Biol. **155**, 193–200
12. Richnau, N., Fransson, A., Farsad, K. and Aspenstrom, P. (2004) Biochem. Biophys. Res. Commun. **320**, 1034–1042
13. Harada, A., Furuta, B., Takeuchi, K., Itakura, M., Takahashi, M. and Umeda, M. (2000) J. Biol. Chem. **275**, 36885–36891
14. Ge, K. and Prendergast, G.C. (2000) Genomics **67**, 210–220
15. Stahelin, R.V., Long, F., Peter, B.J., Murray, D., De Camilli, P., McMahon, H.T. and Cho, W. (2003) J. Biol. Chem. **278**, 28993–28999
16. McLaughlin, S., Wang, J., Gambhir, A. and Murray, D. (2002) Annu. Rev. Biophys. Biomol. Struct. **31**, 151–175
17. Carlton, J., Peter, B.J., McMahon, H.T. and Cullen, P. (2004) Cur. Biol., in the press
18. Schmitz, A., Govek, E.E., Bottner, B. and Van Aelst, L. (2000) Exp. Cell Res. **261**, 1–12
19. Govek, E.E., Newey, S.E., Akerman, C.J., Cross, J.R., Van der Veken, L. and Van Aelst, L. (2004) Nat. Neurosci. **7**, 364–372
20. Salazar, M., Kwiatkowski, A., Pellegrini, L., Cestra, G., Butler, M., Rossman, K., Serna, D., Sondek, J., Gertler, F. and De Camilli, P. (2003) J. Biol. Chem. **278**, 49031–49043
21. Tarricone, C., Xiao, B., Justin, N., Walker, P., Rittinger, K., Gamblin, S. and Smerdon, S. (2001) Nature (London) **411**, 215–219
22. Miaczynska, M., Christoforidis, S., Giner, A., Shevchenko, A., Uttenweiler-Joseph, S., Habermann, B., Wilm, M., Parton, R. and Zerial, M. (2004) Cell **116**, 445–456

Subject index

actin, 27, 112, 120, 121
activator protein 1B (AP-1B), 57
acylation, 153
adaptor-associated kinase 1, 68, 69
adaptor protein 2, 66–69, 74, 75
  μ2 phosphorylation, 66, 67
  μ2 subunit, 66
adenosine A_1 receptor, 160, 161
adhesion molecule, 115
ADP-ribosylation factor (ARF), 32
β-adrenoceptor, 158
Alzheimer's disease, 130–136
amphiphysin, 226, 227
amyloid fibril, 218
β-amyloid peptide, 130, 131, 134, 136
angiotensin II type 1 receptor, 159, 160
anion-exchanger 1 (*see also* kidney isoform of AE1), 48, 50
apical targeting signal, 40
ApoE4, 133
apoptosis, 178
AT_1 receptor (*see* angiotensin II type 1 receptor)
axonal traffic, 144

BAR domain, 224–230
B-cell depletion, 166
B-cell receptor, 167
Bicaudal D (BicD), 6
bilayer translocation, 200, 206–208
blood–brain barrier, 138
botulinum neurotoxin, 140
bradykinin receptors B_1 and B_2, 161, 162
BRAP/Bin2, 226, 227
breast-cancer-associated protein (*see* BRAP/Bin2)

calcineurin, 90
calcium, 167, 168, 192–196
calcium signalling, 167, 168
calmodulin, 192–196
casein kinase, 23
caveola, 151
CD20,
  and lipid rafts, 169–173
  calcium entry, 168, 169
  clinical relevance, 165, 166
  oligomeric complex, 172
  structure, 166
Cdc42, 27
Cdk1, 16
Cdk1–cyclin B, 16, 19, 20
Cdk5, 90, 91
cell motility, 25, 120–124
cell-penetrating peptide, 200, 206–208
cell polarization, 24
cellular prion protein (PrP^c), 112, 113, 211–213
ceramide, 178
  as detergents, 180
  long-chain, 179
  short-chain, 179
  solute efflux, 183, 184
  transbilayer lipid motion, 185
chemokine receptor 5 (CCR5), 157
cholesterol, 131–136, 143
cisternal unstacking, 18, 31, 65, 68, 69
clathrin,
  coat (dis)assembly, 73
  structure, 72
clathrin-coated vesicle, 66, 72, 73
  visualization of, 74
CLEM (correlative live cell and EM), 81
coated vesicle, 144
COPI, 3
COPII, 2, 6–8
cyclin-dependent kinase (*see* Cdk1; Cdk1–cyclin B; Cdk5)

Danio rerio (*see* zebrafish)
distal renal tubular acidosis, 49
Drosophila, 115
dRTA (*see* distal renal tubular acidosis)
dynactin, 5
dynamin I
  de(phosphorylation), 89–93
  domain structure, 89
  in synaptic vesicle endocytosis, 93, 95
  regulation of activity, 92–93
dynein, 4
dynein/dynactin complex, 147

early endosome, 100
early endosome antigen 1 (EEA1), 102

electrostatic potential, 190, 191
endocytosis (*see also* synaptic vesicle endocytosis), 105, 106, 123
endophilin, 226, 227
endoplasmic reticulum, 2–10, 15
endoplasmic-reticulum-associated degradation (*see* ERAD)
endosome, 34, 79–83, 105
endosome–lysosome fusion, 82
epidermal growth factor (EGF), 102, 105, 106
ERAD, 52
ERK2 (*see* extracellular-signal-regulated kinase 2)
exocyst, 123
extracellular-signal-regulated kinase 2, 18,19

fibrillization, of prions, 214
filipin, 156
flip-flop motion (*see* transbilayer lipid motion)
flotillin (*see also* reggie-1 and reggie-2), 124
fluorescence resonance energy transfer, 156
fusion–fission hypothesis (*see also* hybrid organelle hypothesis), 80

GAP43, 124, 125, 195
glycosylphosphatidylinositol, 42, 142, 152, 212
GM130, 16, 17, 18,
Golgi apparatus, 2–10, 15, 34
  disassembly, 18
  fragmentation, 19, 20
Golgi matrix, 16
Golgi matrix protein of 130 kDa, 16, 17
gonadatropin-releasing hormone receptor, 157, 158, 162
GPI (*see* glycosylphosphatidylinositol),
G-protein (*see also* small G-protein), 151
G-protein-coupled receptor, 151
  associated with lipid rafts, 154, 155
  raft localization, 157
GRASP65, 19
GTPase, 88, 99, 100
GTPase-activating protein, 230
guanine nucleotide-exchange factor, 230

hybrid organelle hypothesis (*see also* fusion–fission hypothesis), 80
hypercholesterolaemia, 135

α-intercalated cell, 49

kidney isoform of AE1 (kAE1), 48–58
  effect of dRTA mutants, 51, 52, 53
  expression, 50
  hetero-oligomers, 52, 54
  structure, 50
  targeting, 50, 51
kinesin, 4
'kiss and run' hypothesis, 80

late endosome (*see* endosome)
lipid phase-transition, 181, 182
liposome, 200–208
Lis1, 6
LLC-PK1 cell, 51
lysosome, 78
  delivery to, 78, 79, 80, 81
  secretory, 78
  -related organelle, 78

Madin–Darby canine kidney cell, 40, 42, 43, 51
MAPK (*see* mitogen-activated protein kinase)
MAPK/ERK kinase 1, 18, 19, 20, 21
MARCKS, 191–196
MEK1 (*see* MAPK/ERK kinase 1)
membrane dipeptidase, 40, 42, 43
membrane permeabilization, 183
methyl-β-cyclodextrin, 156, 171
microtubule, 2
mitogen-activated protein kinase, 19
molecular motor, 146
motor neuron, 140
motor protein, 4, 5
  phosphorylation of, 9
multivesicular body, 79
myristolyated alanine-rich C kinase substrate (*see* MARCKS)

N-acylsphingosine (*see* ceramide)
N-BAR domain, 227, 228
nerve growth factor, 148
neuronal development, 113
neurotoxin, 140
neurotransmitter, 87
N-glycan, 40, 42, 43
non-lamellar lipid phase, 187
Nudel, 6

oculocerebrorenal syndrome of Lowe (OCRL), 35

p35, 94
p50/dynamitin, 147
$p75^{NTR}$, 148
$p150^{glued}$, 8
PAK4, 27
paracellular barrier, 51

phosphatidylinositol 4,5-bisphosphate [*see* PtdIns(4,5)P_2]
phosphatidylinositol 4-phosphate [*see* PtdIns(4)*P*]
phosphoinositide, 32
phosphoinositide 4-kinase (PI4K), 32
phospholipase C, 167, 192, 193, 194
plasmalammal raft, 120–124
plasmin, 131–133
plasminogen, 130–132
pleckstrin homology domain, 33, 88, 229
Plk1 (polo-like kinase 1), 16, 17, 18
PM1 motif, 100
polarized kidney cell, 50
polysialoganglioside, 142
prion conversion, 218
prion disease, 211
prion protein (*see also* cellular prion protein)
  α and β-isoforms, 214, 215
  anchorless, 214
  scrapie isoform, 211–213
protein kinase A, 22
protein kinase C, 22, 90, 91, 191, 192, 195
protein kinase D, 24
protein–membrane interaction, 191
PtdIns(4)*P*, 32–34
PtdIns(4,5)P_2, 34, 35, 120–125, 192–196
pyrene, 185

Rab, 99–106
Rab4, 106
Rab5, 101, 102
Rab5a, 105
Rab21, 101, 102, 104, 105
Rab22, 105
Rab22a, 102
Rac, 122, 123
raft patching, 123
Ras, 23
red blood cell band 3, 48
reggie-1 and -2, 110–117
  capping in T-cells, 112
  in *Drosophila,* 115
  in zebrafish, 112
  morphogenetic functions, 114
  neuronal development, 113
  scaffolds, 112, 115, 116
  signal transduction, 112
retrograde transport, 141, 146
RICH/nadrin, 226, 227
rituxan, 166

Sar1p, 7
seladin-1, 135
SH3 domain, 89, 229
SHP-2, 59
SLC4A1, 48
small G-protein, 230
SNARE, 82–84
solute efflux, 183
Southeast Asian ovalocytosis, 49
sphingomyelin, 187
sphinogmyelinase, 183
Src homology 3 domain (*see* SH3 domain)
Src kinase, 23
statin, 136
Ste20 kinase, 25
store-operated calcium channel, 167, 168
synaptic vesicle, 87
synaptic vesicle endocytosis, 87–90
  role of dynamin I, 93, 95
synaptosome, 94
synthetic bilayer, 201, 206, 207

Tat peptide, 200–208
tetanus neurotoxin, 140–148
TGN (*see trans*-Golgi network)
thapsigargin, 169
Thy-1, 142, 143
tissue-type plasminogen activator, 130, 131
transbilayer lipid motion, 185
transfection protocols, 145
transferrin, 106
trans-Golgi network, 15, 24, 31–35, 43, 44, 67, 78
transmissible spongiform encephalopathy (*see* prion disease)
type II PDZ binding motif, 56

vacuole, 78

yeast Sps1/Ste20-related kinase 1 (*see* YSK1)
YSK1, 25, 26
YXXΦ motif, 56, 66
zebrafish, 112